T0313800

Wings for the Rising Sun

Harvard East Asian Monographs 428

Wings for the Rising Sun

A Transnational History of Japanese Aviation

Jürgen P. Melzer

Published by the Harvard University Asia Center
Distributed by Harvard University Press
Cambridge (Massachusetts) and London 2020

Printed in the United States of America

The Harvard University Asia Center publishes a monograph series and, in coordination with the Fairbank Center for Chinese Studies, the Korea Institute, the Reischauer Institute of Japanese Studies, and other facilities and institutes, administers research projects designed to further scholarly understanding of China, Japan, Vietnam, Korea, and other Asian countries. The Center also sponsors projects addressing multidisciplinary and regional issues in Asia.

Cataloging-in-Publication Data is on file at the Library of Congress.

Library of Congress Control Number: 2019948105

ISBN 9780674244412 (hardcover : alk. paper)
ISBN 9780674244429 (pbk. : alk. paper)

Index by Jac Nelson

♾ Printed on acid-free paper

Last figure below indicates year of this printing
29 28 27 26 25 24 23 22 21 20

To Jun

Contents

List of Maps, Figures, and Tables xi

Acknowledgments xv

Preface xix

Note on Sources xxiii

Introduction: A New Perspective on Japanese Aviation History 1

Technology: Transnational Transfer and National Diffusion 2

Mobilizing the Public 4

Aircraft and Statecraft: Aerial Armament and International Diplomacy 5

Part I. Early Japanese Aviation, 1877–1918

1 Powerful Images and Grand Visions 11

Early Gas Balloons: Low Tech, High Risk 13

Japan's First Balloon Launch 14

Balloon Fever Grips Japan 17

Balloons in the Russo-Japanese War and a Technological Dead-End 19

Two Reports about Western Aviation 22

The Road to Japan's First Motorized Flight 24

Conclusion 37

2 The French Decade 40
The Flying Baron Shigeno 41
The Qingdao Air War: Brief Encounters and a Lasting Myth 44
A Technocrat Shapes His Vision: Kusakari Shirō 50
Inoue Ikutarō: The Army Air Force's Mastermind 54
The French Aeronautical Mission to Japan 56
Reconsidering the Exclusive Devotion to French Aviation 64
Conclusion 65

PART II. GERMANY AND JAPAN'S ARMY AVIATION, 1918–37

3 Japan's Army Aviation in the Wake of World War I 69
Early German Influence, 1919–25 71
The Army's Struggle over a New Air Doctrine 80
Squaring the Circle: Disarmament and Airpower Buildup 83
Visions of Internationalism and National Prestige: The "Visit Europe Flight" 86
Conclusion 93

4 On the Way to Independent Aircraft Design 96
Industrialists, Engineers, and Teachers 97
German Airliners into Japanese Bombers: Junkers in Japan 114
The Army's New Aircraft and the Manchurian Crisis 125
Conclusion 132

PART III. BRITAIN, GERMANY, AND JAPAN'S NAVAL AVIATION, 1912–37

5 Navigating a Sea of Change 139
Japanese Observers in Britain during World War I 142
An Early Compromise: Ship-Based Floatplanes 144
A New Launching Technology 147
Redefining Naval Airpower: The Early Years of Carrier-Based Aircraft 149
The Arrival of the First British Aeronautical Engineers in Japan 151

The British Aviation Mission to Japan 154
Conclusion 167

6 Japan's Naval Aviation Taking the Lead 169
Toward an Autonomous Airpower: Large, All-Metal Flying Boats 170
The Next Generation of Japanese Aircraft Carriers 183
A Second Generation of Carrier Planes 186
A New Role for Carrier Aircraft: Preemptive Air Strikes 188
Britain's Waning Influence and a Fateful Legacy 191
Conclusion 194

Part IV. Toward Pearl Harbor and Beyond, 1937–45

7 US Know-How for Japanese Aircraft Makers 201
Late Japanese Interest in US Aviation 202
US Aviation Technology Comes to Japan 205
A Craving for US Machine Tools 213
Know Your Enemy: US Assessments of Japanese Airpower 219
Japanese Perceptions of the US Aviation Industry 231
Conclusion 232

8 Jet and Rocket Technology for Japan's Decisive Battle 234
Early Japanese Experiments 236
German Technology to Japan 238
Japan's First Rocket Aircraft 241
One More Miracle Weapon: Jet Airplanes 248
The Maiden Flight of the Shūsui 252
The Kikka's Maiden Flight 254
Suicidal Cherry Blossoms: The Ōka Attack Aircraft 258
Conclusion 260

Epilogue 264
Technology Transfer: Causes, Conduits, and Consequences 265
The Media and the Public: Anxieties, Exhilaration, and Fervent Nationalism 268
International Relations: From Cooperation to Alienation and Conflict 269

Transwar Continuities and Postwar Disruptions: Japanese Aviation after 1945 271
Conclusion 282

Notes 285

Bibliography 313

Index 329

Maps, Figures, and Tables

Maps

Japan: major airfields and arsenals xxvi

Figures

1.1 Japan's first balloon launch at the Ministry of the Navy's Tsukiji parade grounds 16
1.2 Shimazu Genzō's balloon launch 18
1.3 The Yamada Airship no. 2 21
1.4 Student pilot Tokugawa Yoshitoshi in a Farman biplane 28
1.5 Japan's first pilots 31
1.6 Young Tarō in his Farman biplane bombing the enemy 36
2.1 A Japanese aircraft made in France: Shigeno's *Wakadori* 42
2.2 The navy's Type Mo large seaplane 45
2.3 The "Aviator of Qingdao": Gunther Plüschow in his Taube monoplane 48
2.4 Kusakari's sketches illustrating the new concept of three-dimensional warfare 52
2.5 A triumphal arch saluting Faure and his officers at Gifu Station 59
2.6 An early flight simulator at Kakamigahara 61

3.1 The title page of *Asahi*'s *Memorial Picture Report* on the "visit Europe flight" 91
3.2 The "Great Visit Europe Flight" as a board game 92
4.1 Building the Do N: a look at Kawasaki's shop floor 100
4.2 The Kawasaki Type 87 bomber 101
4.3 Baumann's 1927 Tobi reconnaissance aircraft that became known as the "praying mantis" 106
4.4 The Dornier-Kawasaki team 108
4.5 The proposed bombing run from Taiwan to the US bases at Manila 117
4.6 A large group of army officials proudly posing in front of the Superheavy Bomber 122
5.1 The made-in-Japan version of Heinkel's Hansa Brandenburg W 29 146
5.2 The successful launch of Heinkel's floatplane from a gun turret 150
5.3 Sir William Francis Forbes-Sempill, the leader of the British Aviation Mission 158
6.1 A workman holding a large duralumin spar 172
6.2 The Mitsubishi Type R flying boat 177
6.3 The Nippon Kōkū airline's Kawanishi K-7A seaplane 180
7.1 Japan Air Transport's "Super Express in the Sky" 209
7.2 "DC-4 Off for Japan" 211
7.3 Pratt & Whitney's engine plant and Mitsubishi's Daikō plant 214
7.4 Asahi's record-breaking *Kamikaze* aircraft 224
7.5 High-tech made in Japan: the long-range research airplane *Kōkenki* 226
7.6 A map of the *Nippon*'s 52,860-kilometer around-the-world trip 228
7.7 Front page of one of the Japanese Files Research Project reports 230

8.1 A functional diagram of the TR-10, Japan's first turbojet engine 237
8.2 The BMW 003 jet engine: schematic and blueprint 249
8.3 A functional diagram of the Ne-20, the engine that powered Japan's first jet fighter 250
8.4 The Shūsui rocket interceptor shortly before takeoff 253
8.5 The Kikka jet aircraft preparing for takeoff 256
8.6 The Kikka during takeoff run 256
8.7 An Ōka suicide attacker 259
E.1 A once-proud air fleet crammed in a scrapyard 273
E.2 The five-ring Olympics symbol painted in the sky over Tokyo on October 10, 1964 278

Tables

4.1 Three generations of Japanese bombers 115
4.2 Three generations of the Imperial Japanese Army's aircraft, 1921–35 134
5.1 A list of the aircraft delivery that arrived with the Sempill mission 159
7.1 Japan's annual aircraft production: speculation and facts 222

Acknowledgments

The completion of a book—like the safe completion of a flight—depends in countless ways on the ongoing support and enthusiastic commitment of many people. This book was able to take off only thanks to those who generously offered guidance, intellectual support, and encouragement.

I want to express my deep sense of gratitude to my teachers at Princeton who offered new perspectives, sustained help, and strong motivation for my scholarly pursuits. I am profoundly indebted to my dissertation adviser, Sheldon Garon, whose important feedback and comments masterfully directed my research and helped me put my topic into a wider context. I am immensely grateful to Emily Thompson for her invaluable input during all stages of the project. Susan Naquin's firm advice inspired me to strive for clear and concise writing. She also became a role model for me of a teacher who offers time and effort to students with boundless generosity. Richard Samuels of MIT greatly stimulated my work with his thought-provoking questions, enthusiastic interest, and warm hospitality.

I am delighted to acknowledge the following sources of funding. For their generous financial assistance, I thank Princeton University's East Asian Studies Program and John and Julia Sensenbrenner; for funding my year-long research in Japan, I extend my gratitude to the Japan Foundation. I also express my thanks to Harvard University's Program on U.S.-Japan Relations and its director, Susan J. Pharr. The program's postdoctoral fellowship gave me the best possible opportunity to turn my dissertation into a book.

In Japan I was fortunate to be affiliated with the University of Tokyo, where I benefited from the guidance of Hashimoto Takehiko, whose work on aeronautical research provided considerable input to this book. Kudō Akira and Tajima Nobuo imparted their immense knowledge about German–Japanese relations to me, and Kōzu Masao shared his remarkable experience as an aeronautical engineer and open-handedly entrusted me with invaluable research material.

My research greatly depended on archival materials written in four different languages: English, French, German, and Japanese. I thank the archivists for their professional advice and assistance. Tsubone Akiko greatly supported my research at the Mitsubishi Archives in Tokyo. Hayafuji Hiroshi and Okano Mitsutoshi granted me access to a wealth of primary sources at the Mitsubishi Nagoya Aerospace Museum. Takai Takasumi generously made the archive of the Kakamigahara Aerospace and Science Museum available to me, and Kikkawa Yō carefully prepared the material for my visits to the Mitsui Bunko Business Archives. Kanda Shigeyoshi, the manager of the Japan Aeronautic Association's Aviation Heritage Archive, provided me with some of the most wonderful photographs of Japan's aviation history. I greatly benefited from the resources provided to me by the archive of the Deutsches Museum, Munich; the Mercedes-Benz Classic Archiv Daimler AG, Stuttgart; the Department Military Archives (Bundesarchiv-Militärarchiv), Freiburg; and the Dornier Unternehmensarchiv EADS Corporate Heritage. My work also received important input from the National Air and Space Museum's archivists and curators, especially Evelyn Crellin and Russ Lee, in Washington, DC.

I was fortunate to have access to two of the foremost East Asian libraries in the West. I am indebted to the librarians at Princeton's East Asian Library, especially Makino Yasuko and Noguchi Setsuko, for their sustained library support. I am equally grateful to Kuniko Yamada McVey, who promptly responded to all my requests and made the treasures of the Harvard-Yenching Library readily available for me.

My postdoctoral research on the Harvard campus became an indispensable part of my intellectual journey. Former Lieutenant General Onoda Osamu, JASDF pilot Manabe Seiji, Andrew Levidis, and Subodhana Wijeyeratne engaged with me in a fruitful exchange of ideas on the history of Japanese aviation. Ian Jared Miller spent time and effort to help me with my work.

I gratefully acknowledge Sigrun Caspary and Erich Pauer, who made their expertise on Japanese aviation available to me. I am indebted to Peter Selinger, who repeatedly invited me to his home to share his deep knowledge on German aviation and present key documents to me. I am especially grateful to Christopher Mayo, Seiji Shirane, and Evan Young (aka the "Mayonators"), who for over a decade offered their thoughtful comments on each of this book's chapters. I want to say a cheerful "thank you" to the students in my classes on the history of Japanese technology. I appreciate their keen interest and their probing questions that inspired me to further sharpen and clarify my teaching and writing. Last but not least, I thank Steve Gump, who helped me in the deteutonization of my writing and expertly guided me through the intricacies of the *Chicago Manual of Style*.

The following material has been reprinted in revised form with kind permission from the Weatherhead Center for International Affairs, Harvard University: Jürgen Melzer, "Cooperation, Rivalry, Conflict: America and Japan's Aviation, 1928–1941," USJP Occasional Paper 2014–15, Program on U.S.-Japan Relations, Harvard University.

Finally, words cannot sufficiently express my gratefulness for the unflagging support of the one single person whose patience, understanding, and capacity for encouragement seem to know no bounds. It is with profound gratitude that I dedicate this book to my wife, Jun Shimizu.

Preface

This book reflects my lifelong fascination with flying and my affinity for one of the most intriguing countries I have ever visited. In my previous vocation as airline pilot, I crossed the skies in planes that travel through the stratosphere at almost the speed of sound. While flying these marvels of modern technology, I became more and more interested in the history of flight. The pilots and aircraft makers of the pioneering age especially sparked my curiosity with their courage, vision, and ingenuity. My job gave me plenty of opportunities to travel the world. With every visit to Asia, my interest in the region increased. This was especially the case with Japan, a country that fascinated and bewildered me. What I saw was an ultramodern culture that makes use of the latest technology as a matter of course. Yet a few steps away from the hustle of the main streets, people seemed to cherish the serene tranquility of landscape gardens and religious sites. Maybe, I thought, this fascinating mix of deep-rooted tradition and highly advanced technology was part of my enigmatic experience.

In many ways, my flying career resembled a compressed history of aircraft development. In the spring I turned sixteen, I started flying gliders made of plywood and fabric at a small airfield near my hometown in the south of Germany. At Lufthansa's pilot school, I learned how to fly propeller-driven aircraft, proceeding from single-engine monoplanes and biplanes to faster multiengine aircraft. I then made the transition into the entirely new realm of size, speed, and power of turbojet aircraft. In the

1990s I flew a new generation of airliners that, with their fly-by-wire controls and glass cockpits, offered new experiences and challenges. In a comparable way, the various stages of my flight instruction embodied the step-by-step advance of aviation. My initial training with its focus on "stick and rudder" skills might not have differed much from what Orville and Wilbur Wright, Glenn Curtiss, or Louis Blériot taught at their flight schools in the United States and France. The next phase of my instruction, with cross-country and night flights, took up many of the challenges that Charles Lindbergh and Antoine de Saint-Exupéry describe in their absorbing accounts. A final stage that included polar crossings and zero-visibility landings clearly evinced the more recent milestones in aviation history.

Commercial flight instruction increasingly relies on the safe but artificial environment of flight simulators. Yet in a pilot's life there are, of course, plenty of real-world experiences impossible to forget. When my flight instructor readied me for my first solo flight on a hot Arizona afternoon in 1985, I felt the tense loneliness to take off in and fly an aircraft all by myself and immense release—and pride—after a safe touchdown. I learned the basics of acrobatic flying in an open-cockpit biplane high over the desert. After doing a series of loops and wingovers, I botched the Immelmann turn and ended up in an inverted spin. I still remember the reassuring laughter from the seat behind me when my instructor, John Burris, took over control and brought the aircraft back to level flight.

Some may find that commercial aviation has lost its much of its appeal. But I came to believe that even the most routine flight held the potential for memorable encounters. After takeoff on a foggy day, the aircraft will break through the cloud layer and offer a spectacular view to a remote horizon under a vast blue sky. Toward the end of the flight, the pilot lowers the plane's nose and the aircraft will dive back into the clouds. Within seconds it becomes a "blind flight," where an artificial horizon and a range of other instruments provide accurate guidance to the destination. Among the long-range routes, the night flight from Germany to Japan was my favorite. When overflying St. Petersburg, the city's circular spider web of light would pass by 37,000 feet below. With a little luck—and a few thousand kilometers further—the polar light would shine in all its rainbow colors along the horizon. Several hours later the sublime vastness of Siberia seemed to be all but forgotten when walking through the crowded streets of Tokyo.

Almost two decades ago, my flying career came to an unexpected end. On a sunny day in the Italian Alps, I prepared my parasail for a cross-country flight. Soon after takeoff, the parasail collapsed, and I fell from the sky. All I remember is the noise of the rescue helicopter and realizing that I couldn't feel my legs anymore. After several surgeries and two years of physical rehabilitation, I had to accept the fact that I was no longer fit for flight duty. But I could now walk on crutches and didn't need a wheelchair any longer. This improvement helped me overcome the loss of my pilot's license and encouraged me to reassess my goals.

Eager to learn more about Japan, I enrolled in the Japanese Program at the School of Oriental and African Studies, University of London. While continuing my studies later at Harvard, the idea for this book took shape. During a summer break, I made a trip to Lake Constance. At the Zeppelin Museum, I discovered photos of the German aircraft maker Claude Dornier and his airplanes flying over Japan. After some further inquiry, I learned that during the 1920s nearly all major German aviation companies were involved in the development of Japan's aviation. I began to ask myself how Germany, a defeated nation—and Japan's former World War I enemy—could wield such an influence. With this question in mind, I took up the opportunity to engage with aviation from a different vantage point: that of multinational archives, exhibits, and libraries. As it turned out, I entered an exciting new realm whose many fascinating aspects I set out to explore further during my graduate studies and postdoctoral research at Princeton and Harvard.

Two useful lines emerged in the course of my investigations. I was intrigued by the fact that, from its early beginnings, human flight was a remarkably transnational enterprise. The close—and often unexpected—connections among Japanese, French, German, British, and US aviation proved to be an illuminating field of inquiry. I decided to tell my story from a multinational perspective. Furthermore, I shifted the focus of my research away from the intrepid aviators and toward those who actually devised and built these remarkable flying devices. It is precisely in this respect that the aviation history of Japan becomes especially intriguing. Initially Japanese aircraft designers followed the tracks of their Western counterparts. But they soon embarked on entirely new and uncharted territory. This book pays special tribute to the ingenuity and perseverance of Japanese aeronautical scientists, engineers, and workmen. In their laboratories, in their design offices, and on their shop floors, they not only absorbed new technologies and manufacturing processes at a breathtaking speed but also conceived, designed, and built their own boldly original flying machines.

Note on Sources

Most of the primary documents mentioned in the endnotes originate from the following archives.

Japan

Archives at the Mitsubishi Economic Research Institute (MERI), Tokyo
Asahi Shinbun (Kikuzō II bijuaru)
Diplomatic Archives of the Ministry of Foreign Affairs of Japan, Tokyo
Ishikawajima Archives, Tokyo
Japan Aeronautic Association Aviation Heritage Archive, Tokyo
Japan Center for Asian Historical Records (JACAR), Tokyo
Kakamigahara Aerospace and Science Museum, Kakamigahara (Gifu)
Kobe University Library Newspaper Clippings Collection
Military Archives of the National Institute for Defense Studies, Tokyo
Mitsubishi Heavy Industries Komaki Archive
Mitsubishi Nagoya Aerospace Museum
Mitsui Bunko Business Archives, Tokyo
National Showa Memorial Museum, Tokyo
Ritsumeikan Shishiriyō Center, Kyoto
Tokyo Metropolitan Library
Yomiuri Shinbun Yomidasu Rekishikan

- France
  - Archives du Service Historique de l'Armée de l'Air, Vincennes
  - Centre des Archives Diplomatiques du Ministère des Affaires Etrangères, La Courneuve
  - Musée de l'Air et de l'Espace, Aéroport de Paris—Le Bourget
- Germany
  - Archive of the Deutsches Museum, Munich
  - Archive of the Deutsches Technikmuseum, Berlin
  - BMW Group Archiv
  - Bundesarchiv-Militärarchiv (BA-MA), Freiburg
  - Dornier Unternehmensarchiv EADS Corporate Heritage, Friedrichshafen
  - Mercedes-Benz Classic Archiv Daimler AG, Stuttgart
  - Politisches Archiv Auswärtiges Amt, Berlin
  - Stadtarchiv Heilbronn
- United Kingdom
  - Imperial War Museums, Duxford and London
  - National Archives, Kew
  - Royal Aeronautical Society, London
- United States
  - Glenn H. Curtiss Museum, Hammondsport, NY
  - National Archives, Washington, DC
  - Smithsonian Institution Archives, Washington, DC

Japan: Major Airfields and Arsenals

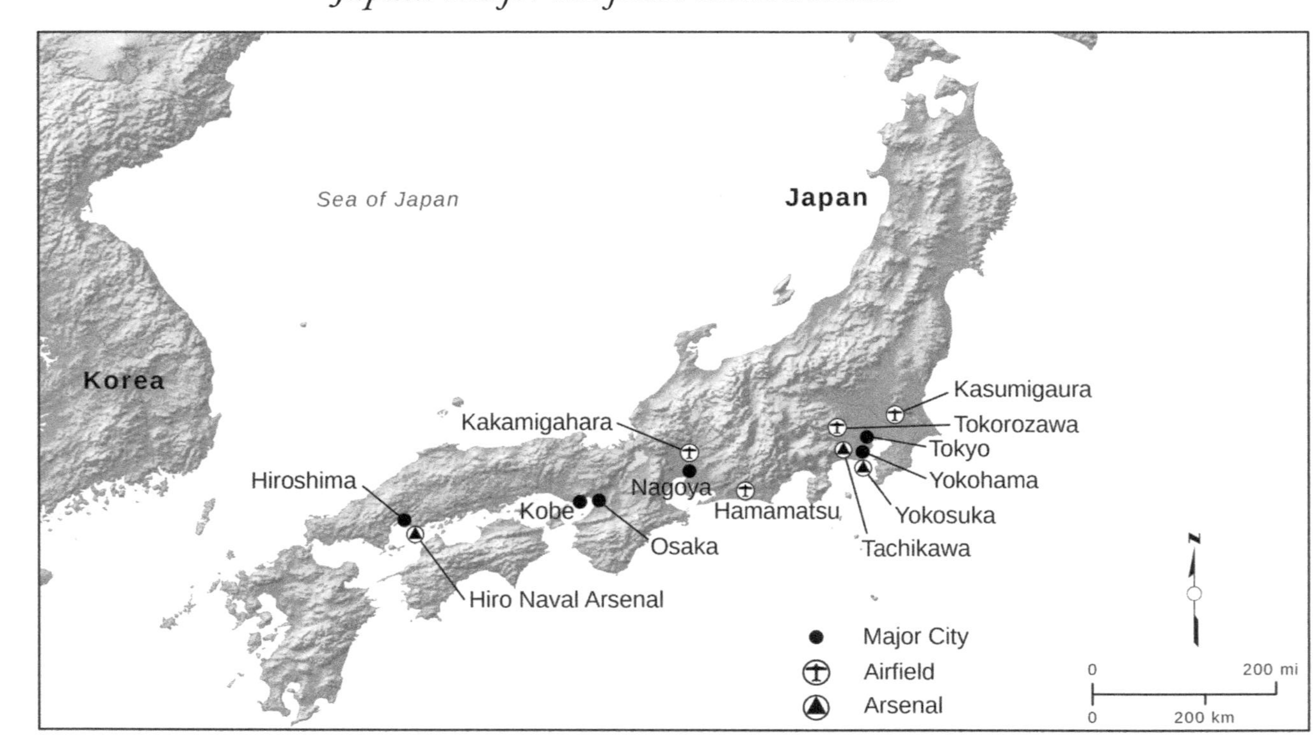

INTRODUCTION

A New Perspective on Japanese Aviation History

This book examines the eventful trajectory of Japan's aviation from its first balloon launch in 1877 until summer 1945, when Japan's first rocket-driven aircraft took off in the final stages of World War II. The history of Japanese aviation offers many stories of heroic achievements and dismal failures, of boundless enthusiasm and sheer terror, of brilliant ideas and fatally flawed strategies. For navigating such a seemingly labyrinthine field, this book offers a bird's-eye view that allows us to look at the history of Japan's aviation from a new perspective. First, instead of presenting the conventional pre-1914, World War I, interwar, and World War II periodization, the narrative arc follows the successive reshaping of Japan's aviation under French, British, German, and American influence. Second, I base my methodology on the assumption that the history of Japanese aviation can only be understood within a broader frame of reference. This approach conceives of aviation as a technological system that is firmly embedded in a wider historical context.

Contextual aviation history lends itself to a transnational approach that directs its focus on "movements, flows, and circulations" of objects, people, and ideas.[1] It allows us to transcend state boundaries and identify transnational actors, connections, and dependencies. It also challenges the assumption of aeronautical advance as a predominantly Western domain or the notion of a purely unidirectional flow of technological know-how and expertise. Indeed, with the Japanese being "the world's greatest transnational actors and learners,"[2] studying the history of Japanese aviation opens up new avenues to analyze how Japan's access to the flow of aeronautical material, experts, and know-how ultimately led to the loss of Western competitive advantage and military supremacy.

Technology: Transnational Transfer and National Diffusion

Technology transfer runs as a central thread in this book. A focus on transnational patterns in the creation, exchange, and diffusion of technological know-how helps us judge their impact on the rapid advance of aviation in Japan. Economists have proposed a three-stage model that begins with "material transfer," such as the import of machinery and production facilities. "Design transfer" is the next step, where the receiving country acquires know-how from blueprints and patents to set up licensed production. The final stage, "capacity transfer," enables an industry to acquire enough insight into methods of research and development to arrive at independent designs.[3]

This book adds several important aspects to this basic pattern. As the development of aviation in Japan demonstrates, the mastery of large-scale production is another defining feature of a country's technological self-sufficiency. Furthermore, the speed and range of a technology's diffusion are critical for its successful transfer. Political scientists have argued that successful diffusion of technology greatly depends on building up organizational structures that support the introduction of a new technology and the dissemination of knowledge about its use and application.[4] Accordingly, this book devotes special attention to the transfer and establishment of an efficient aircraft production process; the organizational buildup of Japan's airpower; the formation of an air strategy; and the training of engineers, workers, ground crews, and pilots.

Japanese concern about the good of the nation often went hand in hand with anxiety about security in a hostile international environment.[5] These worries mounted when World War I exposed the devastating power of advanced weaponry, such as heavy bombers and attack submarines.

Historians have argued that in the wake of the Great War a new ideology closely tied Japan's fate to the domestic advance of science and technology.[6]

Japan's emerging aviation industry greatly benefited from the military's apprehension about the country's backwardness and the fast progress of Western aviation. Japanese aircraft makers capitalized on the aspirations of army and navy officials to "catch up and surpass" (*oitsuki, oikose*) Western powers. As a result, the industry could count on a steadily increasing flow of orders from the military as its main customer. The military's apprehension of being at the mercy of foreign imports further spurred the quest for technological independence and indigenous innovation. Import substitution through domestic production (*kokusanka*) became the new axiom on which the military built its procurement policy.

Japanese aircraft makers could build on a tradition of acquiring foreign technological expertise that was well established long before the first airplane took off from Japanese soil. Already during the Meiji era (1868–1912), the government dispatched an increasing number of students overseas and established a system for recruiting foreign specialists. As a result, between 1868 and 1900 approximately 2,400 of these experts, who became known as honorable hired foreigners (*oyatoi gaikokujin*), were on the government's payroll.[7] The Ministry of Public Works (Kōbushō) became the largest employer for foreign engineers and technicians, who worked in a variety of fields, including railways, shipbuilding, and communication.[8] Among these *oyatoi* instructors were British, French, and German military officers who became instrumental in the buildup of Meiji Japan's armed forces.[9]

Japan's aeronautical experts, rather than being passive recipients of foreign technology, took an active role that soon surpassed the mere improvement of imported designs. Several research centers greatly contributed to this process. The Aeronautics Department of Tokyo Imperial University opened its Aeronautical Research Institute in 1918.[10] The institute's scientists were exempt from teaching duties and enjoyed considerable leeway in the pursuit of their research. The Japanese Navy established its Naval Aircraft Test Laboratory (Kaigun Kōkūki Shikenjo) in 1918, and in 1919 the army followed suit with a new research division at its Tokorozawa Flight School.

By the late 1920s, Japanese engineers could base their designs on scientific evidence, numerical analysis, and systematic wind tunnel experimentation. Their expertise soon reached a level where they designed a new generation of aircraft that successfully demonstrated full independence from foreign technology. By the start of the Pacific War, Japanese aeronautical engineers systematically developed airplanes that matched and often surpassed those of their US counterparts. Very much to the Allied forces' surprise, Japan's aviation technology had taken off.

Mobilizing the Public

Popular aviation enthusiasm is another thread that connects the stages of Japan's aviation history. To be sure, fascination with human flight was and is a transnational phenomenon.[11] Historians have explored how, within a few decades, technology revolutionized our notions of space and time.[12] Aircraft intensified these experiences by providing new perspectives and new thrills. As an artifact that was initially built by enthusiasts for no utilitarian purpose, the aircraft's image was elusive enough to promise both national identity and international brotherhood, quasi-religious redemption, and the triumph of modern rationality.[13] The aircraft also stirred an aggressive enthusiasm for technology, like that of Italian Futurist Filippo Marinetti (1876–1944), who, in his 1909 *Futurist Manifesto*, included the aeroplane in his glorification of war.[14] For architect and designer Le Corbusier (1887–1965) the airplane was "the symbol of the New Age" that can attract the masses with "sensational demonstrations."[15]

Sociologists of science and technology have argued that "the socio-cultural and political situation of a social group shapes its norms and values, which in turn influence the meaning given to an artifact."[16] Yet in the case of Japanese aviation, this process worked in both directions. The lure of the aircraft also shaped social groups. Indeed, matter-of-fact technocrats turned into visionaries, and hard-boiled militaries engaged in visions of Japan as an aerial superpower that enticed them to initiate enormously expensive armament projects of questionable military use. The new flying machines brought a sense of euphoria, mythmaking, and ideology. Time and again, the aircraft's image transcended the logic of practicality and replaced utilitarian thinking with images of military supremacy and invincibility.

The airpower advocates in the Japanese military successfully included the public in the national aviation project. They effectively linked the fate of Japan's airpower with national security and ardent patriotism. Balloon launches, air shows, and patriotic donation campaigns fired the imagination of the general public. Japanese national aviation enthusiasm intensified and ultimately reached epic proportions. It turned out that an advanced technology, considered the embodiment of the rational-industrial age, had the power to evoke intense reactions usually associated with a preindustrial past. Transcending the lure of mechanic novelty or sensationalism, airplanes became effective vehicles to propel national pride and identity.

Japanese mass media played a central role in stirring, bolstering, and intensifying public enthusiasm for aviation. Previous studies have examined the government's influence on Japan's media and how it intensified during the establishment of military rule.[17] This book widens the scenario. It shows how the nation's leading newspapers actively participated in the military's aviation project from its very beginnings. A mutually beneficial relationship evolved in which the military provided the news that the press then enthusiastically presented to the eager public. Close collaboration with the press became the military's key strategy for effectively staging the spectacle of flight and creating an atmosphere of intense popular participation. Most major newspapers engaged themselves in the aviation enterprise. Their efforts culminated in wildly popular international goodwill flights that further stirred air-mindedness at home and propagated the high standard of Japan's aviation abroad.

Aircraft and Statecraft: Aerial Armament and International Diplomacy

As a third unit of analysis, this book examines the reciprocal influence of Japan's aviation policy and international relations. Earlier scholarship on Japan's geostrategic decision making emphasized the quest for economic self-sufficiency and the role of technocratic administrators.[18] This book adds a new perspective by examining how aeronautical advance shaped and was shaped by international diplomacy and military strategy.

The diplomatic and geostrategic ramifications of World War I had a profound impact on the development of Japan's aviation. With the 1919 Treaty of Versailles, the Japanese gained unrestricted access to Germany's aviation technology. Furthermore, because the treaty regulations prohibited Germany from building any military aircraft, nearly all prominent German aircraft manufacturers started licensing negotiations with Japanese companies and began to send their engineers abroad. After the end of the war, the aviation industries of the victorious Allied powers went into a decline as well. To dispose of wartime surplus production and secure large-scale orders for their industries, France and Britain competed for a dominant influence in the buildup of Japanese airpower. Japanese military and industrialists skillfully took advantage of these international rivalries to selectively import aeronautical material and know-how.

After the horrors of World War I, international diplomacy concentrated on peace and arms reduction. During the interwar period, Japan signed two international disarmament treaties that ironically resulted in a significant expansion of the country's airpower. In 1922, the Washington Naval Treaty banned the completion of two Japanese battleships. Instead of scrapping the half-finished hulls of these ships, the navy converted them into aircraft carriers and started an ambitious program for carrier-based aircraft. The 1930 London Naval Treaty limited the number of submarines and light cruisers but did not set any restrictions on airpower. As a result, the Ministry of the Navy redirected appropriations into a large-scale expansion of the Imperial Japanese Navy's air-strike capability.

Paradoxically, the more Japan expanded its airpower to widen its scope of action, the less room for diplomatic maneuvering remained. During Japan's 1931 occupation of Manchuria, the army's new generation of heavy bombers engaged in large-scale bombing campaigns. Chinese media and US officials especially condemned Japan's ruthless air raids on cities behind the front line.[19] Ultimately the controversy over this violation of China's sovereignty led to Japan's withdrawal from the League of Nations in 1933 and the country's decision for an "autonomous diplomacy."[20]

While the navy was expanding its airpower in preparation for the great battle over hegemony in the Pacific, the Japanese Army's dilemma in China intensified. After the outbreak of the Second Sino-Japanese War in 1937, the Japanese military increasingly depended on airpower, especially long-range bombers for air raids on the Chinese hinterlands. The massive bombing of civilians antagonized the US public and accelerated a series of embargoes against Japan's aviation industry that culminated in the oil embargo of August 1941. Thus, the Pearl Harbor attack became emblematic of the breakdown of international relations, a clock-running-out scenario of dwindling resources, and an aggressive military doctrine based on long-range air strikes.

During the initial phase of the Pacific War, the enormous flight range of Japanese bombers, fighters, and reconnaissance aircraft offered a geostrategic advantage in the vastness of the Pacific and the Asian continent. However, beginning in June 1944, large-scale Allied bomb attacks created havoc in Japan's aviation industry. Under these dire circumstances, Japanese engineers responded to the military's demands with increasingly radical designs in attempts to turn the tide of the war. Jet fighters and rocket-powered aircraft were to intercept high-flying B-29 bombers. During the final phase of the war, when an Allied invasion became imminent, jet-propelled flying bombs were to "send the enemy's invasion troops to the ocean floor."[21] Catapulted out of their secret hideouts, these special suicide attackers became the ultimate Japanese aeronautical innovation.

PART I

Early Japanese Aviation, 1877–1918

CHAPTER I

Powerful Images and Grand Visions

A short glance at the many illustrated books about early aviation suggests an almost linear development that begins with balloons, moves forward to clumsy flying wireframes, and culminates in streamlined, powerful, efficient machines. Of course the history of flight did not proceed along a perfectly straight line. This holds especially true for early aviation in Japan.

Historians of technology have examined the means—or "vehicles"—of technology transfer, such as imported machinery, engineers, and technical publications.[1] While acknowledging the importance of these issues, this chapter goes a step further and looks into the motives—or the "motor"—for the transfer of aviation technology to Japan. It deals with the question why the Japanese—without any pressing need—imported a technology that so far had not proven its practical value and for which there existed neither an organizational nor logistical structure.

This chapter sets out to demonstrate that Japan's choice of technology was largely the result of powerful images that heightened the public's perception and sparked the experts' imaginations. Images of a flying machine created visions of national grandeur and military supremacy. An analysis of Japan's early aeronautic imagery can help explain the motives behind the emergence of the country's aviation. The imagery of a spectacular technology proved to be remarkably effective in arousing a widespread aviation enthusiasm among the Japanese public. Popular participation in flight demonstrations, exhilarated press reports about "courageous air battles," and the military's call for a public "aviation craze" built the foundation for the country's new air-mindedness that secured voluntary and even passionate support for the new aviation project.[2]

From its early beginnings, the rise of Japanese aviation was closely linked to the modernization and buildup of the military. As I lay out in this chapter, the Japanese Navy and Army were the main actors in promoting the launch of a manned balloon. Therefore, an excursus on the formation of the armed forces during the early Meiji period and the developing dynamics between them will help us better understand the ups and downs of early aviation in Japan. After their formation in 1868, the Imperial Japanese Army and Navy developed along different lines. Initially the army strongly relied on the help of French advisers, and its organization, doctrine, and training followed the model of the French army. The Ministry of the Army (Rikugunshō), established in 1872, oversaw the military's administration and command. Using the German military as a model, the army established a fully autonomous general staff. With this move, the army leadership could evade any civilian influence.[3]

In contrast to the army, the navy decided to emulate the English naval forces. Initially, the Meiji leaders, who had little fear of a foreign invasion and no immediate plans for imperialistic expansion, were reluctant to build up a naval power under autonomous command. Only in 1893 did the admirals' efforts for an independent navy general staff show success.[4] During several decades of scrambling for power, influence, and funding, the navy crafted a reputation for promoting technical innovation and training elite technicians.[5] Naval shipyards and arsenals turned into centers for developing new weaponry and production machinery.

In spite of their differences and rivalry, the army and navy of Meiji Japan shared a common problem: estrangement from the rest of society. The army struggled with widespread opposition against conscription, which was institutionalized in 1872. The navy was even more isolated from the public, both geographically (by having its few bases in port towns) and culturally (by cultivating its distinct Britishness). It is therefore easy to hypothesize that the officials of the early Meiji military would welcome any opportunity to foster much-needed civilian popularity.

Indeed, during the early years of the Meiji period, the Japanese military took the initiative to build and launch Japan's first flying machine.

A spectacular public balloon launch held the promise of generating positive publicity and improving the image of the armed forces. Thus, while the history of Japan's early ballooning is interesting in its own right, it also reveals much about the emerging dynamics and interaction among the military, the media, and the public that shaped Japanese aviation throughout its turbulent history.

Early Gas Balloons: Low Tech, High Risk

Building a balloon, even with the technology available in mid-nineteenth-century Japan, was astonishingly simple. Balloon technology—unlike other advanced machinery of the early Meiji era, such as locomotives and telegraphs—could be easily indigenized. *Washi* is a paper made from the bark of mulberry trees. A balloon maker would glue several layers of *washi* sheets together to create a gas container that was strong and airtight enough to function as a balloon envelope.

To make the balloon lighter than air, hydrogen was the first choice, despite its high explosiveness. The alternative, a balloon filled with hot air, would require an envelope at least four times bigger. A hot-air balloon also would have to carry either a burner with the necessary fuel or would have a very limited flight time because of the gradual cooling of the air inside the hull. (Helium, the standard gas used in today's balloons, was not commercially available until the twentieth century.)

Hydrogen gas production is a straightforward process that Paracelsus (c. 1493–1541) described as early as the sixteenth century. When metal comes into contact with a strong acid, the metal ions displace hydrogen from the acid, thus releasing hydrogen gas. What may be called a hydrogen generator consists of nothing more than an acid-resistant container that is large enough to accommodate a sufficient quantity of scrap metal over which sulfuric or hydrochloric acid is poured.

Controlling the altitude of the balloon was another challenge. The obvious solution would be dumping ballast for gaining altitude and a delicate device that allowed for the controlled escape of gas to descend. In most cases, the preferred alternative was a simple mooring line that would leave it up to the ground crew to control the balloon's height.

This brief description already hints at two of the most prominent risks a prospective nineteenth-century balloonist had to face. One was the constant hazard of a most violent explosion right above his head; the other was the total loss of control of the aircraft because of a ground-crew failure, leaving him no choice but to be literally gone with the wind.

Japan's First Balloon Launch

The Japanese Naval Academy (Kaigun Heigakkō) played a central role in the launch of Japan's first flying machine. The academy had been established in 1869 at Tokyo's Tsukiji district, a stretch of reclaimed land near the mouth of the Sumida River. In April 1876, one of the academy's engineers, Baba Shinpachi, succeeded in launching a small-scale captive balloon.[6] In the following year, the academy's balloon project came to national prominence as a venture to guarantee the survival of the young Meiji state. The Satsuma Rebellion, an uprising of discontented samurai, turned into one of the most serious challenges to the new Meiji government. In early 1877, Saigō Takamori (1828–77), also known as the last samurai, gathered a rebel militia of about 40,000 men who were determined to march against Tokyo. To counter this threat, the government sent its conscript army, which was to be supported by the most modern technology available: heavy artillery, telegraphs for communication, and a reconnaissance balloon for observing the enemy's activities behind the front line.[7]

Aware of the naval academy's balloon project, the army minister asked for the navy's help. In April 1877, naval academy engineer Azabu Buhei and his team began building Japan's first reconnaissance balloon that featured an advanced silk envelope covered with a thin layer of rubber.[8] By the next month the balloon was ready for its maiden flight. The navy sensed an opportunity to boost its public image and allowed the press to closely cover the launch of Japan's first manned balloon. Japanese newspapers eagerly accepted the offer. Even before the official event they raised the public's expectations with several reports about the impending launch. Finally, on May 22, *Yomiuri Shinbun* reported "the first time ever launch into the sky of a made-in-Japan balloon" in front of the Ministry of the

Navy at Tsukiji.[9] According to the article, five people, one by one, starting with Mr. Azabu, took a ride into the sky. Colored flags were used to pass commands from the balloon basket to the ground crew: red for going up, blue for going down, and white for stop. The article appealed to readers' national pride, emphasizing that the balloon was built and launched "without any foreigner's assistance" and concluding with utmost praise for the successful accomplishment, calling it "splendid" (*appare*).[10]

Newspapers were not the only media to cover the launch. Within a week a large number of woodblock prints by Hiroshige III (1841–94) circulated widely. The prints effectively captured the excitement of the launch (figure 1.1). From a distance, a huge crowd is watching the balloon with its Rising Sun Flag. In the foreground, an excited group in Western clothes cheers at the balloonist, who is courageously waving the red flag, signaling his readiness for further ascent. As a typical example of a "Civilization and Enlightenment picture" (*kaika-e*), the print teemed with national pride and praise of technological progress. The artist could not resist the temptation to put into an already densely packed picture yet another epitome of Japanese modernity, the steam locomotive.

While the navy was still celebrating its successful balloon flight and enjoying its widespread publicity, the army was still fighting a civil war. The army leaders forcefully reminded the naval academy that the balloon's original purpose was to be deployed in the campaign against Saigō Takamori and his rebels. Navy officials expressed their worries that their delicate flying machine might suffer some damage and requested the army not to use it too close to the battlefront. In response to these hesitations, by the end of June the army decided not to use the balloon at all, arguing that its troops had already considerably weakened the enemy's force.

In an obvious attempt to shake off its technological dependence, the army decided to build its own balloon. Already in April 1877 Uehara Rokushirō, a teacher at the army academy, was in put charge of the project. Generously funded with ¥10,000, Uehara received instructions to build an even larger balloon to carry two passengers. Whereas it took the navy only a month to get its new balloon into the air, the army's project proceeded much more slowly. On June 21, 1877, Uehara submitted a construction plan for a balloon that could carry a payload of more than 300 kilograms. He proposed to fabricate an airtight envelope from unbleached silk covered with a paste made from the starch of the *konnyaku*

FIGURE I.1 This 1877 triptych print shows Japan's first balloon launch at the Ministry of the Navy's Tsukiji parade grounds. The center panel gains its dynamic tension from portraying a balloonist who is still reassuringly close to his ground crew, while the upper part of the balloon's hull has already escaped the confines of the picture. (Source © Tokyo Gas Co., Ltd. Gas Museum)

plant. The hull would be covered by a net on which a parachute-type safety device (*anzen-gasa*) could hang. In an emergency, the balloonists could detach this "safety umbrella" together with the basket from the rest of the balloon and land safely. The balloon would also feature a telescope, an anemometer, and a camera.[11]

It took Uehara and his team a year to finish their balloon. The army presented its new flying machine at the formal opening of a new building of the army academy. On June 10, 1878, more than a year after the navy's successful flight, the balloon was launched successfully in the presence of the emperor. The event was duly reported by newspapers and even documented by a photograph that—unlike the colorful, action-packed woodblock print—showed the launch taking place in a rather austere ceremony, a far cry from the carnival atmosphere of the navy launch. Nevertheless, this picture found its way into many textbooks as documenting a milestone of aviation history in Japan.

Balloon Fever Grips Japan

In the late 1870s, "balloon fever" left the military's test grounds and spilled over into the public. Balloonists and their crews could build on a long-standing Japanese tradition of spectacle shows (*misemono*). Such performances became increasingly popular in the nineteenth century, often presenting wonders of Western technology, such as telescopes, panorama theaters, or steam-driven flywheels.[12] Thus, soon after the military's first successful flights, civilians attempted a whole series of public balloon launches with mixed motives and results. To "promote scientific education," the Kyoto city government appointed industrialist Shimazu Genzō (1839–94) to build and launch a manned balloon.[13] In December 1877 a crowd of 50,000 people gathered in front of the former Imperial Palace to watch Shimazu's balloon climb to a height of thirty-six meters (figure 1.2). Soon ballooning became an attraction that promised considerable financial profit. In early January 1878, a public balloon flight was announced in Osaka, and tickets were sold for 3 *sen* each. However, to the great disappointment and anger of the crowd, even after several trials the balloon did not take off. A Nagoya newspaper reported a similar

FIGURE I.2 This illustration presents Shimazu Genzō's balloon launch in Kyoto. The artist depicts in great detail the filling of the gas generator on the left side, from where a pipeline delivers hydrogen to the balloon's hull. He also adds a humorous note with two putti who keep the print title floating in the air long while the launching crew still firmly holds the balloon to the ground. (Source © Shimadzu Foundation Memorial Hall)

event in March 1878, when a flight there also ended in failure; this time the infuriated spectators pressed the organizers to refund their money.

In the following years, the novelty of balloon ascents wore off. In the early 1890s, breakneck aerial stunts became the latest fad. In fall 1890, English balloonist Percival Spencer (1864–1913) came to Japan to perform his daredevil stunts in Yokohama and Tokyo. Accompanied by the blaring music of a band, he climbed with his balloon to an altitude of around 200 meters. He then jumped out of the gondola, holding onto a parachute with one hand while waving his hat with the other. Spencer performed his most notable public flight in front of the Tokyo Imperial Palace, where he even attracted the attention of the emperor.

The aerobatics of American Thomas Scott Baldwin (1854–1923) were even more circus-like. During the balloon's ascent, Baldwin performed various acrobatic tricks on a wooden horizontal bar. To further attract the crowd's attention, he had contrived a special device that discharged a trail of dense black smoke during ascent. Baldwin's balloon performances created a sensation in Tokyo, inspiring colored prints, balloon-shaped crackers, and ornamental hairpins that became bestsellers. The balloon even made it to the stage. Kabuki star Onoe Kikugorō V (1844–1903), who had met Spencer in person, enthralled his audience by impersonating the intrepid Englishman.

Balloons in the Russo-Japanese War and a Technological Dead-End

While the balloon's image became deeply entrenched in popular culture, the Japanese Army further developed its aerial machinery. In 1890 the army purchased a balloon from French manufacturer Gabriel Yon (1835–94), who built a reputation with his high-altitude reconnaissance balloons. However, during the long sea transport from France to Japan, the balloon's hull decomposed, convincing discontented army officials to rely on their own production methods.[14]

The Japanese Army's pursuit of technological independence received a decisive stimulus from a local innovation. In 1897, the inventor Yamada Isaburō (1864–1913) discovered a new way to improve existing balloon technology with the design principles of traditional Japanese kites. Yamada had the idea to modify the balloon's spherical shape into an ellipsoid with a stabilizing fin at its tail. Such a kite-like balloon could deflect the wind and generate additional lift, allowing for a smaller hull with less gas volume. In 1900 Yamada obtained a patent for his new kite-type (*tako shiki*) balloon. Based on this invention, the army built its own kite-type balloon. On December 23, 1901, the test flight in the presence of the Meiji emperor was a spectacular success. Two army officers climbed to a new record altitude of 500 meters. They stayed in direct contact with the ground crew using a telephone cable attached to the mooring rope. This innovative arrangement was Japan's first air-to-ground communication, which greatly enhanced the balloon's effectiveness for reconnaissance missions. During several maneuvers, the Yamada-type balloon underwent further testing where it proved its military value.

In February 8, 1904, the Japanese fleet launched a surprise attack on Russian warships outside Port Arthur at the southern tip of the Liaodong Peninsula. In an effort to secure control over Manchuria and Korea, Japan waged its first war against a major European power. In May 1904, the Japanese Second Army started a series of land assaults on the Russian stronghold at Port Arthur. Soon the need for effective airborne observation became more pressing. However, budget restrictions imposed by the Ministry of Finance along with long transport times precluded the import of foreign-made balloons.[15] Army officials once more turned to Yamada, passing an urgent order to build another two of his "purely made-in-Japan" (*junkokusan*) balloons. In less than two months, Yamada designed, built, and delivered two advanced observation balloons. Their hulls consisted of two layers of light, high-quality silk, and their improved shape provided additional lift.

In July 1904, the army's newly established provisional balloon unit (*rinji kikyūtai*) left for China to participate in the Siege of Port Arthur. On August 17, two days before the second large-scale attack, the balloon unit started its reconnaissance operations, during which the crews successfully obtained information about the location of the Russian command center, troop movements, and mine-laying operations. The Japanese Army was well aware of the vital role that Yamada's reconnaissance balloons had played in the surrender of the Port Arthur garrison: Minister of the Army Terauchi Masatake (1852–1919) granted Yamada the Sixth

Order of Merit "for the development of his balloon . . . which greatly contributed to the success of the Port Arthur Siege operations."[16]

The prestigious award inspired Yamada to take his aeronautical endeavors a step further and build his first engine-powered airship. Such a dirigible balloon's radius of action would no longer be limited by the mooring rope. Its motor would provide much greater control and allow for air travel independent of wind direction. Yamada used a hull shape similar to his earlier kite balloons, to which he attached a new type of a gondola that could house one aeronaut together with a fourteen horsepower engine.

A series of troubles, setbacks, and accidents gave rise to the suspicion that Yamada's latest invention might lead to a technological dead-end. In September 1910, a gas leak ended the first test flight. Within a few months, Yamada completed an improved airship with a much more powerful fifty horsepower engine (figure 1.3). On February 8, 1911, army and navy officers, staff from the Imperial Tokyo University, civil researchers, and

FIGURE 1.3 This photograph, taken in early February 1911, shows the Yamada Airship no. 2 before its first launch. Spectators are peeping through holes; others had made themselves comfortable on top of the fence. On February 23, during a test flight, the airship's gasbag exploded over Tokyo Bay. (Source © Japan Aeronautic Association)

journalists, along with several thousand spectators, came to watch the flight. The airship with its crew of four took off and began its flight. But soon the engine stopped, and the now rudderless airship began to drift. In their distress, the crew lowered a rope; the ground team and a group of soldiers grabbed it, but the airship dragged them along and began to cut several electric wires. Only after more soldiers joined to pull the rope could the airship finally be tugged to the ground.

A few days later, during another test flight, a strong gust smashed the airship against its hangar. The gasbag separated from the gondola; it went up into the sky, drifted toward the open sea, and finally exploded. Its burning remains fell into Tokyo Bay. After this series of tragicomic mishaps and serious accidents, people started to say, "Yamada's airships are hopeless" (*Yamadashiki hikōsen wa dameda*), and the press soon lost interest. Yamada's fall from favor might have been a case of the public's fickleness. More important, it led to what might be called a paradigm shift in Japan's aviation technology: away from lighter-than-air balloons and airships to winged airplanes.

Two Reports about Western Aviation

The year 1909 was decisive for the future development of Japan's military aviation. In March 1909, naval officer Yamamoto Eisuke handed his "Opinion Paper on Aviation Research" to his superior, Captain Yamaya Tanin (1866–1940). The report highlighted the fast progress of Western aviation technology and warned about the "horrible power" of future "battleships in the air" (*kūchū gunkan*).[17] Obviously impressed with Yamamoto's analysis, Yamaya approached the army's general staff office within the same month and proposed joint army-navy cooperation in aviation research.

For their part, the army planners had already started similar investigations. In May 1909, Kawata Akiharu, a captain in the army general staff, submitted a memorandum on "Aerial Weaponry" (*kūchū heiki*) to Minister Terauchi. Using foreign publications and reports from Japanese military attachés based abroad, Kawata had collected and analyzed information about the progress of foreign military aviation. After a detailed comparison of airships and airplanes, Kawata concluded that airplanes were the advanced weapon of the future. Unlike the cumbersome dirigibles, planes combined maneuverability with endurance and speed. Kawata could count on the army minister's attention and support. Already in the previous year, Terauchi had expressed his concern over the poor state of Japan's aviation and convinced the Diet to appropriate ¥600,000 for a "prospective aerial war."[18] Within a few days of receiving Kawata's report, Terauchi consulted with Minister of the Navy Saitō Makoto (1858–1936) about setting up Japan's first organization for aviation research.

The ministers agreed to establish a joint army-navy project, the Provisional Committee for Military Balloon Research (Rinji Gunyō Kikyū Kenkyū Kai), or Balloon Committee for short. The committee's somewhat anachronistic name reflects the dominance of balloons as tools of aerial warfare at the time of its founding. Despite its outdated name, the new organization played a key role in the rapid advance of Japan's military aviation.

Such unprecedented interservice cooperation might come as a surprise—even more so if we consider that the long-standing rivalry between the Imperial Japanese Army and Navy had intensified during and after the 1904–5 Russo-Japanese War. After Admiral Tōgō Heihachirō (1848–1934) and his sailors had won a stunning victory over the Russian fleet in 1905, the Imperial Japanese Navy's popularity and political clout increased dramatically. Military historians have pointed out that the navy's successful political maneuvering and popular support secured an increasingly larger portion of the government's budget for naval armament. At the same time, it fomented strife with the army over appropriations for further expansion.[19]

The navy committed itself to the new Balloon Committee with little enthusiasm, limited manpower, and no funding. Thus, from the outset, the army was in firm control of the Balloon Committee. After the successful deployment of its reconnaissance balloons, the army could easily capitalize on its technological lead in military aviation. Furthermore, the air-minded army minister could secure enough funds to cover the Balloon Committee's entire budget. As a result, Nagaoka Gaishi (1858–1933), the head of the army's Bureau of Military Affairs and well known for his enormous, propeller-like moustache, became the committee's chair.

Eleven of the twenty committee members belonged to the army. Six naval officers and three scholars from Tokyo Imperial University joined the select group.

Still, the navy's restraint from engaging more actively in aeronautical research and in the buildup of airpower needs further explanation—especially if we recall that throughout the Meiji period the Imperial Japanese Navy had built a reputation for openness to technical innovation. In 1909, the same year the Balloon Committee came into being, the Kure and Yokosuka Naval Arsenals laid down Japan's first dreadnought battleships, the *Settsu* and *Kawachi*.[20] These warships incorporated the latest advances in propulsion, armor, and firepower. They clearly show that Japan's naval planners opted for the fast advance of battleship technology rather than for engaging in any exotic machinery. Indeed, one source claims that there was still widespread skepticism among senior naval officers about the value of land-based aircraft in future sea battles.[21] Such a proposition seems plausible, considering that at the time of the Balloon Committee's founding, the aviation world had not yet seen the flight of the first seaplane.[22] In any case, appropriations clearly show that air-mindedness among the navy's leaders had a slow start. Until 1917, the navy's annual aviation-related budget amounted to only a fourth of the army's expenditure.[23]

The Road to Japan's First Motorized Flight

Two central themes that Kawata emphasized in his report deserve close attention. His far-sighted insistence on the superiority of airplanes over airships helped Japan avoid a technological dead-end. As the misfortunes of Yamada's airships would vividly illustrate, airships might not be an avenue worth exploring. Furthermore, in his detailed analysis of Western aviation, Kawata acknowledged France's leading position in the aviation world, with Germany in second place. He supported his assessment with a comparison of each country's number of military airplanes and crews.[24] An updated version of Kawata's report, published under the title "Aerial Flying Devices" (*kūchū hikōki*) in October 1909, introduced the latest airplanes designed by Louis Blériot (1872–1936) and Henri Farman (1874–1958) as exemplars of advanced French aircraft design.

Kawata's focus on French aviation shows that he was clearly aware that the center of aeronautical innovation had shifted. By 1909 France had taken over the leadership from the United States. After the Wright brothers' seminal flight in December 1903, the US military failed to see any military potential for the Wright Flyer. Only in 1908, when the Wright brothers were negotiating with European countries, did the US War Department finally decide to buy one of their flying machines. However, in the same year the Wrights began their protracted legal disputes with Glenn Curtiss (1878–1930) over the design of aircraft flight control. The Wright brothers' patent war lasted for nearly a decade and greatly obstructed the development of US aircraft manufacturing.[25]

In 1909, several milestone events drew the attention of the world's aviation experts to France. In this year the French army was the first military force to introduce airplanes as a weapon of war. It also pioneered creating a comprehensive organization for selecting and training pilots and acquiring and maintaining aircraft. Furthermore, the French general staff outlined a new doctrine and command structure to make best use of the new possibilities offered by the flying machines. These efforts received valuable popular support, whipped up by the aviation pioneer Clément Ader (1841–1925), who published his book *L'Aviation Militaire* (Military Aviation) the same year. Ader's bold vision about airplanes revolutionizing modern warfare fascinated the French public, and the book soon became a bestseller.

The outstanding event of 1909 was Blériot's successful crossing of the English Channel on July 25. With his achievement, the French pilot not only proved the feasibility of sustained powered flight but also aroused an unprecedented aviation craze on both sides of the English Channel. Upon his return to France, an estimated 100,000 people enthusiastically welcomed the man who demonstrated that Great Britain was "no longer an island."[26] Blériot's flight across the channel received extensive news coverage in Japan's newspapers as well. One article reported the "mad enthusiasm" (*kyōkiteki nesshin*) of the French about the feat of their countryman. At the same time, another report raised questions about the backwardness of Japan's aviation research, with the author expressing his hope that "the Japanese rapidly learn this [French] spirit of inquiry."[27]

After Blériot's landmark achievement, the number of French aviation aficionados increased rapidly. In August 1909, half a million paying visitors rushed to the Reims Air Meet, or the Grande semaine d'aviation de la Champagne, as it then was called. With the French president attending and under the watchful eyes of a delegation of the War Ministry, twenty-five aviators from six different countries competed for cash prizes for the fastest, highest, and longest flights. The stunning performances of the new flying machines were celebrated by the general public as a feat of progress. For experts, they proved that aviation was in a transition from a breakneck sport to a substantial means of transport and warfare. To nationalists, the overwhelming success of the French pilots was further proof of the "glory and the honor of our race."[28]

The driving forces behind these developments—even more than the courage of the daredevil pilots—were significant technological improvements conceived and worked out by French aircraft designers. Blériot with his monoplane already anticipated the standard arrangement of motor, wings, and elevator of future aircraft. Farman became one of Blériot's fiercest competitors. Farman gained international fame for establishing a new world distance record of 180 kilometers with his Farman III biplane at the 1909 Reims meeting. The Farman III was a remarkable aircraft that could be steered by means of four ailerons at the two wings' trailing edges. This innovative device was a decisive step toward improved maneuverability and avoided the notorious instability of the Wright brothers' method of wing warping for directional control. Farman built about thirty specimens of this aircraft. It became an international bestseller that soon appeared in the skies of Belgium, England, Hungary—and Japan.[29]

THE DISPATCH OF JAPAN'S FIRST STUDENT PILOTS

The Balloon Committee's members were fully aware of the rapid progress in European aviation, and their work picked up momentum. On April 8, 1910, the army informed Japan's Foreign Ministry about the dispatch of Captain Hino Kumazō (1878–1946) and Captain Tokugawa Yoshitoshi (1884–1963) to Europe. Hino belonged to the army's Technical Research Group (Rikugun Gijutsu Shinsabu).[30] He was a gifted engineer and inventor who specialized in constructing firearms before turning his interest toward designing airplanes. In 1909, Hino began the construction of the Hino Type 1, a monoplane made from bamboo and cypress wood. However, due to its weak eight horsepower engine, the aircraft was never able to take off.[31] Nevertheless, Hino's efforts attracted the attention of the Balloon Committee, which made him a member in August 1909. In March 1910, Tokugawa joined the committee as well. His father was Count Tokugawa Atsumori (1856–1924), the head of one of the three branches of the once-powerful Tokugawa clan.[32] In addition to his illustrious ancestry, Tokugawa was well connected. His superior officer in the army's Balloon Corps, along with committee chairman Nagaoka, recommended that he become a committee member.[33] Some sources speculate that Tokugawa was promised an opportunity to promote the prestige of his lineage by becoming Japan's first pilot.[34]

The two officers were to receive flight training in France and Germany and purchase a variety of the latest aircraft models.[35] The Balloon Committee granted Tokugawa and Hino considerable leeway to choose the most suitable aircraft types with the proviso that the new machines must be easy to operate and combine reliability with high performance. The committee required that the men be able to fly any airplane purchased abroad and that all such planes could be maintained and repaired in Japan.[36]

After his arrival in France in late April 1910, Tokugawa set out for the small town of Étampes, sixty kilometers south of Paris. There he became one of the first students of the École de Pilotage de Henri Farman, Farman's newly founded flight school. Tokugawa described in a detailed account how the training started. Sitting behind his flight instructor he was clinging with one hand to the aircraft. With the other hand, he reached over the instructor's shoulder in order to touch the control column lightly and to get the feeling of how his teacher steered the plane. Several flights later, after Tokugawa had become proficient enough, his trainer allowed him to move to the front seat and steer the aircraft by himself (figure 1.4). Tokugawa described in detail his first solo flight and how he strongly felt the loneliness of the pilot fending for himself. He also mentioned that much to his surprise, the aircraft was much easier to handle without the corpulent instructor on board. After a few more solo flights, Tokugawa was ready to take the final examination, for which he had to buy the aircraft for the steep price of ¥16,000.[37] Apparently Tokugawa

FIGURE 1.4 This photograph taken in summer 1910 near Étampes, France, shows student pilot Tokugawa Yoshitoshi. The Farman biplane's gossamer upper wing stands in striking contrast to the huge pusher propeller behind the airman. (Source © Japan Aeronautic Association)

passed the check flight with flying colors and received his French pilot license with the serial number 289 on October 8, 1910.[38]

The Balloon Committee's decision to send the country's first student pilots to France and Germany was well founded. Even though France established itself firmly as the leading aviation nation, German aerial armament had reached a level that caused considerable French worries. Warnings that Germany would build up its airpower with "Teutonic tenacity" evoked an image of France losing its technological edge to an enemy that would be able to ignore the Rhine River as France's natural defense barrier and strike from the sky.[39] Even though the German military had shown a keen interest in the giant airships of German aviation pioneer Count Ferdinand von Zeppelin (1838–1917), it initially was reluctant to invest in airplanes. Instead, a private initiative promoted the development of made-in-Germany aircraft. In spring 1908, German industrialist Karl Lanz (1873–1921) offered an award of 40,000 marks to the first German pilot who successfully flew a flat figure eight around two pylons, placed 1,000 meters apart. An additional requirement stipulated that the aircraft had to be produced in Germany and powered by a German motor. More than a year later, Hans Grade (1879–1946), an engineer who had specialized in the development of lightweight engines, became the first German pilot to meet this challenge. On October 30, 1909, in front of 2,000 spectators, he completed his flight in less than three minutes. Making use of his prize money and sudden popularity, he founded Germany's first flight school and started producing his prizewinning aircraft.

The Japanese Army assigned Hino to Hans Grade's new flight school at Johannisthal. Located in the southeastern fringes of Berlin, Johannisthal was one of Germany's first airfields and had attracted several aircraft makers who set up their workshops in its vicinity. After Hino completed his flight training, he followed the Balloon Committee's policy and bought one of Grade's aircraft. The Hans Grade monoplane, built with an unusual mix of bamboo and welded steel tubes, was a marvel of lightweight construction. It was also simple and cheap—its price of ¥7,000 was less than half the amount Tokugawa had to pay for his Farman biplane.[40] Designed as a light single-seater, the monoplane had limited military potential. Moreover, the eccentric steering mechanism—the pilot had to operate the airplane with a control stick hanging down from above his head—required a special piloting technique that included a mix of lateral, longitudinal, and twisting movements of the control handle. For a long time, Hino was the only airman in Japan able to fly this aircraft.

THE YOYOGI FLIGHT

In November 1910, two shipments of oversized wooden containers arrived in Yokohama. They contained a precious load from Europe: two French aircraft built by Farman and Blériot and one German Grade monoplane. The dockworkers carefully transferred the giant boxes with their delicate contents to special carts, which were then—in a very old-fashioned way—pulled by oxen to Tokyo. In a balloon hangar that had fallen into disuse, the army's specialists assembled the Farman and Grade aircraft and prepared them for Japan's first motor flight.

The Japanese Army aimed to attract a huge crowd to witness the spectacular event. This was in marked difference to the Wright brothers, who had admitted only a small group of observers to their seminal flight. The Balloon Committee obtained the Ministry of the Army's permission to use the Yoyogi parade ground in the heart of Tokyo for Japan's first flight demonstration.[41] The area provided sufficient space for the flying machines to carry out their ground maneuvering, takeoff, and landing. The open space could accommodate a huge crowd of spectators. Equally important was the fact that, already by the end of the nineteenth century, parade grounds were associated in the public mind with the immensely popular pageantry and ceremonial pomp of military parades, war victory ceremonials, and military reviews.[42] It could be expected that people would come in droves to watch another spectacle there.

Indeed, the forthcoming flight show began to attract increasing attention. It promised the public to participate not only in the country's first motorized flights but also in a sensational competition between Europe's leading aviation nations. German engineering, embodied by a small twenty-four horsepower, 225 kilogram monoplane, was challenging French technology, represented by an impressive fifty horsepower, 600 kilogram two-seater biplane (figure 1.5). On a more personal level, the event could also be advertised as a struggle between the lonely infantryman Hino, who had the reputation of being a more talented technician than a staunch military officer, and the aristocratic Tokugawa, who was a popular figure among his peers and enjoyed the full support of his comrades.

Preparations for the great event began in early December 1910. The parade ground gradually transformed into what can be best described as a quasi-ceremonial space with a short provisional airstrip at its center. Large tents close to the runway served as aircraft hangars. With their enormous white tarpaulins, these hangars cloaked the flying machines in a shroud of mystery. A well-planned layout provided special seating arrangements for the military aviation committee and distinguished civilian guests. The southern part of the parade ground was reserved for military officers and cadets, and the western section was to accommodate various groups and school classes; the remaining space was to be open to the general public.

FIGURE 1.5A & B Japan's first pilots getting ready for takeoff: Tokugawa Yoshitoshi dominating the scene in his French Farman biplane (above) and Hino Kumazō barely visible in his German Grade monoplane. (Source © Courtesy of Makoto Tokugawa and Japan Aeronautic Association)

The final transport of the aircraft from their assembly place to the parade ground occurred at night to avoid obstructions by other vehicles or curious bystanders. The aircraft were put on horizontal ladders attached to wooden poles, which a large number of soldiers then hoisted onto their shoulders. Contemporary observers did not fail to notice the obvious similarity of each airplane to a heavily decorated portable Shinto shrine (*mikoshi*) carried in a similar way during ceremonial festivals.[43]

On December 11, the first test runs of the aircraft engines began. These motors filled the air with a penetrating noise that had never before been heard in Japan.[44] The strange, intermittent sound must have excited the curiosity and expectations of even the more distant bystanders. The next day, spectators could watch the two pilots in their aircraft practicing ground maneuvering. During one of these exercises, Hino's aircraft gained considerable speed, took off, and climbed to a height of about two meters.[45] However, officials declared this short jump of the monoplane an "unscheduled lift-off" that did not qualify as the country's first flight.[46]

In the following days, large crowds gathered at the northern parts of the parade ground. Hundreds of members of the Committee for Balloon Research and military officers were present near the hangars. Moreover, more than 300 spectators, most of them high-ranking military officers and foreign attachés, were already there. But everybody was kept waiting. The only spectacle that front-row spectators could see was the capsizing of Hino's airplane during another ground maneuver that resulted in damage to a propeller. Finally, an official announcement notified the crowd that the historic event was to take place on the afternoon of December 17. One spectator's report shows how, in spite of earlier disappointments, an unwavering crowd continued its daily trip to Yoyogi in anticipation of the spectacle. Under the headline "An Out-of-Season Garden Party," an elderly person gave an even more excited account of the ongoing festival mood:

> I will look at whatever I can see. Then I can die in peace. I came this morning to Yoyogi with my grandson to see the flight performance. Already so many spectators have arrived and still, with the nice weather, trainloads of people just keep on coming. They march like an army of ants from Harajuku Station to the north of the parade ground. A steady stream of people is rushing in, singing military songs. Soldiers try to control the crowd. Small school kids are pushing close. Even though the airplanes do not fly

> the sightseeing crowd encircles the whole parade ground. We wait for one hour, for two hours, but nothing goes up.[47]

Later the organizers issued this notice: "because of strong wind, the flight demonstration has been canceled." Guards dispersed the crowd, and people went home disappointed. On Sunday, December 18, the patience of the crowd was further tested when the flight was canceled again because of high wind speeds. With more than 100,000 spectators making a daily pilgrimage to the parade ground, the pressure on the pilots to fulfill the crowd's expectations regardless of the wind conditions increased.

The next day, presumably to avoid any gusts, as early as 4:30 Tokugawa climbed into his aircraft. He had fastened to his back a heavy lead battery that was connected by two wires to the aircraft's engine to supply its ignition system. According to the somewhat breathless report of the *Tokyo Nichinichi* newspaper, Tokugawa cast a quick glance at the committee members and assured them: "Today I will definitely succeed," and "with these strong words he dashed off westward." Another article continues the story of the heroic flight:

> Captain Tokugawa grabbed the control stick of his aircraft with his left hand and raised his right hand in a magnificent way. A gale-like cloud of dust lifted the big monster and moved it three hundred meters in a westerly direction. At that very moment, the aircraft lifted off the ground and gently began to fly. In front of all our eyes it suddenly climbed straight up to a height of seventy meters.[48]

While Tokugawa was in the air, Hino was desperately working to start his aircraft's engine. Obviously aware of the pressure to get his plane airborne as well, he reportedly said: "Let's see if in the end man or the machine will win."[49] With a spluttering motor running on just three of its four cylinders, he finally took off. Hino successfully fought the winds that tossed his tiny plane up and down and managed to fly a half circle and land safely. As improvised as these flights might have seemed, they did not fail to impress the audience and unleash a wave of national pride. As one excited newspaper report put it, "For those who saw it, it was close to a divine miracle. These first official flights were the dawn of our aviation era . . . [Japan] has finally become a member of the international aviation world."[50]

Within a relatively short time, the proficiency of the Japanese pilots and the reliability of their machines improved. Short jumps of several hundred meters gradually gave way to air trips beyond the airport's vicinity. When the range of these flights widened, the spectacle of the air show gained a new dimension. A report of Captain Tokugawa's first air trip to Yoyogi from the military airport in Tokorozawa, about thirty kilometers northwest of Tokyo, reveals more of the growing appeal of Japan's early flight shows:

> Spectators came with sleepy eyes to Yoyogi to watch the magnificent event. Harajuku Station was completely crowded. Around 5:30 already ten thousand people had gathered together with a large number of military policemen on horse and on foot. Everybody stared at the sky, full of expectations. Then, at 6:15, at the northwest of the parade ground suddenly one black spot appeared. From all directions voices could be heard: "he is coming." The black spot gradually became bigger and looked like a bird. Four soldiers waved with big white flags to show the landing point. Five minutes later the aircraft had made its impressive appearance over the forest and descended with a deafening sound. Any moment now it would come close and fly over the crowd. Everybody shouted "banzai" as often and as loud as possible. The crowd was running all over the place, so Captain Tokugawa could not land as planned. He landed southeast of the parade ground instead at 6:27. He had covered 31.5 kilometers in 32 minutes. After the aircraft touched down everybody made a rush for Tokugawa still shouting "banzai."[51]

What might have seemed like a playful fairground spectacle with a crowd on the verge of getting out of control was actually a performance carefully arranged by the Japanese military. Events like this one did away with doubts about the practical use of the army's new weapon and secured public support for further aviation development. The public's endorsement of the costly project was important, especially when considering that the military's annual budget depended on the Diet's approval. Tokugawa himself was aware of the propagandistic effect of his endeavor. In his memoirs, he expressed his satisfaction that his flight to Tokyo was reported with banner headlines and declared: "I believe that I contributed to the diffusion of an aviation ideology."[52]

Parallel to the military's efforts, several other ways evolved to spark and spread the nation's air-mindedness. Popular songs and children's books reproduced the powerful images of the seemingly magical flying machines. Takano Tatsumi's "Song of the Big Contest between the Two Captains Hino and Tokugawa" joined the chorus of praise for Japan's first two airmen.[53] It was published in 1911 in a lavishly illustrated songbook. The lyrics tell the story of the skillful pilots who—in front of an open-mouthed audience already excited by smoke, explosions, and the daring sound of howling propellers—fly into the sky "like arrows released from a bow." The crowd watches them descend suddenly, dangerously almost touch the ground, then climb up again until they disappear. The two aircraft then enter the realm of the heavenly princesses, a place full of cherry and plum blossoms, peaches, and nightingales. In an unexpected change, the planes fly into the "clouds of war." They succeed in pulverizing the enemy's warships and fortresses with a "shower of air torpedoes and bombshells." Returning to the waiting crowd, the pilots head for the finish line, where tens of thousands of spectators receive them with cheers. The songbook celebrates the two airmen and their aircraft for setting new records in the "civilized world" and "giving extreme joy to the people." The song gives evidence of the popular hero worship that developed in the wake of Tokugawa's and Hino's first flights. It also presents a public united by aviation enthusiasm and is proof of how the aircraft, even when absent from view, spurred fantasies of celestial paradises, military omnipotence, and Japan's place among the advanced nations.

In 1911, Iwaya Sazanami (1870–1933) published *The Young Aviator, an Illustrated Picture Story*. To a much younger audience, his book tells a story similar to Takano's song.[54] In an aircraft with the distinguishing features of Tokugawa's Farman, the young pilot Tarō daringly flies over smoking volcanoes and spouting whales and manages to park his aircraft on the mast of a magnificent Japanese warship. After these achievements, he becomes involved in aerial warfare. When approaching a group of foreign-looking soldiers, Tarō wants to scare them and drops a "toy bombshell" that explodes in a powerful blast. The soldiers are caught by surprise; they topple over and lose their guns (figure 1.6). Even though several artillery soldiers attack Tarō from behind, he manages to escape and continues his aerial adventures. The story ends with cheers for the Japanese empire: "Dai Nippon teikoku banzai." At first glance, the picture book describes a children's fantasy about the appeal of the flying machines. But it also shows how, from the early years of aviation in Japan, the country's youth

FIGURE 1.6 This eye-catching scene is part of a series of illustrations that appear in a picture book for children published in 1911. We see the triumphant young aviator Tarō, who just has bombed the enemy's soldiers from above. (Source: Iwaya, *Hikō shōnen*, 34)

grew up with texts and images that associated the flying machines not only with heroism and adventure but also with military action and patriotic enthusiasm.[55]

It must be noted that such a blend of militant nationalism and aviation was a transnational phenomenon.[56] For instance, in 1907 German author Rudolf Martin presented in his *Berlin-Bagdad* a hawkish vision of 10,000 German airships engaged in imperial conquest. One year later, H. G. Wells published *War in the Air*, conjuring a dark view of Britain's backwardness to defend itself against an imagined attack by the German air fleet, while Frenchman Émile Driant expressed high expectations for a powerful French air force in his *L'aviateur du Pacifique* (1909). In a more ethereal way, poet Gabriele D'Annunzio created an Italian version of the new myth of flight. In his 1910 *Forse che sì, forse che no*, he evoked a promise of moral and spiritual elevation based on danger and the sacrifice of human lives.[57]

TWO DIFFERENT TRAJECTORIES

After his seminal flight, Tokugawa Yoshitoshi became a prominent public figure who appeared in countless publications. He also played a key role in the ongoing development of Japan's military aviation. Visits to the newly opened Tokorozawa airfield became increasingly popular, and

Tokugawa's fame grew more as the master pilot who now also made flight demonstrations with a new Blériot aircraft. Newspapers celebrated Tokugawa for carrying out Japan's first passenger flight, establishing new flight records, and making the first flight across Tokyo that attracted tens of thousands of spectators.[58]

With Tokugawa's assistance, the aeronautical expertise of the Balloon Committee developed rapidly beyond the mere assembly or replication of foreign aircraft. In April 1911, Tokugawa and several committee members began to design and construct an improved version of the Farman aircraft. To make the plane faster, they reduced the span of the lower wing, changed the airfoil's curvature to provide more lift, and modified the undercarriage to increase the propeller's ground clearance. These upgrades resulted in a series of original aircraft, called Kai-Type, that played an important role for the training of the army's new pilots.[59]

While Tokugawa rose to stardom, Hino faded into obscurity.[60] When he decided to work on a new aircraft with a made-in-Japan engine, he received only minimal support, whereas the Balloon Committee fully covered Tokugawa's expenses for the production of the Kai-Type aircraft. Hino soon resorted to his own funds and even sold his house. Eventually the army decided to transfer him to the Fukuoka Infantry Regiment. With the outbreak of World War I, Japan had no more access to imported aircraft engines, and the army became interested again in Hino's engineering knowledge. An offer was made to return him to Tokyo to help build up Japanese engine production, but Hino turned it down.

Conclusion

During its formative years, aviation in Japan already followed a trajectory that combined public air-mindedness with aerial armament and fervent nationalism. Epoch-making events attracted the public's attention and stirred profound emotions. Spectacles, performances, and media reports provided powerful images that shaped the public's attitude toward the country's aviation project. Flight demonstrations could refer to a traditional framework of spectacle shows, religious festivals, and military pageantry that provided a formal structure to mitigate the shock of a new technology and helped channel public euphoria. Japan's aviation pioneers became the focus of public attention. They became national heroes, and press reports, school textbooks, and popular songs celebrated their achievements. The aviators' flying machines transformed from a simple means of transport into a highly efficient propaganda vehicle that incited popular enthusiasm, support, and air-minded patriotism.

At the same time, a pattern for the transfer, diffusion, and indigenous development of aeronautical technology emerged. Early balloon technology lent itself to indigenization. Japan's first flying machine featured a rubber-covered silk envelope filled with hydrogen produced by a rudimentary gas generator. Inventor Yamada Isaburō greatly improved this basic design with his kite-type balloon that allowed for a much smaller hull. The outbreak of the Russo-Japanese War marked the beginning of aerial warfare. This conflict also accelerated Japan's move away from dependence on foreign reconnaissance balloons toward the promotion of domestic production.

Anxiety over the country's relative backwardness turned into a recurrent theme that profoundly shaped the history of Japanese aviation. As a response to the fast progress of Western aviation, the Provisional Committee for Military Balloon Research was established in 1909 to provide an organizational structure and funding for building Japan's airpower. Close observation of Western developments, dispatching student pilots to Europe, and importing foreign aircraft led to the seminal Yoyogi flight. In the same year, Germany and Great Britain started training their first military pilots. Together with the United States and France, Japan now belonged to an exclusive club of five countries that began to fully integrate aerial warfare into their military strategy.

The first years of powered flight in Japan show how French aviation could gain a superior foothold over its rival Germany in terms of technology and prestige. France made its entrée into the Japanese aviation world with a most impressive airplane and with a pilot who was immensely popular and well connected. Hino was arguably the better pilot and definitively a more skilled aircraft engineer than Tokugawa, but Hino could not free himself of the image of the eccentric tinkerer who always came in second. It is easy to imagine that this mental picture almost inevitably influenced the Japanese perception of German aircraft technology as a whole.

The next chapter examines the crucial role of France in the development of Japan's early aviation, which culminated in the French Aeronautical Mission to Japan. We will see how the Japanese learned quickly from their mentors and how the country's military and aircraft makers achieved full independence from French guidance and technology within a decade.

CHAPTER 2

The French Decade

France had a decisive impact on the early buildup of Japan's airpower. In 1909, when Japanese Army officials became seriously interested in powered flight, they were convinced that France was the world leader in aircraft technology. For a period of about ten years, France maintained its position as a producer of superior aircraft. Moreover, the aviation enthusiasm of the French public—culminating in the proclamation that "our future is in the air"—impressed Japanese observers and made them strive for a similar popular movement at home.[1] Most important, the Japanese military looked to France as the first country to demonstrate how to integrate the aircraft as a new weapon into its armed forces. Within the five years preceding World War I, the French military had built up new organizational patterns and strategies to transform the flying machines into a powerful tool of modern warfare.

This chapter examines the French decade of Japanese aviation. It investigates the experiences and visions of Japanese pilots and air strategists who returned from France and proselytized the aviation gospel at home. An account of Japan's first air war over the German concession of Qingdao in 1914 shows how the Japanese military deployed its aircraft for bombing and reconnaissance missions and how an exhilarated press celebrated the exploits of the Japanese pilots in their French aircraft. Finally, the chapter explores the culmination of Franco-Japan aviation collaboration, the 1918–20 French Aeronautical Mission to Japan. Each of these events shaped Japan's military aviation and the country's relationship with France. Ultimately, the end of the French decade and the eventual collapse of the French aviation monopoly in Japan came not as a bolt out of the blue but as the result of missed opportunities, personal animosities, and the emergence of a powerful new competitor: Germany.

The Flying Baron Shigeno

Baron Shigeno Kiyotake (1882–1924) was one of the most colorful characters among the early Japanese aviators. An early dropout of the Hiroshima cadet school, Shigeno entered Tokyo Music School in 1905. In 1910, the same year Tokugawa and Hino were dispatched to Europe, Shigeno went to Paris to study piano. In 1911 he started flight training and began to design his own aircraft. In August 1911, Shigeno presented to French aircraft designer Charles Roux a blueprint of his original aircraft. He chose to name it *Wakadori* ("young bird") in honor of his deceased wife, Wakako. The following year brought Shigeno a double success. He became—after Tokugawa—the second Japanese person to receive a coveted *brevet de pilote*, a French pilot's license. The aircraft on which Roux and Shigeno had worked together during the last few months of 1911 successfully passed its test flights. The *Wakadori*, with its conspicuous Rising Sun Flag on its tail, was praised by the French journal *L'Aviation* as the "world's fastest biplane" that combined the latest Western metalworking technology with Japanese skill and workmanship (figure 2.1).[2] Indeed, the plane had a total weight of only 300 kilograms, and its powerful sixty horsepower Anzani engine could accelerate the aircraft to a speed of up to 130 kilometers per hour. The airplane made Shigeno popular, and with his already extravagant lifestyle (he drove his own splendid car and wore a mysterious wedding ring with a black velvet ribbon), the "flying baron" soon became the toast of Paris.

In spite of the popularity he enjoyed in the French capital, Shigeno decided to return to his home country. In 1912, he arrived in Japan together with the *Wakadori* and soon set a new Japanese altitude flight record for civil aircraft of 300 meters. The Provisional Committee for Military

FIGURE 2.1 This photograph, dated May 1912, was taken at the Issy-les-Moulineaux airfield, known as the birthplace of French aviation. It shows the *Wakadori*, the celebrated aircraft that Baron Shigeno Kiyotake designed during his stay in France. Note the biplane's lightweight design, its landing skids that were to prevent a nose-over, and its circular tail with the rising sun symbol. (Source © Prieur-Branger/ Coll. Musée de l'Air et de l'Espace—Le Bourget)

Balloon Research asked him to become an official flight instructor for training five new pilots. Shigeno agreed, but the committee's hopes of benefiting from his experience as both a pilot and an aircraft designer did not materialize. Several confrontations with Tokugawa, the celebrated doyen of Japanese aviation, made Shigeno's life difficult. Tokugawa had accumulated only one hour of flight time during his stay in France and fifteen hours since his return to Japan. He viewed Shigeno with his flight log of over 100 flight hours as an unwelcome competitor. When Tokugawa refused to assign a proper maintenance crew for Shigeno's aircraft and confiscated Shigeno's Anzani engine for his own aircraft, Shigeno quit his short career with the Japanese Army.

After these bitter experiences, Shigeno made great efforts to promote the advance of Japanese aviation outside the military's sphere. In 1913, he published his book *Tsūzoku hikōki no hanashi* (*A Popular Story about Aircraft*), in which he praised the airplane as a symbol of modernization and convenience. Appealing to the beauty of flying through clouds, wind, and rain, Shigeno even suggested that human flight could become a means to reconcile science and art. He called on his fellow countrymen to secure Japan's place in the international aviation world. Shigeno proposed to expand Japan's civil aviation sector and build a domestic aircraft industry that would make Japan independent from foreign imports.

In April 1914, Shigeno went back to France and joined a small group of Japanese volunteer pilots who enlisted in the French flying corps.[3] During World War I he flew the latest French Voisin, Nieuport, and Spad models. With two confirmed and six unconfirmed downings of German aircraft, Shigeno rose to fame again. He became part of the exclusive league of World War I flying aces and received, for his "daring and skill," the National Order of the Legion of Honor, the highest French decoration.[4]

In the context of French influence on Japanese aviation, Shigeno is interesting for several reasons. As an artist and as a privileged outsider—his family funded his first stay in France—he could develop his own view about aviation, one that was different from that of a military officer or a government technocrat. Far from being simply an eccentric, Shigeno was a skilled pilot and a gifted designer. He had great potential to successfully transfer essential knowledge and expertise from Europe's leading aviation nation to Japan. However, because of quarrels within its hierarchy, the Japanese military missed this unique opportunity, and Shigeno soon left his home country again to pursue his career elsewhere.

Even during his second stay in France, the pattern of petty disputes between Shigeno and the Japanese military continued. In 1915 Shigeno became friends with Kusakari Shirō (1880–1919), a Japanese officer in charge of inquiring about French aviation. However, as Kusakari testifies in his diary, his superiors repeatedly rejected Shigeno's advice and offers to mediate between French and Japanese aviation experts. Even more important, given Shigeno's rise to prominence within the French army, it is astonishing that the Japanese officials did not assign him any role in the setup and implementation of the French aviation mission to Japan. Arguably the mission's fate would have been different with recourse to Shigeno's skill, knowledge, and repute.

The Qingdao Air War: Brief Encounters and a Lasting Myth

In summer 1914, when Japanese volunteer pilots had already flown their first missions over European battlefields, the Japanese military prepared for aerial combat on the other side of the world. For the Imperial Japanese Army and Navy, World War I provided the first opportunity to test their new airpower under battle conditions. As we will see, Japan's first wartime reconnaissance flights, bombings, and air battles seemed ineffectual, even by the standards of the time. They nevertheless made the military and the general public aware of the new weapon's potential to observe and strike from the sky. At the same time, reports about a well-matched opponent cast doubt on Japan's sole reliance on French aviation technology.

Japan entered World War I in response to a request from faraway London. On August 7, 1914, Foreign Secretary Sir Edward Grey (1862–1933) asked for Japan's assistance against armed German merchant ships in the East China Sea. Britain referred to the 1902 Anglo-Japanese Alliance that regulated the two countries' mutual military support.[5] The Japanese government responded quickly to the appeal and even pressed a step further. On August 15, Prime Minister Ōkuma Shigenobu (1838–1922) presented Germany with an ultimatum to withdraw their fleet from Asian waters and hand over the Jiaozhou Bay concession in northern China by September 15. Japan's rush to establish a foothold in China, nearly 1,800 kilometers from Tokyo, went far beyond the scope of the Anglo-Japanese alliance. However, Japanese politicians wanted to seize this "one chance in a thousand years," especially when the Western powers were absorbed with the war in Europe.[6] Furthermore, for the Japanese Navy, the impending conflict was a welcome opportunity to restore a reputation that had been damaged by the Siemens Scandal in early 1914.[7] With the ultimatum unanswered, Japan declared war on Germany on August 23 and dispatched a contingent of almost 50,000 troops to capture the German base at Qingdao, a small peninsula in Shandong province.

For the first time in Japanese military history, airplanes were to be deployed as weapons of war. However, with inadequate equipment, no coherent strategy, and no experience in air combat, it was doubtful that

FIGURE 2.2 Two crew members haul their floatplane ashore against a strong wind. The signs on the biplane's rudder identify the aircraft as one of the navy's Type Mo large seaplanes. The Yokosuka Naval Arsenal built at least fifteen of these early bombers, which were based on a French design. (Source © Japan Aeronautic Association)

these flying machines and their crews would contribute much to the success of the Japanese forces. The seaplane carrier *Wakamiya* became the world's first aviation vessel to be sent into combat. On August 28, the ship left the naval base at Sasebo, about fifty kilometers north of Nagasaki. It carried four biplanes equipped with floats for takeoff and landing on water. All aircraft were French imports: three Farman Type Mo Small Seaplanes the navy had acquired in 1912 and one Farman Type Mo Large Seaplane, purchased in 1914.[8] The latter was a recent acquisition that, due to its strong 100 horsepower engine, could carry three crew members (figure 2.2). Apparently, the navy had high hopes for the new aircraft. One official declared enthusiastically that he was "happier about the timely delivery of one new Farman airplane than about one million additional soldiers."[9] The floatplanes, which for identification purposes carried for the first time the *hinomaru*, a red circle on a white background, were to carry out reconnaissance and bombing missions. The *Wakamiya* arrived at the Chinese coast on September 1; just three days later, the naval pilots flew their first sorties.

Not to be outdone by its rival, the Imperial Japanese Army sent its own transport ship with five landplanes: one Nieuport and four Farman aircraft.[10] Four of these airplanes were made in France; one was built by the Tokyo Army Arsenal. The army vessel arrived on September 2 at the Chinese port of Longkou, 180 kilometers north of Qingdao. Even though the army's and navy's airplanes reached their destination at nearly the same time, their deployment was poorly coordinated. On September 5, when the navy air force reported the first successful destruction of a German minelayer by aerial bombs, the crews of the army air force were still unloading their planes. Spending a considerable amount of time with building hangars, assembling aircraft, and making test flights, the army finally carried out its first reconnaissance mission on September 21.

In the meantime, the activities of the Japanese Navy's airplanes over Qingdao had already received wide press coverage at home. According to one report in the *Yomiuri* newspaper, bombing enemy positions proved that Japan's air squadrons had made "a big leap forward" and their brave pilots deserve the "greatest honor in history." Contrasting the safety of the capital with the faraway heroic aerial combats, the article continued:

> The same airplanes that the citizens of Tokyo could admire a little while ago at the Taishō Exhibition [held from March to July 1914 in Ueno Park] are now causing enormous damage to the German army. If we think of how they are carrying out their unmatched heroism in an actual air battle in the skies over Qingdao, it feels like a dream come true.[11]

The *Osaka Mainichi* newspaper devoted an extra edition to the Qingdao air war. Referring to a detailed army report, the paper vividly described the German artillery's fierce shooting that "welcomed" the Japanese reconnaissance aircraft.[12] The army's account contained further details of a successful bombing sortie from which all participating aircraft returned safely. It concluded that the mission must have left a strong "psychological impact" on the enemy.[13] Clearly the air raid made a powerful impact on the homeland as well, providing the press with headlines like "The Qingdao Enemy Fleet Pressed" and "Our Airplanes' Bold Reconnaissance."[14]

An investigation into the technology available to the Japanese bomber crews in their campaign casts a different light on the military effectiveness of these missions. The Farman bombers carried sixteen shells, eight to twelve centimeters in size, which were attached to either side of the aircraft. The pilot used an aiming device consisting of one horizontal wire in front of him and crosshairs on a celluloid board under his seat. Approaching the target and correcting the angle of his aiming sight for speed and altitude, the pilot had to determine the right moment to pass a loud "*yoshi*!" to the bombardier behind, who would then manually release the bombs.[15] With their limited explosive power and their primitive aiming and release mechanisms, these projectiles were crude and inaccurate weapons that in all likelihood frightened the enemy more with their deafening noise than with their destructive capacity.

Most Japanese commentators agree on the noteworthy maneuverability of the only aircraft available to the German troops.[16] To take off from the short Qingdao runway, the aircraft had to be as light as possible; German naval pilot Gunther Plüschow (1886–1931) even refrained from taking an observer with him (figure 2.3). With the minimized weight of his Rumpler Taube aircraft, Plüschow indeed easily outmaneuvered his enemies. Japanese sources also mention the German aircraft's superior cruising and climbing speed, concluding that it was possible for Plüschow to escape effortlessly into the clouds by climbing to an altitude of 3,000 meters.[17] However, 2,000 meters was the maximum flight altitude of his monoplane, and Plüschow's own version of his escape tactics seems much more likely. In his war diary, he described that due to the poor performance of his aircraft, the only way to shake off his pursuers was by a nosedive out of an altitude of 1,700 meters.[18]

An eyewitness report written by *Asahi*'s war correspondent under the headline "A Courageous Air Battle" gives an idea of the novelty of aerial warfare and how it impressed observers on the ground:

> When the [German] Rumpler aircraft appeared, its silhouette looked like a falcon. In the calm clear sky, it was flying unbelievably high . . . and provokingly fast. After a while it descended in circles to an altitude of five hundred meters. When our Nieuport aircraft flew close to it, the German airplane effortlessly escaped toward the mountains while our aircraft followed it. Our ground troops, getting impatient, started to shoot; their

FIGURE 2.3 Gunther Plüschow, the "Aviator of Qingdao," in his Rumpler Taube monoplane. Note that the fore observer's seat was left empty to make the aircraft lighter and more maneuverable. (Source © Archiv ROBGER, Berlin/Buenos Aires)

> gunfire sounded like beans being roasted. Soon the Rumpler appeared again, brazenly continuing its reconnaissance mission. Now our Farman aircraft joined the fight. The three planes randomly shot at each other and returned the fire. I could see flashes of gunfire in the sky. After a while the aircraft returned to their bases and the air battle was over.[19]

Even with its unintentional humor and inconclusive ending, the report conveys how the German aircraft could inspire a mixture of awe and trepidation. It is also interesting to note that again, the apparent superiority of the Rumpler does not match Plüschow's assessment. In his memoirs, he showed great respect to the enemy's "excellent huge hydroplanes" and their "outstanding, courageous pilots."[20] He lamented that "with the poor climb rate of my Taube I could not get anywhere against the big [Japanese] biplanes with their three-man crews."[21]

The Japanese accounts refer to a machine gun that Plüschow purportedly had installed in his Taube and that enabled him to put the Japanese aircraft under heavy fire. However, Plüschow's armament was far less dreadful than his enemies might have imagined. In addition to his Parabellum pistol, he carried several makeshift bombs he had to drop manually. According to his descriptions, he had put together these devices by filling empty coffee cans with dynamite, horseshoe nails, and scrap metal. On November 6, 1914, the night before the German troops surrendered, Plüschow had already made his escape by air to Haizhou, a city 250 kilometers southwest of Qingdao, where, according to his account, he landed in a rice field and set fire to his aircraft to prevent it from being captured by the Japanese.

The deployment of aircraft under battle conditions provided valuable experience to the Japanese and helped set an ambitious agenda for the future expansion of the nation's airpower. Within less than three months, the Japanese air forces could log an impressive overall number of 135 sorties and 234 bombings.[22] In a newspaper interview, Navy Commander Yamauchi Shirō commented that aircraft had proved their worth for reconnaissance missions and that they would become indispensable for future wars.[23] In a similar vein, another navy official emphasized that the experience gained from the Qingdao air battles would help further develop the military's war strategy and the flying technique of its pilots. However, with the war going on in Europe, the Western countries' aviation technologies advanced rapidly. To keep pace, the officer concluded, Japan needed sufficient funds for further aeronautical study and research.[24]

Referring to previous victories over China and Russia in 1895 and 1905, some historians have called the Qingdao battle the "third great military triumph for Imperial Japan."[25] But Japan's foothold in China was short-lived. During the 1922 Washington Naval Conference, Japan's representatives gave in to US pressure and returned Qingdao to China. There was, however, a longer-lasting legacy of the Qingdao Siege. French airplanes as new weapons of the Japanese Empire had clearly made their debut in the military theater and—equally important—on the public stage. The only cloud cast on the general euphoria was a sneaking suspicion that there might be other countries that, with their own aircraft, might make equally good use of the new invention to strike from the sky.

A Technocrat Shapes His Vision: Kusakari Shirō

In 1914, the same year Japanese military planners began to analyze the lessons learned from the Qingdao campaign, the army sent one of its most knowledgeable experts to Europe. The on-site visits of Kusakari Shirō to European aircraft makers and military facilities were intended to provide the Japanese Army with a detailed picture of the dramatic advances in Western aviation. Kusakari's engineering background—he was a graduate of the army's Artillery and Engineering School—and his experience gained during his prolonged stay in Europe between 1914 and 1917 made him a new type of aviation expert, and he had a profound influence on the development of Japanese military aviation.

The rather earthbound beginnings of Kusakari's career can be traced back to the 1904–5 Russo-Japanese War, when he served in a battalion overseeing railway construction in Korea. In 1907 Kusakari resumed his studies and entered the Department of Mechanical Engineering at Tokyo Imperial University. Following his graduation in July 1910, he was appointed as a member of the Balloon Committee. In April 1914, Minister of the Army Oka Ichinosuke (1860–1916) officially assigned Kusakari to be dispatched to Europe "for aviation research purposes."[26]

The first weeks of his visit profoundly shaped Kusakari's perception of German aviation. After his arrival in Berlin on June 1, 1914, he received a special welcome as a delegate of a country with the potential to become a prospective customer of the German industry. In less than six weeks, Kusakari was able to visit most of the major German aircraft makers, who proudly presented their state-of-the-art technology.[27] He also witnessed German popular aviation enthusiasm when thousands gathered to watch a flight competition at the Johannisthal Airport, in spite of heavy rain.[28] Further visits to aircraft-related companies followed, and Kusakari excitedly recorded in his diary the marvels of German precision optics, bombing devices, wireless telegraphy, and light but powerful aircraft engines.

In mid-July 1914, just two weeks before the start of World War I, Kusakari left for Paris. With his good command of French, he proved to be a tireless information-gatherer. He collected and translated newspaper articles as well as French and German aviation books. Meetings with Japanese war correspondents who had been at the front line further shaped his image of two countries caught up in a modern war fought in trenches and in the air. Kusakari made friends with his fellow countryman Baron Shigeno, who provided his view of aerial warfare. During his frequent visits to the French aircraft manufacturers, some of them arranged by Shigeno, Kusakari witnessed flight demonstrations of the latest Spad and Nieuport aircraft. He experienced firsthand the efforts of France's aviation industry to meet the military's request for ever more and better aircraft with increasing production and continuous technological development.

Toward the end of his stay in France, Kusakari went to the renowned French military academy École spéciale militaire de Saint-Cyr to conduct engine research.[29] What might appear as a technocrat's escapism became a central theme of his future lectures in Japan, where he continually emphasized the importance of technological advancement as embodied by the aircraft engine, the epitome of state-of-the-art technology of the era. Finally, Kusakari accepted an invitation to fly in a Nieuport aircraft. He enthusiastically described his flight over Paris as the "happiest moment during my stay in Europe."[30]

After his return to Japan in April 1917, Kusakari captured the attention of a large audience—both military and civilian. In December 1917, he presented to the Balloon Committee his "Outline of the Current State of Aviation in Each European Country" (*Ōshū kakkoku ni okeru kōkūkai no genkyō yōran*).[31] Even for today's readers, the report holds a certain fascination with its detailed drawings and wealth of information that reveal an engineer's sense of precision and attention to detail. Various charts compare the strength of each country's air force, industry, aviation expenses, and public donations. Organizational patterns of the various air forces supplement detailed maps of the French, German, and Japanese military aviation facilities. In addition, Kusakari devoted several pages to advanced flying techniques for air combat (figure 2.4).

During Kusakari's stay in Europe, three key issues had taken shape that became the leitmotif of his subsequent reports and lectures. First, Kusakari was deeply concerned about the organizational structure of Japan's air force. Given the highly specialized nature of aviation technology and flight training, he argued that a country's air force could not be an integral part of the infantry or any other subunit. Japan should establish an independent air force as a third armed force on the same

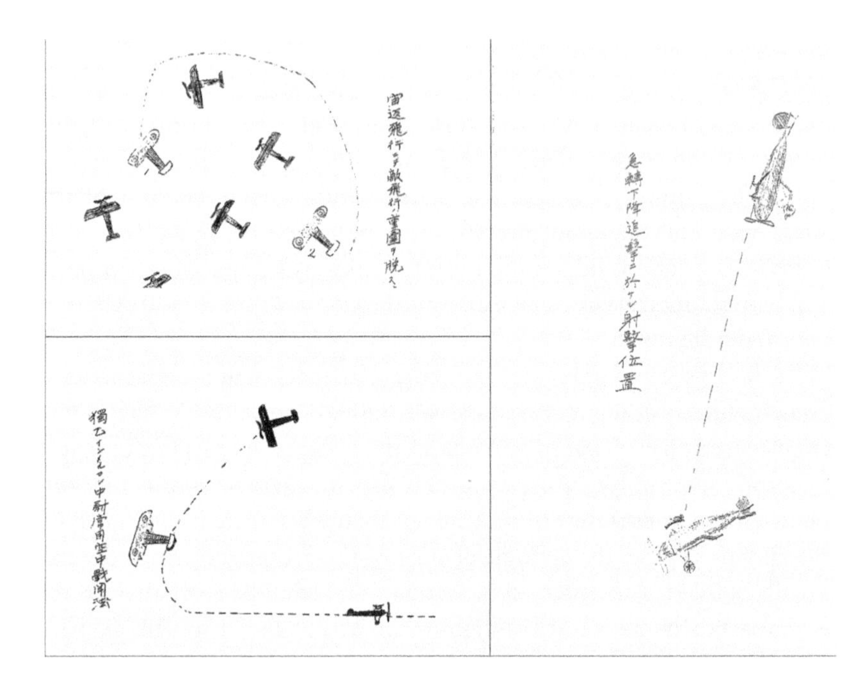

FIGURE 2.4 Kusakari Shirō published these sketches in 1917. Even in their apparent clumsiness, they reveal the author's fascination with the new concept of three-dimensional warfare. The drawings show how to escape from the enemy's line of fire by flying a vertical loop (upper left); the air battle technique of German fighter ace Max Immelmann (1880–1916) (lower left); and an aircraft's best shooting position during rapid descent (right). (Source: Kusakari, *Ōshū kakkoku ni okeru kōkūkai no genkyō yōran*, JACAR ref. C08020891800, 23–25)

organizational level as that of the army and navy. This suggestion was a bold affront to the army's traditionalists who would at best assign an ancillary role to military aviation.

Second, Kusakari emphasized the importance of technology. For him, World War I showed that military success was no longer simply a matter of soldierly spirit on the battlefield. Rather, war had become a struggle of industry and technology. Each country would have to participate in a "war of technology" (*gijutsuteki sensō*) where its manufacturing capability would be of crucial importance. He further argued that securing air superiority would be *the* decisive factor in future wars—an idea that Italian strategist Giulio Douhet (1869–1930) promoted four years later in his book *The Command of the Air*. According to Kusakari, gaining and maintaining air superiority depended almost entirely on the number of high-performance aircraft available. Due to the high attrition and frequent damage from air battle, each aircraft could be expected to last only three months. Kusakari concluded that the technological standard of a nation determined the performance of its aircraft and the growth of aircraft production. It would be of vital importance to strengthen Japanese aircraft manufacturers and advance their production technology.

Finally, Kusakari insisted that fostering public air-mindedness to gain popular support for a country's aviation project was as crucial as developing an organizational and technological framework. He argued that in both Germany and France, popular pressure on politicians and patriotic donation campaigns procured massive funds for the buildup of each country's airpower. He referred to a well-known German donation campaign that emerged in the wake of the explosion of a Zeppelin airship in August 1908.[32] Within a short time, a private initiative managed to raise funds of more than six million marks to finance the construction of new airships. Kusakari emphasized how French and German newspapers and magazines successfully fomented a widespread "aviation craze" and a national "aviation ideology." Being aware of the strong connection between air-mindedness and nationalism in these countries, he lamented that he had never seen such a patriotic spirit manifest itself in Japan. Kusakari implored his audience to engage in a similar all-out effort to support Japan's aviation:

> It is absolutely essential for the aviation project to gain popular support. First the Japanese Diet has to press the government to establish aviation facilities. Then people will patriotically donate the necessary money, property, and buildings. Public aviation organizations shall receive patriotic funding from individual supporters, the imperial family, wealthy citizens, and newspaper and magazine campaigns. Local patriotic aviation clubs shall be established, and private aviation must be encouraged.[33]

Kusakari's lectures were an effective means of conveying his concerns to a broad audience. He skillfully mixed the unsentimental matter-of-factness of an engineer with an ardent patriotic concern for the future of the country. He disseminated his ideas about the prospects of Japanese aviation far beyond a small circle of military experts. Kusakari's prophetic visions of a strong Japanese airpower, large-scale aviation enterprises, and a national aviation ideology became the cornerstones of Japan's aviation project, exhorting the country to catch up with the West and integrate the new weapon fully into the nation's defense.

Inoue Ikutarō: The Army Air Force's Mastermind

Kusakari's dedication to promote the buildup of a powerful air force and a productive aircraft industry received unexpected support from the poor performance of Japan's army aircraft. During World War I, the Japanese military painfully realized how much it depended on foreign technology imports. Once the war began, the fledgling air force could no longer buy aircraft engines from abroad and had to resort to domestic engine production. In 1916 the Japanese Army started a trial production of aircraft engines at its Tokyo artillery arsenal. These motors were based on the design of the renowned 1912 German Daimler 100 horsepower engine that the Imperial Flying Association (Teikoku Hikō Kyōkai) had imported before the war. Due to the low standards of Japanese manufacturing technology and the lack of appropriate metal alloys, these domestic-made engines were notoriously unreliable. Their deficiency became especially manifest during a large-scale military exercise in 1917, when most of the fourteen participating airplanes could not even reach the maneuver area east of Lake Biwa—300 kilometers away from their Tokorozawa base—and had to make emergency landings because of engine troubles.[34]

These alarming incidents brought one of the army's best-educated technocrats to prominence. Inoue Ikutarō (1872–1965) was a graduate of the Army Academy, the Army Artillery School of Engineering, and the Army War College. In January 1918, he was put in charge of an investigation into the poor performance of the army's aviation branch.[35] Inoue submitted a memorandum alerting his superiors that Japan's air force was "conspicuously inferior" to those of the Western countries. He insisted that the main lesson from the ongoing war was the urgent need to "change the present system radically." For this the government had to be ready to provide necessary funding. The progress of army aviation should be given priority over any thoughts of cost saving, Inoue urged.

Inoue's proposed reorganization and expansion scheme included a new flight school with extended facilities for training, research, and material supply. In addition to expanding the Tokorozawa airfield, a new military airport should be built, to which large parts of the present air unit should be transferred. Inoue further suggested establishing an Aviation Department within the Ministry of the Army. This bureau would unify and coordinate all aviation-related activities, which were currently dispersed over various departments, such as engineering, supply, utility, artillery, and the arsenals. Referring to the accidents in the 1917 maneuver, he recommended establishing closer ties between aircraft producers and the military to better coordinate the supply, examination, maintenance, and repair of airplanes. The new Army Aviation Department should be granted full authority to supervise and control Japan's aircraft manufacturing—both civil and military.[36]

In all likelihood, Inoue's far-reaching proposals would have been consigned to oblivion if there had not been a reshuffle in the new Cabinet under Prime Minister Hara Takashi (1856–1921). In September 1918, Lieutenant General Tanaka Giichi (1864–1929) became the new minister of the army. Tanaka's three-year term in office became a decisive period for the development of Japan's military and civil aviation. Bringing together the diverging visions of technical experts and ideologists, Tanaka appointed air-minded army technocrats to influential positions. Building on a 1910 slogan of everybody becoming a "good soldier and good citizen" (*ryōhei ryōmin*), Tanaka insisted that there should be no boundary between society and the military.[37] He forged a strong link between the army and the emerging civil aviation sector.

Tanaka approved the principles of Inoue's memorandum and, in 1919, completely reshaped the structure and substance of the army's aviation program. The minister of the army convinced the Diet to provide the necessary funds for a modernization and expansion of Japan's airpower.[38] As a result the army established its new Aviation Department (Rikugun Kōkūbu) in April 1919. The chief of the department was Inoue Ikutarō, who reported directly to the minister of the army.[39] In August 1920, in a second step, Tanaka promoted the development of Japan's civil aviation sector by setting up an Aviation Bureau (Kōkūkyoku), which, however, he placed under the army's authority. He explained the importance of this move at a meeting of local governors. Whereas in military aviation, "we already can see the first light of dawn," the civil aircraft sector still suffered from a lack of pilots and aircraft manufacturers. Tanaka therefore asked for the "enthusiastic support [and the] united efforts of government and people" to help the Aviation Bureau in its important role of protecting, encouraging, and guiding civil aviation in Japan.[40]

Tanaka's move to install Inoue in the influential position of chief of the new Army Aviation Department had far-reaching consequences. Inoue became one of the most passionate advocates of airpower expansion. In spring 1921, he submitted a proposal for a tenfold expansion of the army's air force to forty-five reconnaissance squadrons, fifty-two fighter squadrons, and fifteen bomber squadrons through 1934, with a proposed total number of 1,164 aircraft.[41] The report contained detailed numbers for the wartime troop strength of reconnaissance, fighter, and bomber squadrons, with a strong emphasis on fighter aircraft that would make up more than half the army's airpower. Inoue planned to assign two-thirds of the fighter force to front-line operations, and he devoted the remaining fighter aircraft to the defense of strategic places and garrisons on the Japanese archipelago, on the Liaodong Peninsula, and in Korea and Taiwan.

One of Inoue's first and most momentous decisions was to equip the Army Air Corps entirely with French aircraft. Rather than selectively import aviation technology from different countries, he chose to strengthen ties with France and take French aviation as the principal model for the army's air arm. He insisted that only after having caught up with the French could Japanese military aviation subsequently achieve independence, at which point it could begin to absorb the technology, strategy, and organizational patterns of other countries as well.[42]

The French Aeronautical Mission to Japan

In March 1918, half a year before Tanaka's appointment as minister, army officials made an opening move for a large-scale import of French aviation matériel and know-how. Army headquarters instructed the Japanese military attaché in Paris to deliver a request for the purchase of aircraft, engines, and related material. Japan also asked for the dispatch of experienced French engineers.[43]

With its request for French assistance, the Japanese Army could hark back to a long-standing tradition of French military missions to Japan. In early 1867, the first French mission arrived in Yokohama to train the Tokugawa shogunate's soldiers. Their efforts came to an unexpected end in November 1867 when Shogun Tokugawa Yoshinobu (1837–1913) resigned and the new Meiji government sent the French advisers back home. The second French mission came to Japan in 1872. With the help of Lieutenant Colonel Charles Antoine Marquerie (1824–94) and his mission members, the Japanese adopted French recruiting laws; established the Imperial Japanese Army Academy (Rikugun Shikan Gakkō), based on the French military academy Saint-Cyr; and set up a large-scale arsenal. A third mission (1884–89), led by the French officers Henri Berthaut (1848–1937) and Étienne de Villaret (1854–1931), instructed Japanese soldiers in French tactics, carried out field maneuvers, and proposed new plans for mobilization and military strategy.[44]

This impressive prehistory could not make up for the fact that the timing of Japan's latest appeal for French support was ill-fated. On March 21, 1918, Germany launched the Spring Offensive with the aim of pushing the Western front deep into French territory. The French high command decided to deploy nearly all available fighters and bombers against the advancing German troops. Accordingly, the French government initially put off the Japanese inquiry. On April 5, 1918, the Japanese renewed their request and added that Japan's legislature had already allocated the budget for a French mission to Japan. Even with funding secured, the Japanese Foreign Ministry had to wait another four months for a French response. In early August, the French military attaché offered to send five balloons and thirty Salmson aircraft to Japan. The proposal included the dispatch of French flight instructors, who would pass along their wartime experience.[45] On August 20, 1918, the French authorities disclosed more details: Artillery Lieutenant Jacques-Paul Faure (1869–1924) was appointed the leader of the French Aeronautical Mission to Japan. In addition, five officers and five noncommissioned officers would accompany Faure as instructors for piloting, reconnaissance, artillery observation, radio operation, and aerial shooting. Moreover, to the great surprise of the Japanese, the French government would cover all expenses.

Several factors might account for this unexpected reversal of the French position. By mid-August the Allied Hundred Days Offensive had resulted in a series of German defeats that relieved the pressure on French airpower and improved the prospect for the dispatch of French aviation experts and matériel. The project of a French aeronautical mission to Japan also received strong support from the French ambassador to Japan, Eugène Regnault (1857–1933). In one of his telegrams, the diplomat urged the French government that "by this means [the Faure mission] we can resume our long-standing influence on the Japanese military and will not allow others to take this place."[46] Finally it is easy to surmise another, even more pragmatic rationale for the Franco–Japanese arrangement. With the war entering its final phase, the Faure mission provided a welcome opportunity for the French to dispose of their wartime overproduction and assign surplus staff to establish an aviation monopoly in Japan.

In the meantime, preparations began in Japan. On December 12, 1918, the Ministry of the Army established the Temporary Committee for Aviation Technique and Practice (Rinji Kōkūjutsu Renshū Iinkai), which would be in charge of the French air team. Two of the army's most air-minded officers led the committee: Inoue Ikutarō and Kusakari Shirō.

After the settlement of all diplomatic, bureaucratic, and monetary disputes, the Faure team left Marseille on November 24, 1918, less than two weeks after Germany had signed the armistice that ended World War I. A group of fifty-nine French officers and noncommissioned officers—twice as many as initially proposed—traveled via Saigon to Shanghai. From there the Japanese government provided a steamer that brought the French mission first to Nagasaki and then Kobe. On January 15, 1919, just after noon, after a train journey of 500 kilometers from Kobe, the Faure team arrived at Tokyo Station.

On their way to Tokyo, the French airmen experienced a first taste of Japanese aviation enthusiasm. Faure's report conveys an idea about the Frenchmen's triumphant reception (figure 2.5):

> Festival mood all over Nagasaki, steamers escort the boat . . . hoorays and fireworks at Moji . . . guided sightseeing tours and a gala dinner at Kobe . . . people thronging each station on the way to Tokyo: "Marseillaise," speeches, presents.

FIGURE 2.5 This photograph shows the triumphal arch, apparently modeled on the French Arc de Triomphe de l'Étoile, that saluted the members of the French Aeronautical Mission at Gifu Station in January 1919. The picture vividly conveys an atmosphere of genuine enthusiasm about the arrival of the French airmen and their leader, Jacques-Paul Faure. (Source © Japan Air Self-Defense Force Gifu Air Base, provided by Kakamigahara City Museum of History and Folklore)

Even more important, his ensuing remarks put the mission's objective in an international context:

> The more than warm, enthusiastic reception shows that for now [the mission] was well chosen, and that the conditions for an extension of French influence in Japan are very favorable. The irritation of the Americans, British, and Italians shows how much our position is envied.[47]

The arrival of the French air team attracted extensive press coverage. Already on January 14, 1919, the *Osaka Asahi* newspaper published a half-page photograph showing Faure and his group surrounded by the mayor of Nagasaki and several high-ranking military figures. Two days later, the Tokyo *Asahi Shinbun* gave an account of the group's arrival in the capital and reported how a big crowd and many dignitaries welcomed them. In more general terms, the paper informed its readers about the enormous increase in the aviation budgets of Western countries and concluded: "By now the crucial role of aircraft in the war is widely understood. Therefore, it is urgent that Japan as well develops its own aircraft."[48]

The blaze of public interest in the French Aeronautical Mission surely inspired an air-mindedness that could provide widespread popular support for the buildup and expansion of Japan's new air force. Furthermore, the impressive number of prominent figures who received Faure testifies to the heightened expectations that Japanese leaders attributed to the French mission. Not only Inoue Ikutarō but also the Minister of the Army Tanaka met the French commander. Tanaka offered Faure free access to all military installations, a rare privilege never before granted to a foreigner. Referring to a long-standing tradition of Franco-Japanese military cooperation, the minister declared that "there is indeed a real affinity between our two armies and I am happy to see it affirmed once more by the arrival of the French aviation mission . . . we consider the dispatch of this brilliant mission just as a beginning."[49] General Uehara Yūsaku (1856–1933), the chief of general staff who had received four years of military training in France, arranged the ultimate honor for the mission's leader: on January 27, Faure was granted an audience with the Taishō emperor.

The French instructors started their work in February 1919 at several locations. Reconnaissance training, aerial shooting, and bombing practice took place in Chiba and Shizuoka prefectures. A school of aerial combat was set up at the Kakamigahara airfield, about twenty-five kilometers north of Nagoya, where, after basic training in a makeshift "flight simulator" (figure 2.6), thirty Japanese student pilots learned how to handle the Nieuport and Morane high-speed fighter aircraft. With the arrival of the French instructors, Kakamigahara became a popular place to visit. Large crowds rushed to the airfield to enjoy a festival atmosphere with scores of street vendors and food stalls—and a chance to see acrobatic stunts performed by the French pilots.[50]

The mission also set up the licensed production of French aircraft and aero-engines. At Tokorozawa, Japan's second major airfield, French instructors trained an impressive 400 workers in the production of Salmson

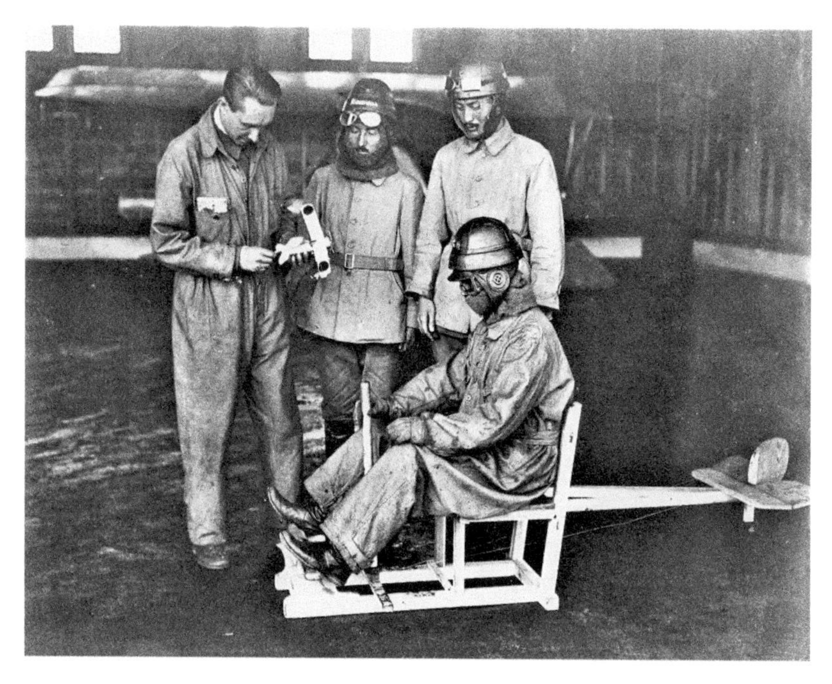

FIGURE 2.6 This wooden contraption functioned as an early flight simulator. The French instructors used it to familiarize the Japanese student pilots with the basic flight techniques of "stick and rudder." In all likelihood the future pilots had to wear their full flight gear to add a touch of reality to an otherwise mundane learning experience. (Source © Japan Aeronautic Association)

and Nieuport aircraft. At the same time, at the Atsuta Arsenal in the suburbs of Nagoya, forty craftsmen received instruction on how to manufacture French lightweight aircraft motors. By the end of 1919, the production plants had reached a combined output of thirty engines and forty-five aircraft fuselages each month.[51]

Within a few months, the French mission seemed to have achieved remarkable results: an exhilarated public in strong support of military aviation, a substantial corps of pilots developing a high degree of proficiency, and an emerging aircraft industry to provide planes and engines to the army's air force. Given these achievements, the Japanese Army asked the French instructors to prolong their work. After some major changes in the mission's contract, the French government agreed to extend the mission until March 1920. Faure's salary was now settled at ¥1,500 a month and those of his subcommanders at ¥1,200—princely salaries considering that the Japanese prime minister had to make do with ¥1,000 a month.[52] Maybe this comes as less of a surprise considering that a practice of generous remuneration had already been established in the early Meiji period, when a significant number of hired foreigners received higher salaries than Japanese generals or ministers; some were even paid more than twice as much as the prime minister.[53]

In summer 1919, the mission's fortunes changed, and a story of high-flying success ran the risk of a hard landing. The army's leaders witnessed how the aviation enthusiasm that welcomed the French mission could easily tip into widespread outrage over the unacceptable behavior of the foreigners and the dangers of a new technology. On August 1, 1919, Faure's deputy, Louis Auguste Ragon, shot himself at the Grand Hotel Yokohama. Although the motives of his suicide remained unclear, rumors began that he took responsibility for a series of scandals that had upset the local population. Things worsened during the stay of the French instructors when four serious accidents took place. The most disastrous crash in Hamamatsu, Shizuoka, killed and wounded several spectators.

Faure's grievances—frequently communicated to his superiors in Paris—provide substantial evidence of the mission's poor organization, insufficient funding, and limited support. The French government repeatedly turned down his requests for more aircraft, machine tools, and specific aviation matériel.[54] For Faure it was obvious that under such circumstances, the mission hardly could meet its goal to establish a French monopoly in Japanese aviation. He feared that "the orders and license acquisitions go abroad" and that British and Italian aircraft manufacturers (like Handley Page or Caproni), which received more generous support by their governments, could easily establish themselves as suppliers to the Japanese.

Slow progress in training Japanese pilots and workmen increased the colonel's discontent. On November 6, 1919, Faure submitted his opinion about the training of the Japanese pilots to the officials of the Balloon Committee.[55] He criticized the Japanese student pilots for focusing too much on theory and neglecting practical training. In general, Faure maintained, the trainees showed a lack of interest, enthusiasm, and discipline. He recommended revising the selection procedure for future candidates. Faure's comments on the Japanese workmen reflect a similar dissatisfaction: "It takes a long time for the Japanese worker, who is very dexterous with his hands, to make a transition from mere tinkering to [serious] mechanical work. The production remains sluggish because of the deficient education of the workers, their lack of efficiency, and their small staff number."[56]

The situation further deteriorated when Faure himself caused considerable bafflement among senior army officers. Asked by the Japanese Army Command about his advice on the future of Japan's air force, Faure insisted on devoting a maximum of funding and matériel to building up sufficiently strong military airpower.[57] In his view, Japan's army would initially need twenty divisions with 824 aircraft and 1,200 pilots at its disposal. To maintain its airpower, the army would have to secure a monthly supply of 323 new aircraft, 600 new engines, and 170 freshly trained pilots.[58] These numbers caused considerable surprise among his audience. Faure argued that these figures were a result from hard-won French wartime experience. However, it is not difficult to imagine that the colonel also wanted to provide a rationale to establish large-scale flight instruction and aircraft production in Japan under direct French control.

Faure elaborated on the air arm's strategy and organization. He argued that a country's airpower should cooperate closely with ground troops. Four different flying corps should be under the direct command of each branch of the ground forces—the artillery, infantry, military engineers, and cavalry. This view clearly reveals Faure's background as an artillery officer who considered airplanes to be ancillary weapons.[59] The French officer opposed the idea of establishing an independent aviation branch within the army. In his view, such an autonomous unit of aviation specialists would soon dissociate from their comrades-in-arms on the ground and, as a result, lack the close connections necessary for effective support. Such a conservative view weakened the position of the army's modernizers, especially at a time when the discussion about an independent air force had reached an advanced stage. The Imperial Japanese Army was very close to making the decision to establish an autonomous air force branch. As a result of Faure's intervention, the matter was put on hold, and the very idea of having an independent air force became less conceivable.[60]

In spite of Faure's outdated views, Inoue Ikutarō found kind words for the French mission. In his "reflections," published in 1936, Inoue put forward: "The Faure mission was an act of sincere kindness of the French government. It has not only been profitable for the flying corps, it also spread knowledge about aviation to both the army and the navy. Moreover, the mission helped to raise the aviation ideology of the whole nation."[61] French evaluations reveal with more clarity some of the underlying motives for the dispatch of the mission. French military attaché Henri Bonnefont de la Pomarède (1872–1957) was one of the first officials to comment on the Faure mission. His report assessed it as "excellent propaganda, not only on the military level, but also for the whole people. This was necessary because of France's long restraint and because of the terrible Anglo-Saxon competition." He drew the triumphant conclusion that "the Japanese military aviation is from now on completely under French influence."[62] In a similar vein, Faure wrote in his final report on April 8, 1920: "from the point of view of instruction, the mission's aim has been brilliantly achieved and has exceeded the most optimistic expectations."[63] Obviously the French diplomat and the mission leader chose grandiloquent words of praise in all their vagueness to submit a success story to their superiors. At the same time, it seems, their Japanese counterpart struggled hard to conceal his disappointment.

Reconsidering the Exclusive Devotion to French Aviation

During the final months of the mission's activities, the Japanese Army's hopes to modernize its air fleet with the latest French technology suffered another serious setback. In January 1919, the month of Faure's arrival, the Imperial Japanese Army had already approved a massive order of 170 French aircraft. The proposed purchase included 30 Salmson 2A2 reconnaissance aircraft, 100 Spad type 13 fighters, and 40 Nieuport 81E2s training aircraft. Nearly twelve months later, these planes finally arrived in Yokohama. To their great disappointment, the Japanese purchasers found that a large portion of the delivery was unusable because of degraded plywood and corroded metal parts. Further discontent emerged when they discovered that most of the damage had not been caused in transit but was the result of slipshod production.[64] The army attaché to the Japanese embassy in Paris sent a damage report to the French government, to the

French Air Force, and to the manufacturers. But lengthy negotiations ended in May 1921 without any loss being recovered.

The Japanese Army's efforts to build up its first bomber fleet led to further frustration about the poor quality of French imports. Several reports about the effects of aerial bombing during World War I had attracted Japanese interest in long-range bombers. Aware that Japan's aviation industry could provide neither the airframes nor the engines for such a heavy aircraft type, the army opted for importing French bombers.[65] After the end of the war, the French had postponed the development of any new bomber type because of military budget cuts, and the Japanese had to make do with the obsolete Farman F.50 bomber. On April 1921, five used twin-engine F.50s arrived at the Tokorozawa Army Flight School. As it turned out, in addition to their poor design and performance, the aircraft had been seriously degraded during their previous use in the French army. The Japanese flight crews soon started to loathe their bombers. In a flight report, one pilot described how part of a bomber's wing had broken apart during flight. He managed to land the damaged aircraft safely, but in outrage the crew destroyed the rest of the wing to avoid having to fly that aircraft again.[66] The flight school soon grounded and finally abandoned all F.50s.

In spite of these problems, the Japanese Army still relied on French technology and placed orders for sixteen Farman F.60 Goliath bombers, the first two of which arrived in Japan in December 1921. As an improved version of its predecessor, the F.60 could carry a heavier bomb load. But the aircraft's increased weight resulted in poor maneuverability. To keep the bomber in the air, the crew had to run the twelve-cylinder Lorraine engine under full power during the whole flight, which resulted in frequent engine damage. Because of these continuing troubles, the army ultimately decommissioned all F.60s in 1928.[67]

Conclusion

Japan's decision to rely on French material and expertise for building their airpower, a domestic aircraft industry, and a national aviation ideology was well founded. Until the beginning of World War I, France enjoyed international recognition as a world leader in aviation. The country could rely on its technological superiority and firmly establish a leading role in the early development of Japan's aviation. France could claim that a French Farman aircraft took off for Japan's first motor flight. French aircraft were the first flying machines that Japan deployed into actual battle over Qingdao, demonstrating the great potential of the new weapon. When Kusakari Shirō returned from France to Japan in April 1917, he disseminated his wide knowledge of chiefly French aircraft and engine manufacturing.

During the French decade, popular support for Japan's early aviation was rising steadily. The seminal Yoyogi flights of December 1910, the exhilarated reports about the 1914 Qingdao air battles, and the exhortations during Kusakari's lecture tours encouraged the public to eagerly support the buildup of a Japanese air force. The triumphant arrival of Faure's air team in January 1919 became a climax of the public's air-mindedness and aviation craze. Thus, the ground was well prepared for the flight instructors and production advisers of the French mission to pursue their declared aim to establish an aviation monopoly in Japan.

In spite of its auspicious beginning, the French Aeronautical Mission to Japan ultimately failed. Exaggerated expectations and the vanishing attraction of the novelty, together with distressing accidents and scandals, eroded enthusiasm and once more proved the fickleness of public opinion. Even more damage occurred in the key area of military organizational reform. After aviation technology finally made it possible to defy gravity, institutional inertia proved to be the much more formidable obstacle. For the traditionalists in the Japanese Army, Faure was a welcome tool to block the far-reaching plans of the reformists like Kusakari and Inoue to establish an independent air force. The world's most advanced weapon had to make do with a subordinate position in an ossified structure.

After the French failure to sustain public support among the Japanese and their undermining of the Japanese reformers' efforts for military reorganization, the only trump card left was France's assertion that it could provide superior technology. The next chapter shows how the arrival of German aircraft as war reparations in Japan seriously challenged the validity of this claim—and how the cards for the ambitious game over gaining a monopoly in Japan's aviation were shuffled again.

PART II

Germany and Japan's Army Aviation, 1918–37

CHAPTER 3

Japan's Army Aviation in the Wake of World War I

World War I (1914–18) was the first total war. Even though Japan sided with the Entente Powers, the country was spared the carnage on the European battlefields. Yet the new and disturbing nature of an all-out war reshaped Japanese assumptions about future armed conflicts. In Japan, as elsewhere, reform-minded technocrats and conservative traditionalists drew their own lessons from the "Great War" to promote their positions.

One group claimed that a country could win the next war only with a powerful national defense ideology that ensured unwavering patriotism and sacrifice from the populace. Another side argued that a modern war would be a war of industry, science, and technology fought with modern tanks, artillery, and aircraft.[1] Japanese technocrats and modernizers could point to a hard lesson the country had to learn. When the European warring powers suspended their exports, Japanese industrialists and military officials became painfully aware of their dependence on Western technology. As a result, the urge for self-sufficiency through import substitution and domestic production (*kokusanka*) gained new momentum, and the pressure on Japan's industry and arsenals to expand their research and development mounted.[2]

This chapter begins with an unexpected consequence of World War I, one that challenged French dominance in Japanese aviation. The Treaty of Versailles granted Japan unprecedented access to Germany's aircraft manufacturers and their products. In 1919, early reports about the advances in German aviation attracted the attention of Japanese experts. After the establishment of the Military Inter-Allied Commission of Control in the same year, the Japanese members of the inspection team gained free access to German factories and military facilities, and their attention turned into intense interest. The arrival of German war trophy aircraft in Japan in 1920 further fueled the desire for access to the latest German aviation technology. The Japanese then took the decisive action to purchase German airplanes and to invite German engineers to Japan.

In the 1920s, Japanese observers brought back from Europe new concepts of air superiority, strategic bombing, and air defense that had profound consequences for Japan's fledgling military aviation. The modernizers in the Imperial Japanese Army proposed a plan of national mobilization that included promoting aeronautical research, building a large-scale national aviation industry, and moving toward a more offensive airpower doctrine. This chapter examines the central role of new Minister of the Army Ugaki Kazushige (1868–1956), who, in an era of international disarmament, managed to provide the resources for a significant expansion of Japan's air force.

Another lesson learned from World War I reemphasized the importance of public support for Japan's ambitious airpower expansion. In an effort to learn more about the reasons for Germany's defeat, many Japanese leaders accepted the stab-in-the-back myth, the *Dolchstoßlegende*. Postwar German right wingers zealously promoted this conspiracy theory, claiming that Germany had been defeated not by its enemies on the battlefield but by a lack of support on the home front.[3] Such thinking influenced Japanese planners, who became convinced that arousing, nurturing, and sustaining public backing was an increasingly vital necessity for the aviation project. The 1925 "visit Europe flight" offered a welcome opportunity to capture the nation's attention with a skillfully staged flight spectacle. The army eagerly provided the funds, logistics, and expertise for an immensely popular project that aimed to stir air-mindedness at home—and to propagate the high standard of Japan's aviation abroad.

Early German Influence, 1919–25

Soon after the end of World War I, the Japanese military began to collect firsthand information about Germany. Technical officers and engineers were eager to gain insight into Germany's latest military technology, access to which had been forbidden to them since 1914.[4] When high-ranking officers, such as Admiral Katō Hiroharu (1870–1939), visited Germany, they were surprised about the progress of science and industry there. In Katō's view, it surpassed that of all other countries. Katō was especially impressed by German bomber technology, which he thought would allow the launching of surprise attacks on Great Britain in the near future.[5]

While Katō was still in Germany, a golden opportunity for direct access to German aviation technology presented itself to the Japanese. Section IV of the Treaty of Versailles stipulated the setup of the Military Inter-Allied Commissions of Control. Commission members, among them seventy-three Japanese, had to ensure that Germany complied with the handover, dismantling, or destruction of all war matériel. The Berlin-based Inter-Allied Aeronautical Commission of Control, consisting of 187 officers, was in charge of Germany's aviation.[6] The commission had to be granted unrestricted access to all German aircraft manufacturers, airports, and depots. The five Japanese members of the inspection committee made good use of this privilege and began to send detailed reports back to Japan.[7]

Starting in March 1920, the commission's reports arrived at regular intervals in Tokyo. To the Japanese observers' astonishment, the German public's aviation enthusiasm remained intact even after the country's crushing defeat. Engineer Lieutenant Colonel Hayashi Masaki, a member of the inspection committee, reported on German postwar efforts to further promote popular air-mindedness. Continuing the activities of prewar local aviation clubs, the German Aviation Association (Deutscher Luftfahrtverband) built on a nationwide network of aviation enthusiasts. The association organized glider and model aircraft competitions, held aviation-related lectures, and published magazines and aerial photographs. The German government supported aeronautical research at universities and funded aircraft exhibitions, while the aviation industry continued its close ties with politicians with invitations for factory visits and sightseeing flights. The Japanese observers also commented on the development of Germany's postwar civil aviation. Regular passenger flights on small hydroplanes at Lake Constance led them to conclude that this kind of safe passenger transport would eliminate the public's fear of flying, whereas in Japan hazardous flight shows with decommissioned military aircraft undermined public confidence in aircraft as a means of transport.[8]

The Japanese did not exclusively rely on intelligence gathered by the Inter-Allied Aeronautical Commission of Control. The army's Aviation Department organized its own inquiry into the state of French and German aviation that proved to be highly instructive. In 1922, Colonel Koiso Kuniaki (1880–1950), an Aviation Department staff member and future general and wartime prime minister, went on a ten-month aviation inspection trip to Europe. During his trip, Koiso became discontent with French aviation technology and developed a high regard for German aircraft manufacturers.[9] While visiting French factories, he noticed that the industry had just begun to experiment with all-metal aircraft. When observing the equipment of the French Air Corps, he concluded that there had not been much progress since World War I. To Koiso's surprise, the French military were still using fabric-covered all-wooden aircraft that, because of underpowered engines, were slow and showed only mediocre climb performance.

When Koiso came to Germany, he gained access to almost all major branches of the German aviation industry. Even though Koiso was a member of the Allied Powers, these companies openly revealed their efforts to evade the restrictions of the Treaty of Versailles. In an obvious hope to do business with Japan, electrical engineering company Siemens showed him a secretly developed machine gun for future heavy bombers, and aircraft designer Ernst Heinkel (1888–1958) explained how civilian passenger aircraft could be easily converted to bombers in wartime. Engine manufacturer BMW presented its new 450 horsepower aero-engine and offered to "bypass" the treaty regulations by exporting the dismantled engine part by part to Japan. Koiso learned that aircraft maker Dornier had already moved part of its production abroad to Italy. During his discussions with the company's director, Claude Dornier (1884–1969), Koiso suggested an improvement for a new fighter aircraft, and Dornier promised to send a revised blueprint to the Japanese embassy as soon as possible.

The Japanese Army also sent Koiso to the factory of Hugo Junkers (1859–1935) in the small industrial town of Dessau, 100 kilometers southwest of Berlin. Welcomed as the representative of a future customer, Koiso visited the company's large wind tunnel and learned about Junkers's unique all-metal technology that replaced all wooden aircraft parts with light metal components. Employees seized every opportunity to impress their Japanese visitor. They explained the company's "scientific method" and highlighted the advantages of their designs over those of competitors Dornier and Adolf Rohrbach (1889–1939).[10] As we will see, the visit to the Junkers factory left an especially long-lasting impression on Koiso.

Koiso was aware of the confidential character of his visits. During his stay in Germany, the Japanese Army asked him to become an official member of the Allied military inspection team, but Koiso refused this attractive offer. He explained that having had access to various "absolutely secret weapons" at the German factories, he could not join a committee that would force these companies to publicly disclose their armament production to the Allied Powers.[11]

Aware of the armament limitations imposed on Germany, the Japanese explored the idea of German companies designing and producing their aircraft abroad. Most of the Japanese commission members conceded that the treaty regulations' damage to the German aviation industry was immense. It seemed quite understandable to them that, for many German aviation manufacturers, cooperating with foreign companies was necessary for survival. The commission members also learned that to evade treaty regulations, several German aircraft makers had already outsourced parts of their production to factories in Holland, Italy, and Sweden.[12]

The commission's reports provided the Japanese military with valuable insight into Germany's latest aeronautical research. A recurring topic that most fascinated the Japanese experts was Germany's pioneering manufacture of all-metal planes. This cutting-edge technology depended on duralumin, an alloy invented in Germany. It was praised by the Japanese media as "the surprise of the world . . . lighter than aluminum, harder than steel."[13] The Japanese became familiar with the superior strength, fire resistance, and durability of all-metal airframes, crucial features that promised to overcome the obvious limits of their wooden counterparts.[14] They also obtained detailed research material on the new technology.[15] The commission members learned about advanced research on giant aircraft, high-performance engines, radio-controlled aircraft, and zero-visibility flights at night or in thick fog. They were even introduced to methods for "stopping engines by electric waves."[16] A 1922 report emphasized the crucial role of Germany's renowned technical universities, which combined theoretical study with practical training and continued to be centers of advanced aviation research. The Japanese observers concluded—perhaps overly optimistically—that due to the aircraft manufacturers' mutual support and open exchange of knowledge, "Germany's whole aviation industry was fully capable of maintaining its leading position in the world."[17]

On several occasions, the inspection trips turned from inter-allied cooperation into inter-allied competition for privileged access to German technology. The Japanese became aware of the extent to which German aviation matériel attracted the interest of the other Allied nations.[18] Obviously, in many cases inspection activities turned into outright industrial espionage carried out under the guise of weapons control. A 1922 report by the Japanese Army members of the Commission of Control emphasized how much every Allied country was keen to import and imitate advanced German aviation technology.[19] The Japanese learned that French and British commission members dispatched their own specialists to German manufacturers and research facilities, each eagerly taking as many notes and photographs as possible. They apparently confiscated aircraft and aircraft parts and seized newly developed materials for their own research purposes.[20]

WAR TROPHIES FROM GERMANY

While the regular reports of the Japanese inspection teams provided valuable insight into all aspects of German aviation, another provision of the treaty allowed for a much more immediate exposure to the products of the German armaments industry. Treaty Article 202 specified that within three months after the coming into effect of the treaty, "all military and naval aeronautical material . . . must be delivered to the Governments of the Principal Allied and Associated Powers." The clause meticulously listed the items in question. Apart from airplanes and hydroplanes, the Germans had to surrender dirigibles and their sheds, engines, nacelles, and fuselages to the Allies. Furthermore, Germany had to hand over all aircraft armament and munitions, including bombsights and other devices for dropping bombs or torpedoes.[21] Finally, the list included aircraft instruments, wireless equipment, and reconnaissance cameras. As one of the victorious powers, Japan was entitled to receive part of the seized material as war reparations. The lion's share went to France and Great Britain, but the treaty regulations granted Japan 5 percent of the confiscated German aircraft and equipment. The rest was to be distributed among the United States, Italy, and Belgium.[22]

For Japan being allotted even a small part of the war trophies from Germany was a matter of immense importance. Minister of the Army Tanaka Giichi urged the Cabinet to fund the shipping expense of 100 aircraft and 500 aero-engines. He argued that receiving this German aeronautical material would advance Japan's prestige by putting the country "on the same level as the other Allied nations" and would be extremely important for the progress and development of Japan's aviation. In September 1919, the Cabinet agreed and granted the substantial amount of ¥2.5 million (the equivalent of US$18.1 million in 2018) for disassembly, packing, and transport of the war trophies.[23]

After payment for the transport costs had been secured, Japanese doubts about the fair distribution of the German aeronautical material arose. In October 1919, Japan expected that the Allies would receive about 6,000 aircraft, of which Japan then would be entitled 300. The Japanese reckoned that among those aircraft around 10 percent would be "very valuable."[24] One clause of the treaty regulations determined that if not enough airplanes of one type were available, each Allied Power was to be provided at least with detailed plans and blueprints of the aircraft type in question.[25] Regardless of this provision—meant to provide equal access to German know-how—a scramble for German aviation technology began.

In the same month, disturbing news from Germany arrived at the Ministry of the Army. Major General Watanabe Kōtarō reported that France, Great Britain, the United States, and others had already begun to purchase—secretly and under false names—the latest aircraft from German manufacturers, who were eager to sell their stock before it was confiscated by the government.[26] With the dwindling number of available aircraft, the Japanese began to worry that they would be left with obsolete material. Adding to this sense of urgency was the fact that the long transport time required to bring the material to Japan would render it outdated on arrival, further diminishing the potential of a transfer of cutting-edge technology.

Confronted with this predicament, the Japanese took a first step toward subverting the treaty regulations. Their negotiators arrived at an implicit understanding with Germany that bordered on complicity and led to the gradual alienation of Japan from its erstwhile allies. In August 1920, Captain Kodama Tsuneo, who was sent to Germany to oversee the distribution of confiscated aircraft, exchanged several telegrams with his superiors in Tokyo. He emphasized the importance of securing accessories for German military aircraft that were soon to go out of production. He urged that "if we miss the opportunity to buy them now, we cannot get hold of them anymore later."[27] It would also be necessary to buy special material for repairing the confiscated aircraft. He implored his superiors to entrust him with the selection and the direct purchase of these items.

With Kodama's proposal, the scramble for German aircraft technology reached a new stage, where it threatened to strain Japan's international relations. The Japanese Foreign Ministry intervened and forcibly reminded the Ministry of the Army that the Treaty of Versailles banned any export of military material from Germany. Therefore, "according to this interpretation of the treaty," the Foreign Ministry could not sanction the army's proposed purchase.[28] Nonetheless, the ministry acknowledged the necessity of carrying out the procurements and proposed to entrust the Japanese trading company Mitsui with the purchase: "Let Mitsui carry out the transaction in complete secrecy and let them have it connected to Japan as little as possible." The ministry continued, "if you proceed like this there will be no strong objection from our side."[29]

The ministry's idea to commission a civil trading company to transfer military armament was not surprising. The Japanese government had already secured the services of Mitsui Bussan for transporting and assembling confiscated aircraft from Germany.[30] The involvement of trading companies is important because it paved the way for future Japanese–German business relations in the aviation sector. It must be also noted that Japanese trading companies, especially Mitsui Bussan and Mitsubishi Shōji, set up their own teams of engineers, who actively collected technical information abroad and provided technical expertise to their Japanese customers. These companies played a crucial, if often underestimated, role in the selection and transfer of foreign technology.[31]

In September 1920, the official reparations arrived in Japan. A massive delivery of seventy German aircraft, nearly 300 engines, and countless pieces of equipment—such as wireless telegraphs, cameras, and aiming devices—was unloaded in Yokohama.[32] Considering that most Japanese aviation experts' knowledge of the progress of German aviation was still sketchy at best—the 1910 Rumpler Taube was still considered the quintessential product of the German aircraft industry—there was widespread wonder and astonishment over this treasure trove of Germany's latest technology. Among the especially impressive items was a giant aircraft, a *Riesenflugzeug*, the Zeppelin-Staaken R.XV. During World War I, this bomber terrorized the skies over Paris and London. Five Maybach engines with 245 horsepower each could propel the Zeppelin-Staaken to an altitude of 4,500 meters. With a wingspan nearly equal to that of the largest World War II bombers, it was designed to fly a 1,300-kilometer mission carrying a bomb load of more than 1,000 kilograms.[33]

Nihon kōkūshi, the authoritative 1936 book on Japanese aviation history, contains a detailed list of the delivered items, emphasizing that "all these German aircraft were powerful, sophisticated, and very well maintained." The authors mentioned that—in an example of German thoroughness—each aircraft was equipped with two interior boxes with repair tools and numbered spare parts. They concluded that "whatever aircraft type you look at they are robust and at the same time elaborate. They incorporate absolute scientific perfection."[34]

During the first months after their arrival, these war trophies attracted more attention with their visual impact and propaganda value than with the potential to advance Japan's aviation. As soon as the aircraft and engines arrived at their final destination of the military airfield at Tokorozawa, they were put on display in a newly built hangar. Between October 1920 and February 1921, army officials could take a close look at the exhibits. The officers even received an illustrated catalog for their orientation and as a keepsake.[35] In April 1921, the general public was given an opportunity to marvel at the German aircraft as well. Then the army distributed the items to the research sections of its flight school and artillery arsenal and to the Aerodynamics Department of Tokyo Imperial University. In the words of an effusive official, "It was a golden opportunity for us to examine [and use] them as research material."[36]

Soon the public witnessed some of the German airplanes in flight. The Japanese military assembled a Junkers CL I all-metal ground attack aircraft and, in the presence of Prince Hirohito, gave an impressive flight demonstration at Tokorozawa. *Asahi Shinbun* praised the Junkers aircraft. The paper emphasized the advantages of the all-metal construction, which eliminated risk of fire. The design would also make the aircraft invulnerable to the enemy's attacks—a claim that reflects the journalist's enthusiasm rather than his aeronautical expertise. *Asahi* surmised that "in the future our all military aircraft may become like this Junkers aircraft."[37]

One trophy in particular epitomizes Japanese enthusiasm for German aviation technology. According to the Allied distribution scheme, Japan was entitled to confiscate one whole airship hangar. The building in question was a 30,000-ton iron structure that the German military had erected during the war at Jüterbog, sixty kilometers southwest of Berlin. After careful consideration, the Japanese exercised their right of ownership and dismantled and shipped the building to Japan at a cost of ¥550,000. Tens of thousands of workers reassembled the hangar at the navy base of Kasumigaura, eighty kilometers northeast of Tokyo. After the construction was completed, the Imperial Japanese Navy could boast of owning the largest building in Asia, a made-in-Germany superstructure that could easily accommodate the new Tokyo Station twice over.[38]

PLANS FOR RECRUITING GERMAN ENGINEERS

After examining the aircraft and engines, the Japanese appetite for German technology increased. The army's aviation section proposed hiring German pilots, engineers, and workmen for the assembly and maintenance of confiscated aircraft and for further research on the engines and aircraft parts. Again, the Treaty of Versailles put a formal limit on Japan's ambitious plans. According to Article 179, Germany had to agree "not to accredit nor to send to any foreign country any military, naval or air mission, nor to allow any such mission to leave her territory." Furthermore, "the Allied and Associated Powers agree . . . not to enroll in nor to attach to their armies or naval or air forces any German national for the purpose of assisting in the military training of such armies or naval or air forces, or otherwise to employ any such German national as military, naval or aeronautic instructor."

It seems that despite these clear-cut regulations, the temptation to violate the treaty became overwhelming. In September 1920, Captain Kodama, who already had been instrumental in purchasing German aeronautical matériel, urged Vice Minister of the Army Lieutenant General Yamanashi Hanzō (1864–1944) to consider hiring German aviation experts. To avoid any suspicions of violating the treaty, the ambitious officer suggested selecting and dispatching German pilots, engineers, and mechanics to Japan as Mitsui employees. The salary of these experts would be declared as transport, packing, and assembly expenses.[39] As Kodama guilefully put it, "If we do so there will be no accounting problems and no diplomatic embroilment."[40] However, the vice minister turned down this clever scheme on the grounds that such a maneuver would both "violate the peace treaty and strain Franco-Japanese relations."[41]

Many army officials could not entirely abandon the idea of inviting German engineers, particularly because the Japanese Army felt increased pressure to catch up with the Great Powers' aviation technology, above all in the field of all-metal aircraft. In 1923 the Army Aviation Department again brought up the topic of all-metal aircraft and claimed that, among all countries, Germany's design and technology for these types of airplanes was the most advanced. The department also felt the need for the advice of German designers and specialists in the fields of aero-engines and wireless communication. The report emphasized the eagerness of Great Britain, Italy, and the United States to employ such German aeronautical engineers and urged that Japan do the same. To bypass the treaty regulations, the Army Aviation Department suggested having civil Japanese companies invite the German engineers. These companies would transfer the German specialists to the army's arsenals, at which point the army would take over their pay.[42] The Ministry of the Army was finally convinced of the urgency and the feasibility of the project and approved the proposal for such an "indirect invitation."

These Japanese maneuvers to obtain German aviation technology did not slip past foreign observers undetected and led to numerous complaints by other Allied countries. Japanese officials took the protests about employing German engineers and importing German military aircraft seriously. Aware that any matériel found by Allied delegates to have come to Japan in violation of the treaty could be confiscated or destroyed, the military strongly dismissed any involvement in illegal imports from Germany.[43] As a consequence, the Japanese used several strategies to obfuscate their breaches of the Treaty of Versailles. These included indirect purchases via third countries. For instance, German aircraft maker Rohrbach received Japanese orders via the company's branch factory in Denmark.[44] Another tactic was to draw a supposedly clear dividing line between military and commercial enterprises. When the French attaché expressed his concerns about the cooperation of the Japanese companies Kawasaki and Aichi Tokei with German manufacturers, the Imperial Japanese Navy declared that it was not involved in the practices of these private firms.[45] Occasionally the Japanese Army would admit knowing about the purchase of sample German aircraft by Japanese civil companies while insisting that these were experimental aircraft used for research purposes only.[46]

The Army's Struggle over a New Air Doctrine

Japanese air strategists' determination to get hold of German technology, and their keen interest in advanced all-metal bombers, was the result of a revolutionary new air doctrine. In the wake of World War I, a major dispute over the future role of the aircraft as a new weapon erupted within the Japanese Army. The more radical among the army's modernizers advocated an "all-powerful air force" (*kōkū ban'nōron*). Referring to the British Royal Air Force, which came into existence in April 1918 as the world's first independent air force, they envisioned a similarly autonomous strike force that could be deployed for strategic bombing missions into future enemies' hinterlands.

During the early 1920s, the Japanese general staff hesitated to adopt such an extreme view. Most officers had little or no World War I combat experience, and it was challenging for them to develop a new air doctrine on their own. As laid out in the previous chapter, the 1918–20 French Aeronautical Mission to Japan under Colonel Jacques-Paul Faure did not foster significant change in the strategists' thinking. Artillery officer Faure emphasized the role of military aircraft to provide ground support and was skeptical about the merits of an independently operating air force. Whereas the French mission had exposed the Japanese Army to a wide spectrum of airpower deployment, it contributed little to the advancement of the army's air doctrine.

The discourse about air strategy was an ongoing transnational discussion. In the early 1920s, radically new concepts about the strategic use of bombers arrived in Japan. In 1921 Italian strategist Giulio Douhet published *Il dominio dell'aria* (*The command of the air*).[47] Douhet became one of the most influential airpower theoreticians shaping the ideas of military planners in the United States, Europe, and Japan.[48] He proposed to deploy airpower as an aggressive weapon to deliver a fast and sweeping victory. According to Douhet, a country's air force had to establish air superiority first. It should then start bombing the opponent's cities and industries. These attacks would defeat the enemy by crippling its aircraft industry and demoralizing its population. Such ideas were attractive to many European military planners who, after the experiences of World War I, were afraid of another war of attrition.[49]

Douhet's doctrine was a close match for the views of Marcel Jauneaud (1885–1947), an instructor at the French army academy. In August 1921, the Japanese Army Aviation Department invited Jauneaud to visit.[50] He gave lectures about French air strategy and the lessons learned from World War I. He attended several Japanese Army field exercises, where he observed combined maneuvers of the army's air and ground troops and long-distance bombing. In August 1922, Jauneaud authored a memorandum for the Army Aviation Department and the Army War College that gave strong support to the army's airpower faction. The French officer recommended a large-scale expansion of Japan's air arm that included establishing a substantial corps of long-range bombers that could operate under an independent command.[51] Jauneaud elaborated on this idea in a book on military aviation that he published shortly after his return to France.[52] Referring explicitly to the German giant bombers deployed against London and Paris, Jauneaud envisioned a long-range strike force of heavily armed bombers. He called these aircraft "genuine flying fortresses" (*de véritables forteresses volantes*) that would burn the enemy's capital, destroy its industry and communications, and undermine public morale.[53]

Jauneaud's memorandum prompted the Imperial Japanese Army to draft a more systematic air doctrine, albeit one that diluted the idea of an autonomous airpower and still focused on the tactical deployment of the army's aircraft. Lieutenant Commander Ogasawara Kazuo (1884–1938), a teacher at the war college, had an influential voice in the ongoing discussion. After attending Jauneaud's lectures, Ogasawara published *Lecture Notes on Air Strategy* (*Kōkū senjutsu kōjuroku*). The book's main chapter on the "General Operational Principles of the Aviation Corps" outlined the specific tasks of the army's reconnaissance, fighter, and bomber corps during distinct phases of battle. Very much in line with Jauneaud's proposals, Ogasawara endorsed the independent deployment of the army's bombers for attacking the enemy's hinterlands before the actual battle started. However, he put forward that during battle the entire army air force should closely cooperate with the infantry.[54] Even with this traditionalist provision, Ogasawara's book was important. For the first time, deployment of the army's aviation arm was officially incorporated into the army's traditional doctrine of ground operations. Ogasawara's *Lecture Notes* also prompted the general staff office to further investigate on "ground troops and the deployment of the flying corps" in the second half of the 1920s.[55]

After Italy and France, Germany was the third country to project an impulse for a revision of Japan's air doctrine. The Japanese members of the Inter-Allied Aeronautical Commission of Control not only made good use of their authority to inspect German manufacturers; access to German military aircraft and air strategy also offered several opportunities to compare the military aviation of France and Germany in terms of technology and air strategy. In 1924, commission member Watanabe Kōtarō evaluated the performance of German and French airplanes.[56] The following year, he compiled a detailed summary of German airpower during the war that included an analysis of various aircraft types and their use. Watanabe focused on the performance and armament of bombers and the new tactics of night bombing and deploying incendiary bombs that the Germans had developed in 1917. He strongly advised that lessons be taken from Germany, suggesting a wide-ranging revision of the army's air strategy that until then had been completely based on the French model.[57]

Squaring the Circle: Disarmament and Airpower Buildup

The high-flying plans of the army's airpower strategists might have been scrapped in the mid-1920s when two major disarmaments sharply cut back Japanese troop strength. However, Japan's military aviation survived arms reductions and severe cuts in the military budget and, after 1925, emerged astonishingly reinvigorated in structure and strength. As an important consequence, modern aircraft technology, domestic manufacturing, and a new focus on research became integral parts of the national mobilization.

Between 1920 and 1929, Japan's military expenses shrank from 49 percent of the national budget to less than 29 percent.[58] Several factors contributed to this remarkable development. Recurring financial crises and the devastating Kantō earthquake of 1923 strained national finances and put the military budget under growing pressure. Moreover, public opinion and a liberal diplomacy under Foreign Minister Shidehara Kijūrō (1872–1951) led to a proactive policy of disarmament. Arguably, the single most important factor was the Washington Naval Treaty, signed by Japan, the United States, Britain, France, and Italy in February 1922. To curb military competition in the Pacific, the treaty prohibited the extension of fortifications and strictly limited the number of capital ships.

The Japanese delegates in Washington also agreed to return Shandong province to China and withdraw their troops from Siberia. With the end of the ill-fated Siberian intervention of 1918–22, Japan's army was confronted with public and political pressure to implement cutbacks similar to those of the navy. In response, the new Minister of the Army Yamanashi Hanzō initiated a major army reorganization in August 1922 that became known as the Yamanashi Disarmament. He downsized the army's infantry, cavalry, artillery, and engineering units, resulting in a reduction of 56,000 troops.[59] However, the army's airpower got off lightly. Even though Yamanashi reduced the budget for flight maneuvers and training, he left the strength of the six air wings untouched. In 1925, another radical reshaping of the army's internal structure took place under Ugaki Kazushige, who had become Tanaka's successor as army minister in 1924. The Ugaki Disarmament—which some historians have called an armaments revolution—led to the elimination of four army divisions, 34,000 soldiers in total.[60] Even while reducing troop numbers, Ugaki substantially enlarged the army's airpower.

With his decision, the minister followed an earlier proposal of Koiso Kuniaki, the young and ambitious modernizer in the Aviation Department who made visits to German aircraft makers and became an ardent admirer of German technology.[61] Koiso's rise within the Aviation Department is perhaps the best evidence of the increasing influence of air-minded technocrats in the Japanese Army. To his conservative critics, who resented the troop reduction and still valued the soldier's fighting spirit over modern technology, Koiso famously countered: "The future war will be a science war. Those people talking about the 'Japanese spirit' [*Yamato damashii*]: will they not die when inhaling poison gas? Will they not burn when hit by incendiary bombs? In this [modern] world we cannot depend on the Japanese spirit anymore."[62] Ugaki laid out the rationale behind his new focus on airpower in an essay addressed to a public audience on "the Imperial Army's new facilities and their connection to national mobilization."[63] In his impassioned appeal for building up Japan's airpower, Ugaki repeatedly referred to World War I, which he described as "a war of applied science and fully mechanized battles." He argued that this new type of war had completely changed the foundations of national defense and demonstrated the urgent need to mobilize the country. Furthermore, he continued, after the war the great powers had made new efforts for their national mobilization, but Japan lagged far behind. Ugaki explained that with the notoriously bad condition of Japan's national finances, it was impossible to wait for an improvement in the fiscal situation. Therefore, he was forced to cut four divisions and at the same time to increase the air corps' preparedness. Ugaki warned that the poor state of Japan's air force made the homeland—especially Tokyo—acutely vulnerable against enemy air attacks. Moreover, he lamented the lack of bomber units for attacking the enemy's strategic points and military resources. The minister concluded that the army must assign its most urgent priority to the improvement of its airpower.

Ugaki soon turned his words into action. His "reduce-and-modernize" doctrine completely changed the size, strategy, and organization of Japan's Army Air Force. To underscore the increased importance of military aviation, Ugaki advanced the Army Air Force to the same hierarchical level as the army's infantry, cavalry, and artillery arms. Accordingly, the Army Aviation Department (Rikugun Kōkūbu) moved up to become the Army Aviation Bureau (Rikugun Kōkūhonbu). It was now under the direct command of the minister of the army and obtained additional authority to supervise military flight training, aeronautical research, and the aircraft production of all civil manufacturers.[64] The Aviation Bureau set up a new policy of research and development to make Japanese aviation less dependent on foreign imports.[65] To catch up with Western aviation technology, the army increased its efforts to collect information about technological developments abroad. To promote domestic research, the Aviation Technology Department of the Tokorozawa Flight School received additional funds and gained the status of an independent institution that reported directly to the Aviation Bureau.[66]

Arguably even more important than the organizational promotion of the army's air arm was a fundamental change in its battle doctrine. The Ugaki reform provided funds to radically revise the army air strategy away from reconnaissance and toward offensive air battle and bombing. The extra resources made it possible to more than double the number of fighter squadrons and establish four new bomber squadrons.

Along with the modernization of its air force, the Imperial Japanese Army increased its efforts to boost domestic aircraft manufacturing. The aim was to set up a self-sufficient industry that would fully cover Japan's demand for aircraft, engines, and equipment during peacetime and in times of war. This move met with harsh criticism from those who suggested that it would be better and faster to import cheap foreign products than to use expensive made-in-Japan products. To counter this argument, the Army Aviation Bureau forcefully referred to the bitter experiences during World War I when the United States and Europe had cut off the Japanese military from importing machines and matériel.[67] The arguments for an independent Japanese aircraft industry prevailed. To overcome the foreseeable initial problems of aircraft production, the army devised a carrot-and-stick policy. The country's aircraft makers could count on guaranteed prices but had to accept strict inspections by the army's specialists.

Visions of Internationalism and National Prestige: The "Visit Europe Flight"

The endeavor to expand Japan's airpower and aviation industry greatly depended on the idealistic and material support of the common people. As we have seen, since the early days of aviation in Japan, the military recognized the importance of popular aviation enthusiasm. An exhilarated public participated in Japan's first balloon launch, celebrated the country's first engine-powered flight, and enthusiastically welcomed the news about Japan's courageous air battle over Qingdao. In the 1920s, a new opportunity arose to have the public eagerly join the national aviation project.

The rise of mass media was one of the defining features of the Taishō era (1912–26).[68] To boost circulation and survive fierce competition, Japanese newspapers had to stimulate and satisfy the appetite of an increasingly demanding readership. Aviation was therefore a welcome, novel source of news. Soon major papers engaged actively in the aviation project with their own aircraft and crews, a move that enabled them to further attract and exploit the public's interest in the flying machines and their heroic pilots. By capitalizing on the drama of flight, Japanese newspapers played an increasingly important role in the diffusion of airmindedness among the Japanese public.

In August 1923, *Asahi Shinbun*, one of the country's biggest newspapers, informed the Imperial Japanese Army of its project to organize and carry out a "visit Europe flight."[69] The paper's representatives argued that such a pioneering flight would be of national and international importance: it would not only demonstrate Japan's international competitiveness to the world but also benefit the military by preparing for any future emergency when it became necessary to mobilize public opinion for national defense. *Asahi* therefore asked the army to support the expensive project, the cost of which the company could not bear alone. In particular, the paper requested the army's assistance in providing the necessary equipment, expertise, and staff. Multilingual pilots, flight engineers, and ground staff would be needed, as well as detailed route information, especially for the parts east of the Ural Mountains. The request included a detailed route description and a comprehensive calculation of the costs involved.

As it turned out, the army was kindly disposed to these requests. Vice Minister of the Army Tsuno Kazusuke (1874–1928) instructed Vice Minister of Communication Kuwayama Tetsuo (1881–1936), to appoint a civilian aviation commissioner and secure the assistance of the Japanese attachés based in the countries along the proposed flight route. Additional support included a detailed survey of the flight route and en-route weather conditions, as well as providing rescue boats in case of an emergency landing on water. An intense exchange of notes among the Ministry of Communication, the army, and the Foreign Ministry testifies to the close cooperation of these three bureaus to ensure the necessary overflight permits and ground support. The army dispatched four officers to the Soviet Union to secure fuel supply, means of communication, and the provision of adequate airfields. To disguise their identity, the small group entered the Soviet Union as aviation experts from the Ministry of Communication or as *Asahi* employees. Thus, in addition to their preparations for the flight across Soviet territory, the Japanese Army officers could gather valuable military information about Siberia.

The press campaign for the decade's most ambitious aviation project was well timed. On January 1, 1925, *Asahi Shinbun* announced the "visit Europe flight" in its New Year's edition.[70] Several Western nations had already successfully completed long-distance flights to Japan, which underscored the national significance of *Asahi*'s endeavor. In 1924 alone, four overseas flights had arrived in Japan, causing worries about more "black ships from the sky" and the backwardness of Japanese aviation.[71] *Asahi*'s project, as promising as it was to promote Japan's national pride at home and its international reputation abroad, could expect great public attention.

To further stimulate the readers' interest in the "visit Europe flight," *Asahi* intentionally chose a route none of the Western aviators had tried before. *Asahi*'s aircraft were to reach Western Europe by a high-risk Siberia crossing, where the obvious obstacles of limited fuel supply, inadequate maps, and climate extremes had to be overcome. The newspaper officials had to solve another problem that went beyond the manifest aeronautical challenges. During the previous two decades, memories of the Russo-Japanese War of 1904–5 and the disastrous Siberian intervention of 1918–22 vexed Japan's relations with Russia and the Soviet Union. Only in January 1925, with the signing of the Soviet–Japanese Basic Convention, did the relations between the countries normalize.[72] *Asahi* started negotiations with the Soviet government in the same month and received approval for overflying Siberia five months later. The prospect that the proposed flight would help improve Soviet–Japanese relations and provide an opportunity to overcome the revolutionary government's international isolation obviously soothed initial Soviet worries about Japanese spy activities.

Along with its diplomatic efforts, *Asahi* started a publicity campaign to sell the "visit Europe flight" to the public. As a first step, the paper presented the four airmen that it had chosen for the project. Abe Hiroshi, a skillful pilot and elite career officer, and Kawachi Kazuhiko, an experienced pilot of *Asahi*'s air fleet, were to be accompanied by engineers Katagiri Shōhei and Shinohara Shunichirō, who had both been trained in Europe and were now temporarily employed by *Asahi*. When *Asahi*'s publicity machine gained momentum, the airmen increasingly had to cut back their flight training and exchange their flight suits for dinner jackets. *Asahi* organized meetings with politicians, bankers, members of the Imperial Court, and notables of the aviation world. Further detailed coverage of the aviators' participation at numerous Shinto ceremonies, receptions, and "encouragement conventions" soon turned them into national celebrities who were closely associated with the newspaper.

Moreover, *Asahi Shinbun* launched a countrywide competition for a popular song to accompany and promote the project. The winning lyrics were chosen from among over 3,000 entries and set to music by popular composer Tamura Torazō (1873–1943). The "Visit Europe Campaign Song" (*Hōōhikō Ōenka*) was released as a phonograph and continuously played on the radio. It celebrated *Asahi*'s "silver wings bringing peace," and its refrain—"Go! Go! Heroes to the far away sky!"—soon became a popular tune. To further boost the popularity of the project, the paper published a variety of posters and material for elementary schools and distributed them all over Japan.

Asahi's promotion campaigns showed amazing results. Beginning in March 1925, donations from all over the country rushed in. Japan's Military Reserve Union, youth associations, students, and teachers eagerly supported the "visit Europe flight," and the paper was glad to announce further generous donations from railway companies, Kabuki actors, and amusement theaters. *Asahi* even ran reports about people who abstained from tea drinking and smoking in support of the company's project. When the newspaper settled the final account of the "visit Europe flight," it turned out that actual costs far exceeded the planned budget, but it also became clear that donations could cover more than half of the flight's expenses.[73]

On July 6, as a prelude to an even bigger event, *Asahi* held an ostentatious naming ceremony at the Tachikawa airfield in the western suburbs of Tokyo. The two "visit Europe" Breguet 19 aircraft—French imports assembled by the Nakajima Aircraft Company—were at the center of the function. *Asahi* had invited one of Japan's most prominent dignitaries, Prince Yamashina Takehiko (1898–1987), who after graduating from the Yokosuka Naval Flight School in 1920 became known as the "Prince of the Skies" (*sora no miyasama*).[74] The prince named the two planes *Hatsukaze* and *Kochikaze*, or "Early Autumn Wind" and "East Wind."[75] According to *Asahi*'s report, 10,000 elementary students attended the event, and 150 students of the Tachikawa Youth Choir cheerfully sang the campaign song. Then, from July 15 onward, *Asahi*'s small fleet of aircraft set out for the country's major cities, over which they dropped leaflets with the latest details of the "visit Europe flight."

Once more, the Yoyogi Parade Ground became the venue where Japanese aviation history was written. On July 25, 1925, a day with perfect flight weather, it was impossible to escape the sights and sounds of a landmark event. All of Japan was caught up in a festival fever. Already in the early morning, signal flares, speaker cars, and radio broadcasts informed the people in and around Tokyo about "today's decisive action." Tramways and buses were decorated with flags, and their conductors issued special "visit Europe flight" tickets. In every corner posters advertised the flight, and the streets were filled with children singing "Go! Go! Heroes!"

At the parade ground, 200,000 people—including hundreds of government officials and military and business leaders—assembled in an order where nothing was left to chance. The two "visit Europe" aircraft were at the center of a ceremonial arrangement, facing the baldachin of the imperial family. In between, a small canopy with a microphone on a desk had been set up. A wide half-circle of dignitaries surrounded the planes from the rear. At the beginning of the departure ceremony, the master of ceremonies (*shikaisha*) stepped to the microphone. When he announced the opening of the performance, his voice was carried by numerous large speakers that had been installed all over the grounds. Then Prince Kan'in Kotohito (1865–1945) delivered a congratulatory address, followed by a farewell speech by the minister of communication and a message from Prime Minister Katō Takaaki (1860–1926). After three cheers of *banzai*, a chorus of 3,000 elementary school students sang the campaign song, accompanied by a military band (figure 3.1). A formation of army aircraft crossed the ceremonial space and dropped celebration messages. Finally, the "visit Europe" crews climbed into their planes, started the engines, and—with the whole crowd watching, waving, and cheering—took off into the "far-away sky." Actually, their initial flight took them to the Jōtō parade ground in Osaka, where another 150,000 exhilarated people welcomed them.

Finally, the two planes left Japan for their long journey of more than 17,000 kilometers. *Asahi* reported proudly how locals received the aircraft and crews enthusiastically along their route. Reports of welcome ceremonies and receptions at each scheduled stopover provided the paper with a reliable source of new headlines. These accounts reassured readers that the flight crews were safe and Japan's international prestige was on the rise. Mixed with tales of adventure and exoticism, the reports did not fail to keep the public's interest alive.

In conjunction with its relentless coverage of the "visit Europe flight," *Asahi* used a whole range of marketing tools. Free extra editions, festivals, aviation-related crossword puzzles and board games, and contests for guessing the flight time sustained the enthusiasm for the mission's success (figure 3.2). The strategy worked: with the rising popularity of the project, *Asahi*'s circulation and sales figures shot up. On October 27, the day the two aircraft reached their final destination of Rome, the paper was flooded with congratulatory telegrams and received no fewer than 5,600 submissions in its competition for a welcome song. When the flight crews returned to Japan, an enormous crowd of 100,000 greeted them at Tokyo Station. The nation celebrated, and pilot Kawachi Kazuhiko received the Order of the Rising Sun, Japan's second most prestigious decoration.

Asahi's flight was a remarkable aeronautical achievement. It also was a classic example of how to secure public support—both idealistic and material—with mass events that went hand in hand with repetitive, all-out coverage. The project appealed to the masses because it was an open

FIGURE 3.1 Rosy cheeks and a sea of brightly colored flags. This colorized photograph is the title page of the *Osaka Asahi*'s 1925 *Memorial Picture Report of the Great Visit-Europe Flight.* In spite of all the hyperrealism, the picture has an abstract note. Depicting no aircraft but with all eyes cast upward, the photo brilliantly conveys the aviation enthusiasm stirred by the paper's "visit Europe flight." (Source: *Asahi* Shinbun, *Ōshū hōmon)*

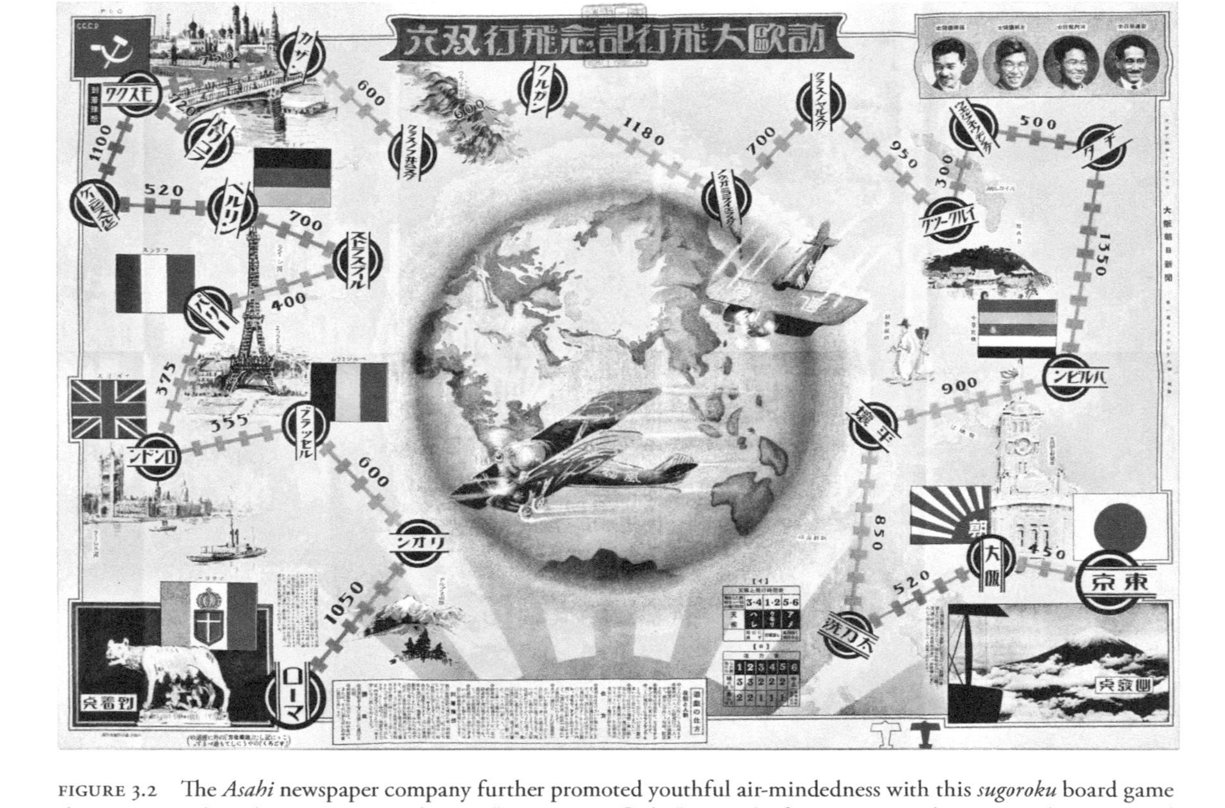

FIGURE 3.2 The *Asahi* newspaper company further promoted youthful air-mindedness with this *sugoroku* board game that invites its dice players to reenact the 1925 "visit Europe flight." Note the four airmen in the upper right corner and their two aircraft whizzing around the globe in the center. The newspaper proudly shows its Osaka head office in a prominent position along with Pyongyang's Moran Hill, the Kremlin, the Eiffel Tower, and the Houses of Parliament in London. (Source © Tokyo Metropolitan Library Special Book Collection Room)

invitation to participate in the intense drama of flight and celebrate national grandeur and international friendship. The songs composed during and after the flight testify to the hopes for a more peaceful coexistence, and pilot Abe Hiroshi's appreciation—"I will never forget in my whole life, how everywhere all along our flight the Soviet people enthusiastically and kind-heartedly welcomed us"—conveys a genuine gratitude for the citizens of a country that long was considered Japan's archenemy.[76]

On the other hand, it is important to note that the success of *Asahi*'s flight nurtured an enduring air-mindedness that had a peculiarly nationalistic note. The paper's report, published shortly after the completion of the "visit Europe flight," emphasized the important consequences of a flight that "still in thirty and fifty years will influence the development of Japan's aviation world and the development of the Yamato race in the world."[77] The more immediate effect was continuing aviation enthusiasm, especially among young Japanese. The story of the "visit Europe flight" made it into school textbooks; schools organized "visit Europe flight" dances and relay competitions and engaged in a boom of building model aircraft.

Finally, the flight to Europe was a national project carried out for the most part by the army but presented to the public as a civil enterprise. Considering that the newspaper reaped substantial benefits to its image and circulation, it is not difficult to surmise that the army's cooperation came at the price of increased state influence on the press. After their generous support, the army could expect the paper's future cooperation to help instill in the public the military's own version of an aviation ideology.

Conclusion

In the wake of World War I, Japan's army aviation underwent dramatic changes. A historic opportunity for importing aviation technology opened up, plans for an unprecedented expansion of the nation's airpower emerged, and national air-mindedness reached a new high. All this happened in a broader context of international diplomacy, transnational studies of a new air doctrine, and a wildly popular transcontinental flight.

The architects of the Treaty of Versailles unwittingly promoted German aviation technology by granting the Japanese unprecedented access to Germany's aircraft manufacturers and their products. Exposure to German aviation technology finally broke the French monopoly in the Imperial Army's air arm. It intensified Japan's interest in German aircraft and engines to the extent that it enticed the country to violate the treaty and turn to complicity with its former enemy.

One particular weapon especially fascinated the Japanese. The army's strategists became aware that during World War I, bombers' fields of operations had expanded beyond the support of ground troops to attacks on the enemy's hinterlands. Germany had been the first country to implement this new strategy with newly developed four-engine bombers equipped with incendiary bombs. When these aircraft attacked Paris and London, they not only brought about widespread destruction but also caused mass panic and heavily undermined public morale. Detailed accounts of these air raids, along with an examination of German war trophy bombers, attracted Japanese interest in the technology for a new type of aircraft capable of carrying a heavy bomb load over a long distance.[78]

The new technology intensified the dispute between traditionalists and modernizers about a new airpower doctrine. Italian, French, and German air strategists had a strong influence on the reform-minded officials of the Imperial Japanese Army, most notably Ugaki Kazushige. Ugaki successfully carried out a disarmament policy while promoting a new doctrine of aggressive airpower deployment. His efforts initiated a significant buildup of the army's air arm, an expansion of the bomber fleet, and the development of a large-scale domestic aviation industry.

Finally, the decisive importance of public support for the military was another lesson learned from World War I. The army's increasingly close collaboration with the press culminated in the spectacular "visit Europe flight" of 1925. The ambitious aviation project excited the nation and imbued the people with a sense of shared purpose and national self-esteem. At the same time, it gave rise to an aviation ideology that was as important for Japan's armed forces as its hardware, like bombers and fighter aircraft. This new air-mindedness made the military confident that it could count on the voluntary and even passionate support by the populace for the project of nationalism and military expansion.

Such was the atmosphere of the mid-1920s, when access to German technology held the promise of raising the level of Japanese aviation to meet or even surpass that of the world standard. It also nurtured hopes that in time, a national aviation industry could become established to make the country independent from foreign imports and know-how. The next chapter follows the arrival of German aeronautical engineers in Japan during the latter 1920s and examines whether their activities lived up to the heightened expectations of their hosts.

CHAPTER 4

On the Way to Independent Aircraft Design

This chapter explores the development of Japan's army aviation in the second half of the interwar period, a time of volatile international relations and fast-changing defense policies. The army's demand for state-of-the-art aviation technology prompted Kawasaki and Mitsubishi, two of the largest aircraft makers in Japan, to invite German manufacturers, engineers, and scientists to Japan. These specialists played a pivotal role in the most advanced form of technology transfer: transmitting the capacity for independent design. German expertise had a remarkable impact on the progress of Japanese airpower. When the Japanese Army went to war against China in 1931, it was able to deploy some of the world's most advanced fighters, bombers, and reconnaissance planes.

In 1925 the newly established Army Aviation Headquarters (Rikugun Kōkūhonbu) arrived at an important decision. The army's aviation research department at the Tokorozawa Flight School would no longer design and build new prototypes. Instead, developing and manufacturing the army's future planes would be entirely contracted out to civil companies. Consequently, without receiving any technological assistance from their government client, these manufacturers had to lay the foundation for their own research and development. At the same time, the army introduced the new concept of "prototype competition," which led to intense rivalry within the aviation industry and increased the pressure for innovative designs. As a result, the major aircraft makers turned to foreign experts for help designing airplanes that could meet army requirements and win lucrative contracts for large-scale production.

The army's strategists were aware of recent developments in airpower deployment. In the mid-1920s, they began to envision a fleet of strategic bombers that could carry a war deep into the territory of a faraway enemy. However, the rise of the ultra-conservative Imperial Way faction (*Kōdōha*) within the army weakened the position of these modernizers. Furthermore, the deployment of army aircraft during the 1931 Manchurian crisis provided little incentive for the army's planners to further modernize their strategy. The progress of the army's air strategy stagnated, and by the early 1930s it fell into what I call a "doctrinal slumber."

Industrialists, Engineers, and Teachers

By the early 1920s, Japanese specialists had become increasingly interested in German aviation technology. As laid out in the previous chapter, reports by the Japanese members of the Inter-Allied Aeronautical Commission of Control—along with war reparations from Germany following the Treaty of Versailles—effectively propagated the recent advances in German all-metal construction. The news about a revolution in aircraft design arrived just in time. In 1923 the Army Aviation Department decided to modernize its bomber fleet. A new generation of heavy bombers would replace the increasingly obsolete French Farman F.50 and Farman F.60 Goliath—both biplanes with fabric-covered wooden frames.[1]

TURNING FLYING BOATS INTO BOMBERS: KAWASAKI'S TIE-UP WITH DORNIER

Shipbuilding company Kawasaki Dockyard played a major role in the army's modernization project. Kawasaki had started building aircraft in 1918 at its Kobe factory. In 1922, their aircraft production expanded and moved near the army's Kakamigahara Airfield, about twenty-five kilometers north of Nagoya. In October 1923 Colonel Sugiyama Gen (1880–1945), section leader at the Army Aviation Department, met Takezaki Tomokichi, the head of Kawasaki's aircraft department, and asked him to find a German aircraft maker who could design a new all-metal bomber for the Japanese Army.[2] Takezaki promptly arranged a meeting with the German trading company Illies, which helped Kawasaki successfully establish contact with aircraft designer Claude Dornier.[3]

Cooperation with hydroplane maker Dornier was promising for several reasons. The Kawasaki engineers were convinced that Dornier's design and production methods for flying boats suited them because they allowed the Japanese company to transfer its well-developed shipbuilding technology to its aircraft section.[4] Furthermore, Dornier offered considerable experience in international cooperation. In 1920, the Allied commission had ordered the destruction of the planes at his German factory. In the following year Dornier decided to outsource large parts of his production to Marina di Pisa, a small port town about 300 kilometers northwest of Rome.[5] Italian bankers and industrialists provided the funds, facilities, and workforce for a German–Italian joint venture, the Costruzioni Meccaniche Aeronautiche. In July 1922, the company began the production of the Wal ("whale") flying boat, which soon became a top seller worldwide and propelled Dornier to international fame. As it turned out, the political, financial, and technological aspects of this international cooperation provided a tried-and-tested pattern for the Dornier-Kawasaki tie-up.

The Dornier-Kawasaki license contract, concluded in February 1924, reveals the remarkable extent and intensity of knowledge, manpower, and material transferred to Japan.[6] The license permitted Kawasaki to build nine different types of aircraft, including small trainers, fighters, reconnaissance aircraft, and bombers. Dornier agreed to provide a sample airplane of each type, along with blueprints, design calculation sheets, specifications of all materials, and instructions for maintenance and operation. In addition, Dornier's technical staff would give instructions for manufacturing and make available the necessary machine tools, templates, drilling jigs, and gauges. In turn Kawasaki agreed to pay to Dornier the sizable sum of ¥875,000, equivalent to nearly a sixth of Kawasaki Dockyard's annual profit.[7] The effect of this payment on Dornier's financial situation cannot be overestimated: the net profit from the license agreement alone more than tripled Dornier's 1924 revenues. It is important to note that only in November 1923, a few months before the contract was signed, had German hyperinflation ended. It had wiped out all financial assets denominated in marks.[8]

To avoid any allegations of violating the Treaty of Versailles, cooperation between Kawasaki and Dornier began as a top-secret arrangement. However, government officials of both countries were well informed.[9] In August 1924, Kawasaki notified Minister of the Army Ugaki Kazushige about the cooperation with Dornier and the invitation of "highly skilled engineers from Germany."[10] Two months later, German ambassador Wilhelm Solf (1862–1936) reported similar news to the German Foreign Ministry: Kawasaki, "with the support of the Japanese government," had bought the license to build Dornier's aircraft. Production would start soon with the help of German experts, who had already arrived.[11]

In the same year the contract was concluded, Dornier delivered the first three aircraft to Kawasaki. Satisfying his obligations under the contract, Dornier himself set off for Japan. He arrived in November 1924, just in time to supervise the reassembly of his planes in Kobe. The public flight demonstration of his flagship was scheduled for December 3. Army and navy officials, along with journalists and several hundred invited guests, attended the event. Dornier's test pilot, Erich Just (1898–1955), climbed into his narrow, open cockpit, and the presidents of the two companies—Claude Dornier and Takezaki Tomokichi—took seats in the more comfortable passenger cabin. At eight o'clock sharp, Just pushed the engine throttles forward. The 7.4-ton Wal flying boat accelerated and smoothly took off for its 500-kilometer nonstop flight from Kobe to the Kasumigaura naval base. Wavering between fright and admiration, the Japanese press praised the hydroplane as a "monster in the sky" (*kūchū no kaibutsu*) and excitedly quoted Dornier's comment that his aircraft would make it possible to fly from Tokyo to Berlin in only thirty hours, an impressive improvement over a sea-and-land travel time of more than two weeks.[12]

After this successful publicity event, a unique technical cooperation began over a distance of 10,000 kilometers. In his Friedrichshafen design office, Dornier redesigned the Wal passenger hydroplane as a land-based bomber and renamed it the Do N.[13] He equipped the plane with three machine guns for defense against enemy aircraft, one camera for

FIGURE 4.1 Building a new heavy bomber for the Japanese Army. This look at Kawasaki's shop floor shows three workmen attaching the casing for the two 450-horsepower engines to the aircraft's fuselage. They are being watched by two white-collared men in Western attire. In the foreground, the tools and workpieces of the laborers and the half-finished wing are all characteristic features of the new all-metal construction that made the airplane a major technological breakthrough. (Source © Airbus Corporate Heritage)

reconnaissance purposes, and one bomb-releasing device. In addition, he modified the wings and fuselage to allow easy disassembly for rail transport.[14]

In early 1925, Kawasaki began the production of a prototype (figure 4.1) under the technical direction of Dornier engineer Richard Vogt (1894–1979). The Japanese Army sent six officers to Kawasaki to closely observe and study the production process. In February 1926, the Do N completed its acceptance flights in strict secrecy.

After the successful trials, Kawasaki composed an enthusiastic message of congratulations in English and cabled it to Dornier:

> Test flights completed within three successive days and to clients' fullest satisfaction STOP Characteristic performances and mechanical

FIGURE 4.2 The Kawasaki Type 87 bomber. Note the size of the aircraft and how the shape of its fuselage reveals its flying-boat origins. The aerodynamically awkward opening at the very front of the airframe was to accommodate the nose gunner with his twin machine gun. (Source © Airbus Corporate Heritage)

> fulfillments of machine proved perfect and excellent even beyond our expectation STOP Not single adjustment nor repairing was necessary during trials STOP Clients and ourselves all take our hats off and salute you for unparalleled success of your authentic design mutual congratulations.[15]

To any contemporary observer, the bomber must have evoked a mixture of awe and trepidation (figure 4.2). Gone was the graceful fragility of earlier constructions. With a wingspan of nearly twenty-seven meters, nose and rear machine guns, and one huge bomb attached to its fuselage, the aircraft had a takeoff weight of nearly eight tons and deserved to be called a heavy bomber.[16] In 1927 the Japanese Army officially adopted the Do N and introduced it under the name Army Type 87 Heavy Bomber. It was Japan's first bomber able to carry a bomb load of 1,000 kilograms. According to one source, the army treated the new aircraft as its priceless treasure or "tiger cub" (*tora no ko*).[17]

While Western powers still relied on wooden biplanes for bombing missions, Japan's fleet of all-metal monoplane bombers was clearly the world's technologically most advanced strike force. Between 1926 and 1932, Kawasaki delivered a total of twenty-eight Type 87 bombers to the army. From 1927 on, the army assigned them to the Heavy Bomber Squadron of the Hamamatsu bomber base. In light of the army's large-scale order, Kawasaki took a favorable view of its cooperation with Dornier. The company was proud of having become the leading designer and producer of all-metal aircraft that had also mastered such advanced design features as the cantilever wing, which did not require external supports, and a streamlined monocoque airframe that integrated the base frame into the aircraft's body.[18] Considering that the United States introduced their first all-metal monoplane bomber in 1931, and the UK Royal Air Force's first all-metal monoplane bomber entered service in 1936, Kawasaki had clearly taken the lead in advanced bomber design.[19]

Kawasaki's achievements also helped soothe the Japanese public's worries about the backwardness of the country's aviation technology, a recurring issue covered in the Japanese press. In a 1932 article, *Kōbe Yūshin Nippō* praised Kawasaki's cooperation with Dornier.[20] The paper pointed out that while Japanese had "swallowed Western technology without chewing" (*u-nomi*; literally, "gulp down like a cormorant") for a long time, Kawasaki was now playing "an active role in the front lines of the Japanese aviation industry." The newspaper congratulated the company for its foresight in contracting with Dornier and adopting the all-metal technology. This move, the article emphasized, enabled the company to build "exactly the type of heavy bombers the army air force was so urgently looking for."

THE THEORY AND PRACTICE OF AIRCRAFT DESIGN: ALEXANDER BAUMANN AT MITSUBISHI

Building a modern strike force of heavy bombers was only one aspect of the Japanese Army's aerial armament initiative. In the mid-1920s the army placed orders for advanced fighters, reconnaissance aircraft, and light bombers. The aim of this ambitious project was twofold. First, it was to provide the Japanese Army Air Force with the necessary equipment to support ground troops and intercept enemy aircraft. Second, the program would also promote an autonomous national aviation industry. As it turned out, two foreigners played crucial roles in leading Japanese aviation engineers toward independence not just from imports but also foreign design.

Mitsubishi was another Japanese company that used its technological proficiency gained from shipbuilding to diversify into aircraft manufacturing. Initially Mitsubishi had joined the aviation industry as an engine maker. Following a navy order for aircraft engines, the company established its Internal Combustion Engine Section (Nainenkika) at its Kobe shipyard in 1916. In May 1920, Mitsubishi founded a new company, Internal Combustion Manufacturing (Mitsubishi Nainenki Seizō) and put it in charge of all aviation-related activities. Mitsubishi Nainenki set up its factory on a stretch of reclaimed land at the port of Nagoya that the company's submarine section had abandoned earlier.

In the mid-1920s Mitsubishi intensified its efforts to not depend on foreign license agreements any longer. A board meeting in 1924 addressed the company's concern to move past mere "copying of Western products." Referring to the fast progress of aircraft design in the West, board members emphasized the importance of upgrading the company's research and prototype development. The board decided to invite Alexander Baumann (1875–1928) to come to Japan and introduce a "new design for the aircraft of the future."[21]

Baumann was well known to Japanese aviation specialists. In 1910, he became the first scholar in Germany to be appointed to a professorship in aircraft engineering. During World War I, he designed a giant bomber, the VGO I, which made its maiden flight in 1915. Mitsubishi approached Baumann in 1924. An offer to pay a monthly salary of more than ¥4,000—about nine times more than the head of Mitsubishi's Nagoya factory—convinced the professor to come to Japan. In his two-year contract, he agreed to "instruct the workmen and engineers, draughtsmen and designers of the employers in all details in relation of [*sic*] aircraft and motor construction."[22]

In 1925, the year of Baumann's arrival in Japan, the army decided to start its first "prototype competition" (*kyōsō shisaku*). In these competitions, several aircraft makers would receive orders to submit newly designed prototypes for competitive evaluation by the Army Technology Council (Rikugun Gijutsu Kaigi), the chief of the general staff (Sanbō Sōchō), and the inspectorate general of military training (Kyōiku Sōkan).

Upon the recommendation of the three departments, the minister of the army would officially adopt one aircraft and make an order for large-scale production to the winning company.[23]

In spring 1925, Japan's three biggest aircraft makers, Mitsubishi, Kawasaki, and Nakajima, received the order to participate in the first prototype competition for a new light bomber. Mitsubishi put Baumann in charge of the project and granted him considerable latitude in designing a new airframe. Baumann decided to use a distinctive mix of materials. The aircraft's fuselage consisted of a fabric-covered duralumin skeleton, and the wings were made from a mix of light metal and wood. W-shaped struts connected the lower and upper wings, eliminating the need for extra bracing wires. The aircraft's unusually long and narrow upper wing followed the aerodynamic theorem that such a shape reduces drag and thereby increases performance.[24] Although designed as a light bomber, Baumann's aircraft was heavily armed with four machine guns and could carry a bomb load of 800 kilograms.[25]

In spite of its high bomb load and a remarkable top speed of more than 200 kilometers per hour, the aircraft did not satisfy the army. The evaluation team decided that the plane's structure was too complicated and (in sharp contrast to Mitsubishi's claims) too expensive to build, with a production cost more than twice that of its competitors' designs. Nevertheless, especially with its advanced wing design, it served as valuable "teaching material" for Mitsubishi staff.[26] Baumann arrived at a positive evaluation as well, stressing the instructional aspect of his endeavor: "I am most delighted that this is the first aircraft that was designed, calculated, and built by Japanese engineers under my guidance. . . . I am sure that after I have left Japan those people will build excellent aircraft without me."[27]

In September 1925, the Ministry of the Army launched its next prototype competition. It ordered the three invited companies—Mitsubishi, Kawasaki, and Ishikawajima—to develop a new reconnaissance aircraft to replace the obsolete Type Otsu-1, which Kawasaki had been producing under French license since 1922.[28] To observe enemy forces well behind the front line, the new plane was to have a flight range of 1,000 kilometers, twice that of its predecessor. In addition, the plane should carry one flexible and two forward-firing machine guns. Once more, Baumann envisioned a very unconventional arrangement. To maximize the pilot's field of vision—an obvious requirement for reconnaissance missions—he increased the distance between the upper and lower wings as much as possible. This design feature reduced the aerodynamic interference between the two wings, resulting in improved flight performance. Baumann also minimized the use of struts and wires to reduce air resistance. The result was the Tobi ("black kite"), a fragile-looking two-seater with a half-size lower wing and a staggered wing arrangement that earned the aircraft the nickname "praying mantis" (*kamakiri*) (figure 4.3).[29] However, during the official test flight in 1927, the aircraft's weak point was not its unorthodox wing construction but its undercarriage design. During landing, one of the Tobi's undercarriage struts collapsed, and the aircraft suffered heavy damage. The army did not adopt the aircraft.

Despite this series of setbacks, Mitsubishi was eager to continue cooperation with Baumann. Even after the professor's return to Germany in autumn 1927, the company signed a three-year follow-up contract. For a handsome annual remuneration of ¥16,000, Baumann agreed to teach a number of Mitsubishi engineers who were to be transferred to Germany. He agreed to return to Mitsubishi's factory in Japan each summer for a period of four to six weeks. But what could have turned into a unique type of technological counseling did not take place. Baumann died of a heart attack in March 1928 at the age of fifty-two.

All of Baumann's designs resulted in costly failures. Why, then, was Mitsubishi still interested in his advice? The answer can be found in Baumann's unique approach to passing on his knowledge. While emphasizing the importance of research and theoretical investigation, he entrusted the Japanese designers with the practical implementation of their studies. As one Mitsubishi engineer recollected:

> The abilities of the Japanese staff had already reached a relatively high level. Therefore, Baumann taught us the [theoretical] fundamentals of aircraft design but left the details of strength calculations and specific design features to us. . . . Baumann drew all his conclusions with a typically German rigor based on [aerodynamic] theory.[30]

This quotation illustrates how Baumann's instructions played a key role in transitioning Mitsubishi's engineers from mere empiricism to scientifically based design. As an academic who was under no obligation to

FIGURE 4.3 This photograph shows Mitsubishi's Tobi experimental reconnaissance aircraft designed by Alexander Baumann and his Japanese team. The German professor's unusual wing design was to improve the pilot's sight and the aircraft's overall aerodynamics. The unorthodox arrangement also gave the airplane its nickname, "praying mantis." (Source © Japan Aeronautic Association)

promote the sale of German aircraft or manufacturing licenses, Baumann could encourage and enable Mitsubishi's engineers to arrive at their own independent designs. Engineers like Horikoshi Jirō (1903–82), who rose to fame as the designer of the Zero fighter, remembered Baumann as both a strict teacher and an instructor who devoted his time "from dawn to dusk" to training Mitsubishi's engineers.[31] Baumann, in turn, held the Japanese in high regard. Shortly after his return from Japan, he cautioned his German audience against underrating the Japanese. He asserted that the "diligence of the Japanese was more than a match for their German counterparts."[32]

TEN YEARS IN JAPAN: RICHARD VOGT

The designs of a different young German engineer propelled the Kawasaki company into the league of Japan's biggest aircraft makers. As a high school student, Richard Vogt had made himself familiar with airplanes at a small airfield near Stuttgart, where aviation pioneers like Ernst Heinkel, Hellmuth Hirth (1886–1938), and Baumann tested their early inventions. In 1916 Vogt joined the Zeppelin factory at Friedrichshafen; the next year he accepted Dornier's offer of becoming head of the company's department of aerodynamics. Though he had no formal training as an engineer, Vogt managed to lead the section with Dornier's help and advice. After World War I, he studied under Baumann's supervision at the Technische Hochschule Stuttgart, where he earned his PhD in 1921. Then, in 1924, Dornier promoted Vogt to head engineer of a small group of experts to be dispatched to Japan. Together with two other engineers, two master craftsmen, and one test pilot, Vogt arrived in Yokohama in September 1924 (figure 4.4).

As soon as Vogt began his work at Kawasaki, he realized his job involved much more than supervising the assembly of aircraft delivered from Germany. Five army officers had their desks close to his office. Kazumi Kensuke, a former member of the Inter-Allied Aeronautical Commission of Control, led the group that had been dispatched to Kawasaki for "research on the experimental production of the heavy bomber."[33] Vogt soon learned that his salary was paid by the army and the main purpose of his stay was training these officers. They expected the young engineer to answer all their questions, even when they went beyond the

FIGURE 4.4 A group photo of the proud Dornier-Kawasaki team. We can see the six Dornier employees in the two middle rows, with the youthful Richard Vogt in the center wearing a gray suit (no hat, no moustache). Note the two men in uniform whose presence clearly reveals the military's involvement in the German–Japanese cooperation. (Source © Airbus Corporate Heritage)

scope of his experience.[34] Vogt also became aware that Kawasaki ultimately aimed for independent design of their aircraft. When he brought up this topic to his superiors at the Dornier company, he received the instruction to "delay such a situation as much as possible."[35] This short but revealing directive already hints at Vogt's conflict of loyalties. He had to cope with Claude Dornier's insistence to stay in business by keeping Kawasaki in a semi-dependent position. Kawasaki, on the other hand, expected Vogt to lead its designers toward technological self-sufficiency.

Vogt soon had the opportunity to demonstrate his talent and originality—and his determination. In September 1925, Kawasaki joined the prototype competition for a new reconnaissance aircraft and appointed Vogt as chief designer for the project. The German engineer decided to use an all-metal "stressed-skin" structure, an advanced design feature that Hugo Junkers had introduced in 1923 and Adolf Rohrbach had put to practical use the following year.[36] The stressed-skin design concept allowed for designing a light yet robust aircraft by using the plane's casing to improve its overall structural strength. Vogt devised an unusual airframe with two vertical struts connecting the upper and lower wings of the biplane. Even though this arrangement minimized drag and reduced the overall weight of the airplane, Dornier let Vogt know from Germany that he did not approve of the design and suggested a more conservative approach. Vogt nevertheless convinced Kawasaki to accept his ideas—proof of the authority the young designer was already enjoying at his new employer.

In 1927 three aircraft manufacturers presented their prototypes to the army. The competitive review turned out to be an all-German design contest: Vogt's aircraft had to compete with Mitsubishi's Tobi, designed by Baumann, and Ishikawajima's T-2, built under the supervision of Gustav Lachmann (1896–1966). In the employ of Ishikawajima, Lachmann was a German aeronautical engineer who had already become famous for his innovative design of slotted wings, which improved an aircraft's low-speed capabilities. In January 1928, in spite of the vaunted expertise of his competitors, Vogt's prototype won the competition on account of its superior air speed and flight range.[37] This remarkable success not only led to the adoption of the aircraft by the army but also earned Kawasaki a cash prize of ¥200,000. Apart from carrying considerable prestige, the prize was a much-needed financial windfall for the aircraft department. Kawasaki still suffered from the fallout of the 1927 Shōwa Financial Crisis that had resulted in the layoff of more than 3,500 employees and the provisional seizure of assets at the parent group, Kawasaki Dockyard.[38]

Even more important, the Kawasaki Type 88, as the new reconnaissance aircraft was called, became one of Japan's most successful planes of the late 1920s. Japanese historians have called it "the pioneer of made-in-Japan aircraft" (*kokusanki no sakigake*).[39] The Type 88 was manufactured over a period of more than ten years with over 1,000 units built.[40] The plane made Kawasaki one of the leading aviation manufacturers in Japan. Vogt's proud statement naming the plane "my first big success" fits with the comment of his Japanese apprentice engineer, Doi Takeo (1904–96).[41] Doi, who later became one of Japan's most influential aircraft designers, declared in an interview that with the Type 88, the "obscure aircraft designer Vogt had become world-famous."[42]

FOREIGN EXPERTISE PUT TO THE TEST: THE COMPETITION FOR THE ARMY'S NEW FIGHTER AIRCRAFT

In the mid-1920s Japanese worries about Soviet airpower began to intensify. Reports about the increasing flight range of Soviet bombers emphasized the need for a new, powerful fighter aircraft to intercept the enemy's bombers before they could launch an attack on Japan's major cities.[43] In November 1926, Inoue Ikutarō, the head of the army's Aviation Bureau, and Minister of the Army Ugaki Kazushige discussed the matter. They agreed that the army's present fighter, a licensed production based on a 1918 French design, had clearly become "inferior to the new fighter aircraft of the Western Great Powers" and was unfit for such a task.[44] Inoue reminded the minister that the development of a new aircraft would take almost three years and asked for permission to direct the three largest aircraft manufacturers—Mitsubishi, Kawasaki, and Nakajima—to develop a new interceptor. Three months later, the minister approved Inoue's request.

By that time the army had already set up the specifications for its new fighter. To engage successfully in combat with enemy aircraft, the new plane had to be exceptionally fast and maneuverable. In addition, it needed superior climb performance, an obvious requirement for effectively intercepting high-flying enemy bombers.[45] The experts of the army's Akeno Flight School, who were also involved in the initial planning, furthermore emphasized the importance of an unobstructed downward view and insisted that the new fighter had to be a high-wing monoplane.[46]

The army had developed a new evaluation procedure that was no longer based on rudimentary trial and error methods but on numerical analysis and scientific evidence. Before giving the order to build any prototype, the army instructed each manufacturer to provide detailed performance and strength calculations along with a scale model of the aircraft for evaluation in wind-tunnel tests. The three companies had to submit all the necessary blueprints and data on the proposed aircraft's structure, the wing arrangement and profile, and the engine and propeller.

To meet the army's challenging guidelines, all participating companies relied on the expertise of foreign designers. Nakajima had contracted

French designers André Marie and Maxime Robin to work in the company's design section from April 1927 to April 1929.[47] Mitsubishi's chief engineer, Nakata Nobushirō, began designing the Hayabusa ("falcon") under Baumann's supervision, while Kawasaki put Vogt in charge of the fighter's design. Vogt and his team of thirty Japanese engineers could refer to a Dornier monoplane that Kawasaki had purchased as part of the 1924 license deal.[48] To provide more sturdiness to their experimental KDA-3 fighter, they attached diagonal wing struts and reinforced the fuselage. An improved flight control mechanism was to increase the airplane's overall maneuverability.

In June 1928, each company's prototype was ready for required flight tests. They all ended in fiascoes. Kawasaki's test pilot had to return to the airfield after the engine of his KDA-3 failed. Mitsubishi's Hayabusa showed more perseverance, and pilot Nakao Sumitoshi (1903–60) managed to climb to 5,000 meters. This altitude was necessary to perform the most demanding maneuver that the army required each manufacturer to demonstrate: a nosedive at an angle of sixty degrees under full engine power, during which the aircraft was to accelerate to a top speed of 400 kilometers per hour. The pilot had instructions to throttle the engine only in the extreme case of dangerous airframe vibrations. Most designers deemed this maneuver to be an absurd request, considering that the maximum speed of advanced Western fighter aircraft at that time was 300 kilometers per hour.[49] However, Nakao was eager to fully demonstrate the Hayabusa's high-speed capabilities and forced his aircraft into a vertical nosedive. The airplane could not sustain the maneuver and disintegrated in midair.[50] Nakao had the presence of mind to jump out of the free-falling cockpit and open his parachute. He survived the accident unharmed and became famous as Japan's first pilot to successfully bail out with a parachute from a crashing airplane.[51]

Nakajima's aircraft suffered a similar fate. During the test dive, the plane entered a spin that could not be recovered, and it crashed. Sources state that the obvious reason for the accident was the plane's insufficient structural strength. Records do not indicate whether the pilot survived the crash. According to Vogt, who still harbored hopes that his aircraft would be the winner, the army canceled the competition "without further explanation" and did not adopt any of the three prototypes.[52]

Rather than being discouraged by this setback, Kawasaki decided to develop a new fighter on its own initiative. This was an exceptional move that clearly challenged the army's authority over the fighter program. Vogt's supervisor, Takezaki, gave him free rein to build a fighter aircraft entirely according to his own ideas. Vogt opted for a biplane that was to be powered by a powerful BMW 500 horsepower engine. Doi Takeo, who had been promoted to Vogt's chief assistant, closely studied the results of previous experiments carried out in wind tunnels in Great Britain, Germany, and the United States. Doi proposed to use the NACA M-12 airfoil, which he thought would be especially suitable for fighter aircraft.[53] This wing shape had been devised, tested, and made publicly available by the foremost US aeronautical research institute, the National Advisory Committee for Aeronautics (NACA).[54]

Vogt tried to avoid the fate of Mitsubishi's Hayabusa, which broke apart during the required high-speed test. He took great care to build a robust aircraft that could withstand the most extreme flight maneuvers. An extra-strong main wing spar and special triangle-shaped duralumin struts connecting the upper and lower wings significantly increased the airplane's sturdiness. As a result, during destruction tests the new wing could easily hold out against a load factor of fifteen, exceeding the army's requirement by two points.[55] In July 1930, the first flight tests began. The new KDA-5 impressed test pilots with its outstanding maneuverability and velocity. With maximum speed of 320 kilometers per hour, the aircraft even outstripped the US and British fighters of that time—only in 1932 did the US fighter aircraft Curtiss P-6 Hawk in its most powerful version reach a comparable performance. Dornier's pilot Just compared the KDA-5 with its predecessor and commented, "It is as different as night and day."[56]

In an effort to convince army officials as well, Takezaki launched a public promotional campaign. On November 5, 1930, *Asahi Shinbun* ran the headline "A World Record Set by a Military Aircraft," reporting that test pilot Tanaka Kanbei had climbed with his "made-in-Japan aircraft" KDA-5 to an altitude of 10,000 meters, withstanding extreme cold and thin air. According to the report, Tanaka continued the climb even after the altimeter stopped working at 10,000 meters and might well have reached 12,000 meters.[57] The publicity stunt worked, and the army allowed Kawasaki to present its aircraft at the Tachikawa Airfield. In April 1931, in the presence of the minister of the army and high-ranking officers, Tanaka skillfully demonstrated the high performance of the KDA-5.

Japan's occupation of Manchuria in September 1931 hastened the adoption of Kawasaki's aircraft. With immediate need for more fighter planes and a significantly increased budget, the army officially accepted the KDA-5 as the Army Type 92 Fighter one month after the outbreak of the conflict. In an atmosphere of nationalist war fever, the Japanese press welcomed the army's decision to choose an aircraft that "flies like a bullet" and held the prospect of turning Japan into a major airpower.[58] Two months later, the army also adopted Nakajima's remodeled fighter aircraft. This unusual move testifies to the great demand for more planes. At the same time, it subverted the original idea of prototype competitions that were meant to determine a single best manufacturer. Yet papers nevertheless presented this uneasy compromise to the public as a success. The *Yomiuri* newspaper called the two new aircraft types "Japan's pride before the world." By perfectly complementing one another in speed and maneuverability, these fighters would "protect the skies of Japan in a time of crisis."[59] In early 1932, both Kawasaki and Nakajima began large-scale production of their fighter aircraft, of which they built a combined total of 700 in under two years.[60] These numbers reflect the belief in the increasingly important role of airpower in defending Manchukuo, the territory that, according to a popular catchphrase, had become Japan's "lifeline" (*seimeisen*).[61]

AN ENGINEER AS TEACHER: VOGT'S LEGACY

Among the large group of foreign aviation specialists coming to Japan, Richard Vogt stood out. His focus on independent design significantly raised the status of the Japanese designers. Before Vogt's arrival, Kawasaki relied completely on imported foreign technology. Research and native design were nearly nonexistent. Minor design details were carried out by technicians who had graduated from technical schools or just finished junior high school and were paid on a daily basis. The work of these "draftsmen" was called *zukō*, which referred to a mixture of drawing and handicrafts.[62] When Kawasaki decided to embark on the construction of metal aircraft, the design section increasingly hired university graduates, who began their careers under the guidance of Vogt.

Doi Takeo turned out to be one of Vogt's most gifted apprentice engineers. He joined Kawasaki in 1927, directly after his graduation from Tokyo Imperial University's new Department of Aeronautics, which was founded in 1920. During his close cooperation with Vogt, he fully developed his design talent and, after Vogt's return to Germany in 1933, became Kawasaki's chief designer for the ambitious Ki-5 fighter project. In the early 1940s, Doi secured his place in Japan's aviation history with his masterpiece, the Hien ("flying swallow"). This outstanding fighter aircraft showed the design influence of Vogt and is considered superior in many aspects to Mitsubishi's much more famous Zero-sen.[63]

Engineer Andō Nario (1899–1987) assisted Vogt with drawing and making calculations. Andō became an important mediator between the army and Japan's aircraft makers.[64] As a designer and examiner in the Army Aviation Headquarters, he was involved in aircraft design at Mitsubishi and Nakajima and in research at the Army Aviation Technology Research Center at Tachikawa. With his technical knowledge and close insight into the capabilities of the civil manufacturers, he was one of the few influential aviation experts who could counter the army's often unrealistic specifications for new experimental aircraft.[65]

German Airliners into Japanese Bombers: Junkers in Japan

In the early hours of October 26, 1931, a giant aircraft made its first takeoff from Japanese soil. With substantial help from Germany, Mitsubishi had built the Type 92 Superheavy Bomber, Japan's first long-range bomber designed to fly non-stop to US bases in the Philippines, where it would drop its bomb load of 5,000 kilograms. Praised by the Japanese press as the "world's largest super-heavy bomber,"[66] the plane had a wingspan of forty-four meters, wider than the US Army Air Forces' Boeing B-29 Superfortress, which made its first appearance in the skies over Japan in June 1944. The successful production of the Type 92 Superheavy Bomber

Table 4.1. Three generations of Japanese bombers

	Farman F.60 Goliath (1919)	Do N (1926)	Type 92 Superheavy Bomber (1931)
Wingspan (m)	28	26.8	44
Max. weight (tons)	5.5	7.7	25.4
Max. speed (km/h)	120	180	200
Flight range (km)	400	660	2,000
Max. bomb load (kg)	800	1,000	5,000

Notes. All technical data are from Nohara, *Zukai sekai no gun'yōkishi: 6*, 25, 32, 43. For comparison, the US Boeing B-17 that made its first flight in 1935 had the following characteristics: wingspan 31.6 m; max. weight 29.7 tons; max. speed 462 km/h; flight range 3,219 km; max. bomb load 2,700 kg.

was a major achievement for Mitsubishi and a significant advance for the Imperial Japanese Army, comparable to the technological leap from the wooden Farman bomber to the all-metal Do N in 1926 (table 4.1).

The Type 92 Bomber, also called "the Superbomber," was the materialization of a dramatic change in US–Japanese relations. To better understand this important development, it is helpful to go back to 1907, when the Japanese military first based its operational planning on an Imperial Defense Policy (Teikoku kokubō hōshin). The document drafted by the army's general staff office and the naval general staff identified Russia as Japan's primary adversary. After the 1917 October Revolution, czarist Russia collapsed, and the Imperial Russian Army and Navy ceased to exist. This event led to a profound shift in Japan's military strategy. For the first time, military planners began to worry more about rising US influence in the Pacific than about Russian territorial expansion. In 1918 the Japanese military revised the Imperial Defense Policy and set up a strategy of first defeating the US Asiatic Fleet, whose main task was to protect the Philippines, and then occupy Luzon, the main island of the archipelago. Next the Japanese Navy would encounter the US Pacific fleet on its way to the western Pacific and destroy it in a single decisive battle.[67] Five years later, in the wake of the Washington Naval Treaty, the Cabinet under Prime Minister Katō Tomosaburō (1861–1923) passed another far-reaching revision of the Imperial Defense Policy. Now the United States was promoted to "Japan's number-one hypothetical enemy" for the army and the navy.[68] This move reflected a perceived increasingly anti-Japanese stance of the United States—and a growing discontent with the arms limitation imposed by the Washington Naval Treaty, which many Japanese saw as a threat to their nation's interest.[69]

The anti-American defense policy still included the preemptive occupation of the Philippines. In 1925–26, the Japanese Army and Navy began drafting a plan for an invasion of the archipelago.[70] The navy would destroy the US vessels in Philippine territorial waters, while the army would invade the island and occupy Manila. The invasion plan included air attacks on US bases on the island. However, it soon became obvious that such an operation was beyond the flight range of any of Japan's military aircraft. Initially the army planned to use the navy's carriers for launching aircraft.[71] Another proposal was to disassemble the planes and transport them by ship to a prepared airfield on Luzon, from where they could participate in the battle.

The invasion plan went through several stages until Koiso Kuniaki, who had been promoted to chief of the Army Aviation Headquarters' General Affairs section in 1927, proposed a new line of reasoning. Koiso argued that, considering the still weak state of Japanese air defense, the only way to protect the homeland from an air attack would be the complete destruction of the enemy's bomber bases immediately after the outbreak of a war.[72] He pointed out that recent developments in the design of civil aircraft showed the feasibility of fast, long-range airplanes carrying increasingly heavier payloads. Well aware of recent developments in Germany, Koiso also referred to the latest German civil aircraft that could be easily converted to long-range bombers.[73]

In early 1928, Koiso's proposal gained momentum. Inoue Ikutarō, chief of the Army Aviation Headquarters, suggested developing an "ultra-heavy bomber" (*chōjū bakugekiki*) for a bombing run from Japan's colony of Taiwan to the Philippines. Such an aircraft should be able to take off from an air base in southern Taiwan, cross the Luzon Strait, and attack Manila (figure 4.5).[74] In February 1928, the army minister ordered Inoue to start investigations into an aircraft that could fly nonstop to such a faraway battlefield for reconnaissance and bombing missions.[75]

The same month, the Army Aviation Bureau set up the specifications for an aircraft that could fulfill such a challenging task. These requirements tested the limits of aviation technology at that time. The new bomber should have an operational flight radius of 1,000 kilometers plus

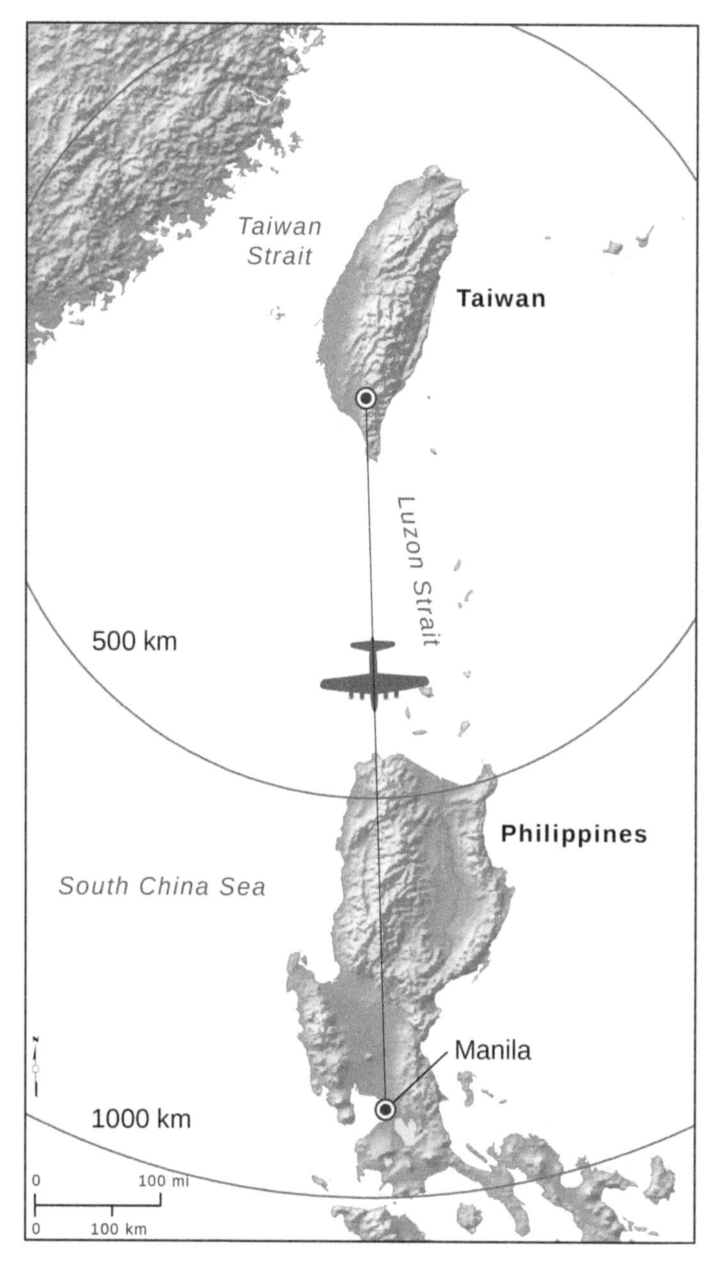

FIGURE 4.5 Heading south: the "ultra-heavy bomber's" proposed 900-kilometer bombing-run from Taiwan to the US bases at Manila.

an additional 500 kilometers for carrying out its tasks over the target. It should be able to carry a bomb load of 2,000 kilograms and be heavily armed for self-defense to perform its mission independently, that is, without the protection of extra fighter planes. Furthermore, the aircraft should be equipped for night flight and long-distance radio communication.

The army made a total sum of ¥800,000, to be distributed over three years, available for the bomber project. This bold decision earmarked 40 percent of the total the army spent on aviation material between 1928 and 1930.[76] The Ministry of the Army's decision to fully support Koiso and Inoue's ambitious project clearly testifies to the unprecedented rise of the airpower faction. Initially the Aviation Bureau insisted on employing only Japanese aircraft manufacturers for designing and constructing the long-range bomber. However, it turned out that no Japanese aircraft maker was prepared to undertake such an ambitious project without foreign help.

The aircraft that the Japanese Army envisioned was already taking shape on German drawing boards. After Charles Lindbergh's successful solo Atlantic crossing in 1927, the German Ministry of Transport envisioned a "Transocean Program" where a new generation of long-range passenger aircraft would expand Germany's aerial network all the way to the Americas. In March 1928, German aircraft maker Hugo Junkers received a massive subsidy of more than two million Reichsmarks from the Ministry of Transport to design and build a new airliner, the G 38, for opening a regular passenger service across the Atlantic.[77] The airplane was to become the world's largest land aircraft. Inside its enormous wings it could accommodate four engines and six passenger seats with a grandstand view.[78]

When the representatives of the Japanese Army showed their interest in the giant airplane, it was easy for Junkers to convince them that the G 38 was the ideal aircraft for long-range bombing missions. Although designated as a passenger aircraft, the G 38, with an all-metal structure and wide fuselage, lent itself to conversion into a bomber with large bomb bays and machine gun turrets. The aircraft's advanced wing layout and fuel-efficient engines made it possible to carry a heavy bomb load over a great distance. In a rare move—the G 38 still existed only on the drawing boards, and its first flight was scheduled for late 1929—the Japanese Army's technical experts recommended buying the license for the Junkers G 38, remodeling it into a bomber, and producing it in Japan.[79]

For the next step, the army had to find a suitable Japanese manufacturer for the licensed production of the new bomber. In early 1928, Inoue proposed to use Mitsubishi for this project, and Minister of the Army Shirakawa Yoshinori (1869–1932) approved the plan in August 1928.[80] The proposal was attractive to Mitsubishi: the Army Aviation Bureau agreed to cover the cost of the experimental production and continuously place orders for the aircraft. For his part, Junkers increasingly depended on revenue from Japan when the growing political turmoil of the Weimar Republic endangered the flow of government subsidies to his company.[81] Furthermore, a successful conversion of his largest passenger aircraft into a bomber would qualify Junkers to become a major supplier for the secret buildup of Germany's Luftwaffe.

LICENSE NEGOTIATIONS AND PREPARATION FOR PRODUCTION

Despite these mutual benefits, license negotiations between Junkers and Mitsubishi lasted nearly a year. For Mitsubishi, there was much more at stake than the purchase of the aircraft. What really counted was the transfer of know-how and production technology. Junkers, for his part, was wary of illegal copies and patent infringements and preferred to sell his own finished products to Japan. However, when he handed over his first tender in March 1928, he gave in to Japanese demands and agreed to set up a production site at Mitsubishi's Nagoya plant.

The two companies finally signed the license contract on September 20, 1928. The agreement, valid for twelve years, involved fifteen aircraft types and three different engines, for which Junkers would provide all necessary data, drawings, and manufacturing information. In addition, the German company committed itself to dispatch "fully skilled and trustworthy staff" to Nagoya. The contract entitled Mitsubishi to have its own employees trained at Junkers's Dessau factory. In return, the Japanese company agreed to pay ¥600,000 (the equivalent of US$4 million in 2018). According to Mitsubishi, this was the largest amount ever paid to any airplane manufacturer in Europe or America.[82]

In April 1929, preparations for the construction of the Superbomber began. The director of Mitsubishi's aircraft section, Shibuya Yonetarō, approved nearly ¥1 million for the construction of an entirely new factory and the purchase of machines, tools, and other production facilities.[83] Training for Mitsubishi personnel in Germany played an equally important role in setting up Superbomber production and advancing Mitsubishi's production methods. The Japanese trainees, among them the future designer of the Zero-sen, Horikoshi Jirō, received detailed instruction on material examination; on designing airframes, wings, and flight controls; and assembling and testing engines.[84]

As a clear sign of the high importance attributed to the bomber project, Hugo Junkers sent his personal assistant, Benno Fiala von Fernbrugg (1890–1964), to Japan. The former World War I fighter ace had studied engineering after the war and had been working closely with Junkers since 1925. Fiala's detailed daily progress reports provide a clear picture of the complex preparations for the production of the bomber.[85] As a first step, he assessed Mitsubishi's share in the production process to decide which tools and machinery Junkers had to send to Japan. Then, with Mitsubishi's engineers, he set up a schedule for wind tunnel experiments and arranged the construction of special earthquake-proof scaffolding. Fiala also examined the options for the thirty-five-kilometer transport of the giant aircraft from the Nagoya factory to the military airport at Kakamigahara. When it became clear that narrow roads and countless overhead wires would make an overland transport unfeasible, he prepared for shipment via the Kiso River. Finally, Fiala gave instructions to compile a German-Japanese dictionary with all the necessary vocabulary to communicate the "Junkers construction method."

The conversion of Junkers's passenger aircraft into a heavily armed bomber was an even more challenging task. Fiala suggested equipping the fuel tanks with extra fire protection. He gave detailed instructions for the layout of the bomb bay; the number, size, and types of bombs; and their release mechanism. Furthermore, he determined the location and firing range of machine guns and cannons that were to protect the bomber against enemy fighters. To use short runways, the aircraft was to be equipped with additional wing flaps, which would create extra lift and lower the takeoff and landing speed. Efficient wheel brakes would decrease the required landing distance. Floating devices ensured that the aircraft could sustain an emergency water landing. For carrying out air raids at night or under poor weather conditions, the cockpit was equipped with instruments for night flights and operation in low visibility. Fiala even successfully navigated the dangerous waters of Japanese popular beliefs. Being aware of Japanese sensitivities toward the number forty-two, whose pronunciation, *shini*, also could be understood as "to die," he assured his Japanese partners that the wingspan of the new aircraft was forty-four meters and not forty-two, as had been reported earlier.[86]

While Mitsubishi was still waiting for the bomber production to begin, project costs began to escalate. In July 1929, the Army Aviation Bureau had to nearly double the initial budget of ¥800,000 to ¥1.5 million.[87] In October 1930, when production of the bomber was already a year behind schedule, Junkers suggested delivering one prefabricated aircraft to Japan. This proposal for the wholesale import of the plane was a delicate matter, as it went against the military's declared aim of a made-in-Japan aircraft. Junkers succeeded in convincing his Japanese business partners that a sample aircraft would "accelerate Mitsubishi's production and advance the schedule by at least one year." It would also allow tests of the extra equipment, the training ground, and flight crews.[88] Mitsubishi agreed to import the main parts for the first two bombers from Germany, and subsequent airplanes were to be built entirely from made-in-Japan components.

The New Bomber's Maiden Flight

Finally, in autumn 1931 the first Army Type 92 Superheavy Bomber (92-shiki chōjū bakugeki ki) was ready for its maiden flight at the Kakamigahara airfield (figure 4.6). Even though Junkers had sent experienced chief test pilot Wilhelm Zimmermann, the army insisted that a Japanese pilot had to fly the new bomber because the aircraft had been "paid for by the Japanese taxpayers."[89] On October 26, 1931, Captain Katō Toshio took off with the new bomber and successfully carried out a first test flight. The next day, Junkers informed the Japanese military attaché in Berlin about the important event. Junkers also sent a letter expressing his "joy and satisfaction" along with a model of the bomber to General Watanabe Jōtarō (1874–1936), the new head of the Army Aviation Bureau.[90]

FIGURE 4.6 A large group of army officials posing in front of the Army Type 92 Superheavy Bomber. The picture conveys an idea of the aircraft's full size. Four 800-horsepower engines, mounted inside the wings, drove the huge 4.5-meter diameter propellers. The flaps visible at the wings' trailing edges are part of the Junkers "double wing" design that significantly reduced the landing speed of the aircraft. (Photo courtesy of Kōkū Jōhō; source: *Nihon no kōkū runesansu*, 82)

Army engineer Kariya Masai, who participated in the flight test, gave a vivid inside picture of the Type 92 Bomber's operation that leaves little room for the glamour of flight.[91] The extraordinary large flight crew consisted of three pilots, five flight engineers, five gunners, and three operators who were responsible for radio communication, bombing, and photographing. According to Kariya, during the long flights at high altitudes, pilots frequently collapsed over their controls because of a lack of oxygen. The engineers who continuously had to attend the aircraft's four engines worked half-naked in their noisy and overheated engine compartments. For their part, the gunners suffered from the cold and the rough movement of the gun turrets. Another curious feature was the landing technique of the giant aircraft, which required the cooperation of the whole crew. Shortly before touchdown, a ringing bell signaled to the flight engineers and operators that they had to run to the back of the aircraft cabin so that the bomber's rear part could be lowered and a three-point landing made.

Even though the bomber program was classified as top secret, the Army Aviation Bureau failed to hide it from a curious public. In March 1931, the first press reports about the bomber project appeared. As a countermeasure, the vice minister of the army ordered the chiefs of staff in all divisions and the commanders of the military police not to call the aircraft a "superheavy bomber" but to refer to it in more neutral terms as a "special experimental aircraft." The vice minister also insisted that the bomber's performance and all other related information were to be kept absolutely secret. He notified the Home Ministry's Newspaper Censorship Bureau (Shinbun Ken'etsu Kyoku) to prohibit journalists from publishing any articles about the aircraft.[92]

The combined efforts of the army and government censors showed little effect. Newspapers reported with a mixture of sensationalism and national pride about the new aircraft. Several months before the bomber's first flight, *Asahi Shinbun* announced the production of a "Superbomber of which we can be proud before the whole world."[93] The article emphasized that Mitsubishi already had spent "an enormous amount of money and time" on a "made-in-Japan machine" that was an improved version of the Junkers G 38. *Asahi*'s competitor *Yomiuri* ran the headline "The Army Air Force's new and powerful machine, the world's largest super-heavy bomber."[94] In astonishing detail, the paper presented the production and examination process and impishly reported that the military now called the heavy bomber a "special experimental aircraft." After revealing all the technical details about the dimensions, performance, and bomb load of the all-metal monoplane, the article concluded confidently that "our army, which until recently had been sluggish in the development of heavy bombers, could finally strengthen its [air]power." Despite this ambivalent praise, the army continued its efforts to crack down on press reports about flights of the heavy bomber, a seemingly impossible task considering the monstrous size of the plane.[95] Only in 1938 was the aircraft officially declared no longer a secret.[96]

THE SUPERBOMBER'S EFFECT ON MITSUBISHI'S AIRCRAFT MANUFACTURING AND DESIGN

The cooperation with Junkers had a deep impact on Mitsubishi's aircraft manufacturing. After building the first two bombers from imported prefabricated parts, the Japanese company managed to take control of the entire production process. Considering the size and complexity of the four-engine aircraft, this transition was a remarkable achievement. It testifies that Mitsubishi could quickly absorb Junkers's innovative design and production technology.[97] According to one commentator, introducing unified production standards and efficient blueprint management became the foundation of Mitsubishi's aircraft production after 1932.[98] Another Mitsubishi engineer recounted that the "massive amount of technological information and manufacturing technology became one of Mitsubishi's most precious assets for the further development of the company's technology."[99] Even their competitors became aware of Junkers's pervasive influence on Mitsubishi's production technology. In 1933 Kawasaki engineer Senba Tadashi paid a visit to Mitsubishi's Nagoya factory. After seeing how Mitsubishi used the Junkers methods of standardized parts and efficient riveting, he enthusiastically commented: "This is a factory we [other] aircraft makers can only yearn for."[100]

Mitsubishi's interest in Junkers's technology continued, and the Japanese company designed two more bomber types for the army that were based on a Junkers model. Rather than long-range strategic bombing, these airplanes were intended for tactical support of ground troops over the battlefield, as advocated by the army's traditionalists. In September 1930, Mitsubishi imported one Junkers K 37 built in Junkers's Swedish factory. The aircraft was about half the size of the Superbomber. It combined the nimbleness of a fighter with the load-carrying capability of a bomber and was one of the few twin-engine airplanes fully certified for acrobatic flight. Demonstration flights started in early 1931, and the aircraft impressed the Japanese Army with its superior maneuverability.[101]

Mitsubishi used the Junkers K 37 as a model for the design of two bomber types. To accomplish this task, the company's all-Japanese team of engineers drew heavily on their experience from the Superbomber project. They completely redesigned the Junkers K 37 and presented two vastly different aircraft types in 1933: the Army Type 93 Heavy Bomber, an enlarged version of the German model that could carry a bomb load of 1,500 kilograms, and the Army Type 93 Light Bomber, a highly maneuverable, lighter version designed for air battle and bombing missions. The army adopted both aircraft types; starting from 1933, Mitsubishi produced them in large numbers.[102] The success of these bombers clearly testifies that Mitsubishi's expertise had reached a level, one where its engineers could significantly alter German design into custom-made aircraft that closely followed the army's specifications.

The Army's New Aircraft and the Manchurian Crisis

On September 18, 1931, a bomb exploded on the tracks of the South Manchuria Railway near the city of Mukden (Shenyang) in northeast China. The sabotage was part of a plot carried out by field officers of Japan's Kwantung Army, who acted without the authorization of their commander-in-chief, Army General Honjō Shigeru (1876–1945). The Kwantung Army blamed the Chinese military for the bomb attack and used it as a pretext to start a takeover of the whole territory. For many historians this event, which became known as the Manchurian Incident, marks the beginning of Japan's Fifteen Years War.

THE OPEN SKIES OF MANCHURIA

For the staff officers of the Army Aviation Headquarters, the Manchurian crisis provided the first opportunity since the 1914 Qingdao air battle to test and deploy their new equipment. However, when the Kwantung Army started invading Manchuria on September 19—approved by General Honjō after the fact—it did not have any air squadron under its command. Just one day after the incident, several of the Sixth Air Wing's reconnaissance aircraft were dispatched from Korea over a distance of more than 350 kilometers to the Mukden area to gather information about the Chinese forces. Another squadron of fighter aircraft followed. After it became clear that Japanese ground troops had already destroyed all of the enemy's airplanes, the fighters were sent back to their Pyongyang base.[103] In November 16, 1931, the Kwantung Army finally set up its own Flying Corps. It consisted of twenty-seven reconnaissance aircraft, nine fighters, and nine light bombers, all built in Japan.[104] The small fighter contingent and the lack of heavy bombers clearly reflected the army's traditional air doctrine of limiting the use of aircraft to support for ground troops.

Considerable disagreement within the top ranks of the Imperial Japanese Army led to an unusually late deployment of the heavy bomber squadrons. The staff officers of the Army Aviation Headquarters wanted to send their heavy bombers to Manchuria right after the outbreak of the conflict to get practical experience under actual battle conditions, which they considered crucial for training their crews and an investigation of bombing techniques. However, in fear of adding to the mounting international criticism of the invasion, the army's general staff office was reluctant to make full use of the bombers. Only on December 27 did the general staff office in Tokyo officially approve and assign four Kawasaki-Dornier Type 87 Heavy Bombers to the Kwantung Army Flying Corps for the "training of cold weather operations."[105] On December 30, 1931, for the first time Type 87 Heavy Bombers entered actual combat in Manchuria. Soon a full-fledged bombing campaign against splinter groups, armed trains, and military bases began. Bomb attacks on enemy troops continued in spite of their retreat into remote mountain areas. During the conflict, the Japanese gained valuable experience in the tactical deployment of bombers that included long-range navigation, using advanced bomb sights for precision bombing, and the importance of a bomber's armament for self-defense. Furthermore, the army was impressed by the bombers' apparent capability to carry out "intimidation flights" (*ikaku hikō*) that terrified an already demoralized enemy. The psychological effects of these missions seemed to confirm Giulio Douhet's assumptions about the moral impact of bombing raids.[106]

Any assessment of the Japanese Army Air Force's operations during the Manchurian takeover must consider the fact that within the first few days of the conflict, all Chinese aircraft in Manchuria were already destroyed or captured before they could engage in air battle with Japanese planes. This allowed the Japanese Army Air Force to operate with the advantage of complete air superiority. Under such conditions, their aircraft could watch and strike from the sky without any Chinese resistance. The army could adhere to its traditional air doctrine of carrying out typical tactical operations that supported the ground troops with fire power, search and rescue missions, and reconnaissance reports.

THE HOME FRONT AND THE OUTBREAK OF "DONATION FEVER"

Immediately after the Manchurian Incident, the press began massive coverage of the Japanese military's activities.[107] A steady stream of news from Manchuria molded the image of the army's new aerial weapon as a highly efficient constituent of military power that fully deserved the public's support. At the end of September 1931, the first articles about the bombing of the Chinese army remnants appeared, but—as the reports pointed out—in most cases Japanese pilots dropped leaflets before the attacks to warn the civilian population of the air raids. At the same time, the Japanese public received reassuring news about the air force's successful operations to protect civilians in the hinterland, where Japanese and Korean families lived in fear of attacks by splinter groups of the retreating Chinese army. However, as the papers assured their readers, their safety improved since Japanese aircraft regularly patrolled the region.

For the Japanese on the home front, reassuring press reports were one way to deal with the crisis. Another, more active course of action was providing direct financial support for the Japanese troops. The military soon realized the potential of collections and donation drives. The armed forces were convinced that raising funds especially for their hugely expensive aircraft needed to be unified and organized from above. At the end of November 1931, a new committee for managing donations was established.[108] The committee members came up with an appropriate name for the aircraft they expected to be purchased with the donations: *aikokugō*, or "wings of patriotism."[109]

For the Japanese press, the emerging donation campaign offered yet another inexhaustible source of news. Not surprisingly, the continuous reports about the victorious air force were soon presented to the public in conjunction with intensive coverage of the donation campaign. In a mutually reinforcing process, the press publicized and shaped a movement that gradually developed into "donation fever." A *Yomiuri* article of December 5, 1931, praised the Tokyo Boy Scouts for collecting money to buy two new aircraft. Their efforts to "protect the skies" should be an inspiration for people all over Japan to donate money for aircraft to the army. The paper did not forget to mention the splendid performance of "our strong Army Air Force in Manchuria," which had crushed the brutal Chinese enemy so easily. Two days later, a similar article congratulated the 50,000 members of the Miyagi Veterans Association for donating one heavy bomber, duly named "Miyagi-gō." More reports about a fundraising campaign in the streets of Tokyo undertaken by junior high school students informed the public about the sudden rise of a "flood of contributions," reaching totals of ¥130,000 on December 8 and ¥150,000 on December 19.

The campaign received additional support with an appropriate song. The "Aikokugō Song" became immensely popular and is still remembered by some Japanese today. Its refrain, "Go! Go! Our Aikokugō!," and its reference to "silver wings shining in the morning sun" is strongly reminiscent of *Asahi*'s 1925 "visit Europe" song. For today's listeners, the innocuous voice of child star Kawamura Junko (1925–2007) singing the Aikokugō Song might stand in bizarre contrast to the bellicose text that celebrates the heroic Aikokugō bombers for heading straight for the enemy's sky and "defeating the enemy in one blow."[110]

When the donation craze spread across Japan, it reached even the fringes of society. A newspaper headline mysteriously announced, "The Caged Birds Fly." The text related that Tanigawa Tori and Katō Kiyo, two Yoshiwara prostitutes, took off for a short flight over Tachikawa airport.

After landing, they suffused their exhilarated comments with praise for the Japanese troops: "Even we, the caged birds, are aware of what is going on in the world. We especially admire the remarkable success of our Army Air Force in Manchuria. Even though we are women of ill repute, we are happy to help Japanese women appreciate the merits of our aircraft."[111] More strange news was reported from the Osaka Prison. Under the headline "The Heart of a Demon Is Moved," readers learned that a criminal on death row donated his lunch money, to which he attached a "letter of tears and blood." Moved by the news from Manchuria, the prisoner wrote:

> Now I am just waiting for the day when I will be hanged. I was born as Japanese, but I have never done anything for our country, far from it, I caused only trouble to people. But now young Japanese soldiers in the cold of Manchuria are fighting for our country and sacrificing their own lives. They are the same age as me. I cannot help feeling ashamed of myself. . . . I have only two sen at hand. Please add this money for buying new aircraft. This will be the first and last honorable action in my life.[112]

While the press reports steadily intensified the donation fever, the military came up with another propaganda tool to further include the public: Japan's first Aikokugō donation ceremony. Ironically, the army chose two foreign imports, a Junkers K 37 bomber and a Dornier Merkur transport aircraft, to become the "wings of patriotism" Aikokugō 1 and Aikokugō 2. The business newspaper *Chūgai Shōgyō Shinpō* strongly supported the army's decision to buy the planes with the proceeds of the donation funds to encourage Japanese soldiers who were "fighting in the severe cold of Manchuria."[113] The article provided detailed information about the planes. It praised the Junkers K 37 as "the most advanced aircraft of all Western countries" with superior armament, high cruising speed, and a wide radius of action. The Dornier Merkur was considered equally outstanding for becoming Japan's first ambulance aircraft. Its special cabin design, laid out by a Japanese army surgeon, made it possible to treat injured soldiers already during their flight from the battlefront to the hospital.

All praise notwithstanding, the grand moment—the official presentation of the first two Aikokugō aircraft—was yet to come. On December

22, *Yomiuri* prepared the public for the event by announcing that the planes would fly to the Yoyogi Parade Ground in Tokyo, where they would be donated to the Imperial Army as "manifestations of the nation's air-defense fever" in a solemn naming ceremony. Then Aikokugō 1 and Aikokugō 2 would take off for a cross-country flight to seven major cities, where they planned to drop millions of flyers to promote the cause of Japanese air defense.

On the morning of January 10, 1932, the grand naming ceremony took place. The whole nation could follow the event via live radio broadcast. His Imperial Highness, the minister of the army, and the Imperial Aviation committee, along with countless veterans, schoolchildren, and Boy Scouts, attended a celebration that followed Shinto ritual. After the Shinto priest's prayer, high-ranking military officers delivered congratulatory addresses. Then the aircraft took off, greeting the audience with a "flight of gratitude" before disappearing on their way to the Manchurian battlefield. After the two airplanes were gone, the army continued to entertain the spectators with a large-scale flight show for which forty military aircraft and several private planes performed maneuvers over the parade ground.

The aircraft-naming ceremonies soon became regular, institutionalized events—within one and a half years, the army registered a total of ninety-two aircraft as Aikokugō. During a typical naming ceremony, the airplanes were named with their Aikokugō number and carried the names of their donors clearly visible on their fuselages.[114] The planes would take off to drop flyers thanking the generous donors. These simple measures added immensely to the popular appeal of the army's air force as they created a strong identification of the contributors with "their" aircraft. After dropping their tokens of gratitude, the Aikokugō aircraft would take the names of prefectures, schools, miners, wealthy brokers, and insurance companies up into the sky and to the far-away battlefront.

THE SUPERBOMBER'S FATE AFTER THE MANCHURIAN CRISIS

The new wave of aviation enthusiasm did not sweep over to the army's Superbomber project. By autumn 1931, shortly after the first successful flight of the Type 92 Superheavy Bomber, Japan's political, strategic, and military environment had fundamentally changed. A radical shift in the military's strategy in the wake of the Manchurian Incident, a reshuffle of the Aviation Bureau, and a new minister of the army's revised battle doctrine were fateful not only for the Superbomber program but also for the whole army air branch.

In 1929 two central figures of the army's airpower faction left the Army Aviation Bureau. Koiso Kuniaki, the mastermind behind the bomber project, was transferred to the Ministry of the Army Economic Mobilization Bureau (Rikugunshō Seibi Kyoku), and Inoue Ikutarō, a staunch supporter of the army's aviation branch since 1911, became a member of the Supreme War Council (Gunji Sangikan).[115] Inoue's successor as the head of the Aviation Bureau was Watanabe Jōtarō (1874–1936). Watanabe seemed to be hardly qualified for promoting the buildup of a strategic bomber fleet against the opposition of the army's traditionalists. According to Koiso's memoirs, Watanabe confessed to him in 1929 that he was a "total amateur in all matters concerning aviation" and asked if there was "any book to make [him] understand it easily."[116]

When the invasion of Manchuria began in September 1931, the army's general staff recognized the Kwantung Army's move as fait accompli, ignored the 1923 Imperial Defense Policy, and made the advance into China its new battle doctrine. This fundamental shift rendered obsolete the army's earlier plans to bomb and occupy the Philippines. As the Kwantung Army had annihilated Chinese airpower during the first few days of the conflict, the need to deploy the Type 92 long-range bomber in northeast China did not arise. As a further fallout of the Manchurian Incident, new Prime Minister Inukai Tsuyoshi (1855–1932) appointed Araki Sadao (1877–1966) minister of the army in December 1931.

Araki's rise to prominence was the result of a momentous shift in the Imperial Army's ideological alignment. The Ugaki disarmament led to growing discontent among the army's young officers, who denounced Ugaki's belief in a technology-driven modernization. Many of them grouped around Araki to form what became known as the Imperial Way faction (*Kōdōha*). In the view of these young radicals, Japan had neither the technological nor economic potential to equip its troops with a sufficient amount of modern weaponry. They advocated a turn away from Ugaki's pragmatism and reliance on expensive equipment toward a revival of an aggressive military spirit and morale (*seishin*). The Imperial

Way faction played a major role in plotting and carrying out the Manchurian crisis. It was also with the backing of this group that Araki could rise to power.[117]

Soon after Araki became minister of the army, he revised Ugaki's policy of enlarging the army's airpower and reducing its troop numbers. Following the tenets of the *Kōdōha*, Araki emphasized the army's fighting spirit and spiritual mobilization over modern technology and material strength. He showed little interest in a technologically advanced army air force that would carry out missions beyond the tactical support of ground troops.

Under these circumstances, there was little prospect for the Army Type 92 Superheavy Bomber to open a new chapter in Japan's military history. Nevertheless, from December 1931 to May 1932 the new bomber underwent a thorough test program. The aircraft was transferred to the army's 7th Air Wing at Hamamatsu, where training for the bomber crews started in June 1933.[118] Whereas in 1933 the bombers made nearly 300 flights over Japan, mainly for training their crews, they mostly stayed on the ground in the following years.[119] With the bomber's raison d'être becoming increasingly doubtful, its production stopped after the sixth aircraft in 1935.

In a half-hearted attempt to assign a role to the bomber in future wars, the army set up a special task force (*tokubetsu ninmu butai*) that, in the case of a war with the Soviet Union, could fly long-range missions that included preemptive air strikes in the Soviet hinterland.[120] However, the task force was established only as a reserve unit that would be activated on a mobilization order. This order never came, and the unit's bombers were disassembled and stored in a hangar at the Army Aviation Headquarters arsenal in Kakamigahara.

Conclusion

From 1925 on, Japanese aircraft manufacturers seized the opportunity for a massive transfer of know-how from Germany. They were willing to invest heavily in employing aeronautical experts and had the foresight to include training for Japanese specialists in the license contracts. This training, along with the delivery of aircraft parts and machine tools to Japan, introduced the new all-metal technology from Germany and exposed the Japanese aircraft industry to an entirely new set of design concepts and production methods. Japanese engineers and workmen gradually built up their knowledge, experience, and skills. They became familiar with radically new design features like the monocoque fuselage, the stressed-skin design, and the cantilever wing. Their newly acquired expertise paved the way for advanced aircraft to emerge and increasingly narrowed the gap with Western aviation technology.

A new generation of Japanese aviation engineers made fast progress toward their ultimate aim: independence from foreign design. They learned eagerly from German specialists when formal training in aeronautical engineering was still in its infancy. Alexander Baumann received an enthusiastic welcome from Mitsubishi's engineers. Responding to the company's high expectations, he built a sound theoretical base for engineering at Mitsubishi and was held in high esteem even after his return to Germany. Engineer Richard Vogt came to Japan in the wake of a license agreement between Dornier and Kawasaki. While he was supervising the experimental production of a new bomber for Kawasaki, he gained the respect of his Japanese engineer colleagues and earned the trust of his Japanese employer. Vogt seized the opportunity to develop his own original designs while having his Japanese colleagues fully participate in this process. Baumann and Vogt significantly contributed to the development of Mitsubishi and Kawasaki; as a result, the technological competence of these two companies—and their employees' confidence in their own technical skills—substantially increased.

Through its cooperation with Junkers, Mitsubishi reached new level of aviation expertise. The company became familiar with the advanced design of large, multi-engine aircraft. Mitsubishi could venture into a three-step process that began with assembling imported parts, led to the licensed production of entirely made-in-Japan aircraft, and ultimately resulted in independent design and manufacturing capabilities. Equally important was the buildup of a production system that allowed the construction of increasingly complex military airplanes.

As a result, German expertise allowed the Imperial Japanese Army to equip their air squadrons with some of the world's best military aircraft. By the early 1930s, the state of the army's airpower had risen to a

Table 4.2. Three generations of the Imperial Japanese Army's aircraft, 1921–35

	Bombers	Reconnaissance aircraft	Fighters
First generation	Farman F-60 Goliath (1919)	Salmson 2A2 (1917)	Nieuport 29 (1918)
	16 imported 1921–26	600 built by Kawasaki 1922–27	608 built by Nakajima 1923–32
Second generation	Kawasaki-Dornier Do N (Army Type 87 Heavy Bomber)	Kawasaki KDA-2 (Army Type 88 Recon)	Kawasaki KDA-5 (Army Type 92 Fighter)
	28 built 1926–32	710 built 1927–31	385 built 1931–33
			Nakajima Army Type 91
			350 built 1931–34
Third generation	Mitsubishi Ki 20 (Army Type 92 Heavy Bomber) 6 built 1931–35		

new level. The army air force's obsolete bombers, fighters, and reconnaissance aircraft, all based on French World War I designs, had been replaced with a new generation of advanced all-metal aircraft (table 4.2). With the only exception of Nakajima's Army Type 91 fighter, these airplanes were all based on designs by Claude Dornier, Richard Vogt, and Hugo Junkers.

In the wake of the Manchurian Incident, the Japanese Army could send its state-of-the art airplanes into battle and secure massive public support. A nationwide effort to further strengthen the air force turned into a donation fever that showed remarkable results. Within a short time, massive funds were raised that enabled the army to substantially reinforce its airpower. Against the backdrop of a looming war, the tried-and-tested tools of mass events, popular songs, and inciting press reports gained a new efficiency. The armed forces and the Japanese press exploited popular aviation enthusiasm as an instrument for the support by the populace—voluntary and at times enthusiastic—for a nationalism and military expansion that initially met with widespread reticence and doubt.

Paradoxically, neither technological advance nor overwhelming popular support led to a breakthrough in the army's air strategy. The doctrinal slumber of the army's planners continued even when the Japanese Army became the world's first armed force to use reconnaissance aircraft, fighters, and bombers in an international conflict after World War I.[121] Rather than giving rise to new concepts, the Manchurian Incident led to a consolidation of the army's traditional airpower doctrine of limiting the field of aerial operations to the close support of ground troops. Without the need to engage in air-to-air combat or in strategic air raids on the enemy's infrastructure, the conflict did not press the army's strategists to map out and implement advanced concepts of airpower, such as strategic bombing or the fight for air superiority.[122]

In a similar way, the army's efforts to revise its perfunctory engagement in aeronautical research ended in a dismal failure. In 1935 the army established the Air Technical Research Institute (Rikugun Kōkūgijutsu Kenkyūjo) at Tachikawa. The new facility received generous funding, especially after the start of the war with China, when its budget nearly tripled to ¥5.1 million. Yet the institute never earned the respect of the army's pilots. The army's flying corps looked down on the Tachikawa engineers as a group of incompetent technologists who had no idea about military affairs.[123] Furthermore, the army still left the design and testing of new aircraft types entirely to civil aircraft makers, a practice that further undermined the status of the Air Technical Research Institute. Even more, the army's leadership high-handedly decided about future aircraft developments without involving the institute's experts. This lofty attitude had devastating consequences, as most career officers of the Army Aviation Headquarters had little or no aeronautical knowledge. Their numerous unreasonable requests for new aircraft with superior flight performance often puzzled and frustrated the engineers of the civil manufacturers, especially when their clients insisted on conflicting requirements that were impossible to implement.[124]

In the end, rather than becoming the epitome of technological advancement to spur new developments, the Superbomber became a symbol for the fall of the Japanese Army's airpower faction. The army resumed its plan for a large four-engine bomber only in 1943 when it ordered Kawasaki to start the design of the Experimental Long-distance Bomber Ki-91. As one of history's bitter ironies, US superbombers began their own air raids on Japan in the summer of the following year. In early 1945, US B-29s dropped their bomb loads over the Kawasaki factories, bringing to a halt the Japanese Army's last Superbomber project, which was by then in its prototype phase.[125]

PART III

Britain, Germany, and Japan's Naval Aviation, 1912–37

CHAPTER 5

Navigating a Sea of Change

This chapter explores the rise of Japan's naval airpower from the navy's first successful seaplane launch in 1912 up to the mid-1920s, when aircraft carriers began to reshape Japan's naval aviation. Covering roughly the same time span as chapter 3, the parallel narrative presented here reflects Japan's well-known interservice rivalry. The Imperial Japanese Army and the Imperial Japanese Navy inhabited their own universes. They jealously shielded their air arms against any attempts to create an independent air force.[1] The branches' proverbial unwillingness to cooperate led to entirely different technical standards and parallel developments. It even resulted in double license purchases, as was the case with the German dive bomber Heinkel He 118 and the Daimler-Benz aero-engine DB 601.[2] Furthermore, the only two companies engaging in large-scale production for both the army and the navy, Nakajima and Mitsubishi, were required to keep their army and navy production lines strictly separated. This policy of bifurcated production effectively prevented any relevant technological transfer between the two services.

The Japanese Navy actively promoted the growth and technological advance of its air arm. To implement a massive airpower expansion, the navy chose a two-pronged approach of both internal and external research and development. In 1925, the army had abandoned its aeronautical research, but the navy substantially increased aircraft development and production at its arsenals.[3] After the 1923 Kantō earthquake destroyed its Aircraft Test Laboratory at Tsukiji, the navy resumed its research activities at the Kasumigaura Air Base. In 1926 the navy operated the country's largest wind tunnel, with a 2.5-meter diameter. It had been built under the supervision of German engineer Carl Wieselsberger (1887–1941), an assistant of the eminent aeronautical engineer Ludwig Prandtl (1875–1953).[4] In 1932 the navy centralized its basic scientific research, the development and testing of experimental airplanes, and aircraft production at the new Yokosuka Naval Air Arsenal (Kaigun Kōkūshō), about twenty kilometers south of Yokohama. At the same time, navy officials and specialists cooperated closely with civil manufacturers. These companies received generous funding and technical assistance from the naval arsenals, which supplied them with technological know-how, advanced designs, and the latest production technology.

In the fast-changing economic, political, and strategic environment of the early 1920s, military innovation became a dominant element of national security. To catch up with the Western powers, many Japanese military planners felt the pressing need to modernize the country's armament and revise long-established strategies. However, in Japan as elsewhere, these reform-minded military innovators encountered various challenges. They had to offer a realistic vision of future warfare that would not get lost in utopian fantasies or pure theory. They had to meet the challenge of budget cuts. Most important, these officers had to prevail in an environment that valued discipline and strict adherence to established procedures—an environment that showed little inclination to consider unorthodox ideas or experiment with new concepts.[5]

Historians have pointed out that the success of military innovation depends on a balanced interplay among strategic, organizational, and technological transformations.[6] The development of early naval aviation in Japan provides a touchstone for such interplay. As we will see, the proponents of naval airpower had to convince their superiors of the value of a new doctrine. They had to demonstrate how their ideas could be implemented with new technology and new organizational patterns. This process of persuasion was difficult, and it comes as no surprise that the modernizers' obvious choice was to turn to foreign knowledge and experience.

Situating Japanese aviation history in a wider strategic context, this chapter demonstrates that importing British and German technology provided decisive stimuli for far-reaching changes in the Imperial Japanese Navy. A whole range of different imported aircraft types offered new answers to the probing questions of how to overcome the huge distances of the Pacific Ocean and how to challenge a powerful enemy. The new technologies helped bolster the navy's modernizers and seriously challenged the "big-ship, big-gun policy" (*taikankyohōshugi*) of the traditionalists. At the same time, British and German hardware and know-how transformed the character of Japan's airpower from a largely defensive force to an air fleet with aggressive first-strike capabilities.

The competing doctrines of "battleship first" versus "aircraft first" (*kōkūheiryoku chūshinshugi*) dominated the development of naval aviation in Japan right from the start.[7] The Japanese Navy had carefully studied Alfred Thayer Mahan's (1840–1914) seminal book, *The Influence of Sea Power upon History*, which was published in the United States in 1890 and translated into Japanese in 1896. The US admiral forcefully argued that only a strong navy could guarantee a nation's great power status and emphasized the central role of battleships for a country's defense and international status. After the Japanese Navy's spectacular successes in the First Sino-Japanese War (1894–95) and the Russo-Japanese War (1904–5), the big-ship, big-gun policy prevailed among navy officials. This doctrine of naval warfare was based on the belief that large ships and superior firepower would be decisive for victory at sea. The battleship-first proponents received more encouragement when World War I provided a boost to Japan's economy. After the European powers suspended their international trade, Japanese exports expanded rapidly. The boom made substantial funding available for entertaining the navy general staff's vision of an ambitious fleet expansion program that would culminate in an "eight-eight fleet" (*hachihachi kantai*) of eight battleships and eight armored cruisers.[8]

Under these conditions, Japan's emerging naval aviation had a slow start. The career of naval officer Nakajima Chikuhei (1884–1949) illustrates this point. In 1912 Nakajima, then a lieutenant, spent four and a half months in the United States. His order was to become familiar with the construction and maintenance of airplanes at Glenn Curtiss's factory at Hammondsport, New York. Nakajima also took flying lessons, on both sea- and landplanes, and gained the Aero Club of America pilot's license. Upon his return, Nakajima's superiors harshly criticized him for transgressing his order, which was limited to factory visits only.[9] Nakajima revived his image as a misfit in 1914 when he presented his "Request for the Allocation of the 1914 Budget" to his superiors. In his memorandum he opposed the navy's focus on large-scale battleships. He argued that only rich countries could afford such a policy. For the price of one battleship, Japan could build 80,000 aircraft. Such a large air fleet, equipped with bombs and torpedoes, could inflict "extreme and unimaginable damage to the enemy."[10] Historians have pointed out that the doctrine of aircraft as the principal weapon and the ineffectiveness of battleships (*kōkūshuheisen kanmuyō ron*) made Nakajima one of the pioneers of air strategy, long before Giulio Douhet and William Mitchell (1879–1936) came up with similar ideas.[11] Yet it seems that the young officer's ideas fell on deaf ears among navy officials, who still clung to the image of the aircraft as an immature technology with questionable military value. In 1917 Nakajima left the navy to start his own company, which ironically became one of Japan's leading aircraft suppliers.

Almost two years passed between the pioneering flights of the Japanese Army and the navy's successful flight, when Lieutenant Kaneko Yōzō (1882–1941) took off with his Farman floatplane on October 6, 1912. In the following month, the navy showed its new air arm to the public for the first time when 2 floatplanes, together with more than 100 other vessels, participated in a naval review at Yokohama.

In summer 1914, the Japanese Navy dispatched its seaplane carrier *Wakamiya* along with four seaplanes to participate in the siege of Qingdao. On September 4, 1914, Lieutenant Commander Kaneko again made aviation history when he flew the Japanese Navy's first wartime sortie. On countless occasions, he experienced firsthand the cumbersome procedure of hoisting seaplanes from their carrier down to the water surface to allow them to take off. Kaneko therefore repeatedly emphasized the need for a device that would enable aircraft to take off directly from military vessels. Soon this vision materialized.

Japanese Observers in Britain during World War I

During World War I, the Japanese Navy took its first steps toward a systematic expansion of its airpower. In 1916, it decided to establish within the next five years three flying squadrons (*hikōtai*), each equipped with six aircraft. At the same time, the navy dispatched several officers, including

Kaneko, to Great Britain to study all aspects of British military aviation and assess the state of the country's aviation industry. In September 1916, Kaneko left Japan. After his arrival in Great Britain, he became especially interested in a pioneering project of the Royal Navy. The HMS *Furious* was the world's first vessel to use a modern carrier design. Initially planned as a battle cruiser, the *Furious* was modified during construction with an aircraft hangar and a flight deck that extended over the forward half of the ship. Kaneko visited the National Physical Laboratory in Teddington, near London, where he observed wind tunnel tests for the new carrier. In May 1917, Kaneko submitted his Investigation Report on British Aviation Technology (*Eikoku kōkūjutsu shisatsu chōsahōkoku*) to Vice Minister of the Navy Suzuki Kantarō (1868–1948).[12] Kaneko urged Suzuki to follow the British lead and strengthen Japan's naval airpower by building carriers and training future aircraft carrier pilots. The report included important material on design and layout that laid the foundation for the construction of Japan's first aircraft carrier.[13] In August 1917, Kaneko's advocacy of the new technology grew even stronger when he learned that Squadron Commander Edwin Dunning (1892–1917) had successfully landed his Sopwith Pup biplane on the flight deck of the British carrier *Furious*. This aeronautical feat made Dunning the first person to land an airplane on a moving ship.[14]

Lieutenant Commander Tosu Tamaki (1877–1949) was another naval officer who became a strong proponent of transferring British naval technology and strategy to Japan. Tosu spent February 1916 to July 1917 as a military observer of the British fleet. In his November 1917 "Outline of the Royal Navy's Strategy" (*Eikoku taikantai sakusen no taiyō*), he offered a concise analysis of the Royal Navy's wartime aviation strategy with a view to its adoption by the Japanese Navy.[15] In the chapter "Use of Aircraft in the Navy's Strategy," Tosu took the discussion about the strategic value of naval airpower to an entirely new level. After explaining how aircraft could be launched from cruisers for reconnaissance missions, he argued that bombers able to take off and land on aircraft carriers would play an even more important role in increasing a fleet's fighting strength. These aircraft could deliver a fatal blow to the enemy by bombing its naval bases and attacking its fleet during a sea battle. It is important to note that a number of early airpower advocates like Tosu and Kaneko were later promoted into influential positions within the Japanese Navy, where they could continue to spur the development of naval aviation. Tosu became vice admiral in 1939, and Kaneko rose to the rank of rear admiral in 1940.

In addition to these groundbreaking reports, directly purchasing aircraft provided another opportunity to investigate the latest advances in Britain's naval aviation. Even though the British aircraft industry struggled to meet the country's wartime demand, the Japanese Navy was able to import a whole range of British planes.[16] In 1916 it purchased a Grahame-White trainer, an airplane that could carry five crew members, making it the largest aircraft in Japan at that time. The navy also acquired one Short 184 torpedo bomber, famous for being the world's first aircraft able to launch an aerial torpedo attack, and a Sopwith Schneider plane that became Japan's first fighter seaplane. In addition, with money donated by ship owner Yamashita Kamesaburō (1867–1944), the navy was able to buy seven Short 320 seaplanes. These were the Royal Navy's largest hydroplanes, each designed to carry a 450-kilogram torpedo. These aircraft arrived in Japan between 1918 and 1919.

An Early Compromise: Ship-Based Floatplanes

Airplanes with floats, like the imported Short and Sopwith seaplanes, were an obvious technological choice during the early years of naval aviation in Japan. Even though the floats (also called pontoons) added weight and air resistance, they enabled a plane to land on water and thus made the aircraft independent from airfields or landing strips. Ordinary merchant ships could easily transport these seaplanes by using cargo hoists to lower and recover the aircraft. Even though most of these early naval aircraft lacked the engine power or structural strength to carry heavy armament or bombs, floatplanes became an invaluable tool for navy operations. Already during World War I, surveillance aircraft in combination with wireless telegraphy and aerial photography revolutionized military reconnaissance. Long before radar was available, airplanes that could act as the "eyes of battleships" to make out a distant enemy were of paramount importance. Ironically the development of "big-gun" technology had reached a point where battleships became even more dependent on aerial support. By the end of World War I, the increasing firepower of naval artillery began to extend beyond the visual horizon. For the effective use of their large guns, battleships needed spotting aircraft to provide accurate artillery guidance for directing their gunfire.

In addition to British imports, the Japanese Navy became acquainted with the latest German floatplanes. Between October 1920 and February 1921, twenty-two seaplanes arrived in Japan as war reparations from Germany.[17] Among these aircraft, the Hansa Brandenburg W.29, a World War I floatplane, interested the navy most. The advanced monoplane, designed by German aircraft engineer Ernst Heinkel, was well known for its superior performance and impressive combat record. Especially when compared with the French floatplanes that the Japanese Navy had deployed in 1914 at Qingdao and was still using in 1920, the German aircraft was nearly twice as fast and could carry double the load. The navy adopted the W.29 as Japan's first ship-based monoplane and ordered aircraft makers Nakajima and Aichi Tokei to mass-produce copies. As yet another consequence of the Treaty of Versailles, German companies had lost the protection of their patent rights, so Japanese aircraft manufacturers did not have to worry about buying an expensive production license. Between 1922 and 1925, the two companies built a total of 310 Hansa Brandenburg planes, which became the navy's main water reconnaissance aircraft (figure 5.1).[18]

The navy's decision to entrust former watchmaker Aichi Tokei with such a massive order for warplanes is not as uninformed as it seems. Aichi Tokei was different from most major Japanese aircraft makers like Mitsubishi, Kawasaki, and Ishikawajima, whose origins can be traced back to shipbuilding. The Nagoya-based company started its business as a clock manufacturer in 1898. During the 1904–5 Russo-Japanese War, Aichi Tokei diversified into armaments production and began to produce torpedo fuses and detonators for naval mines. With increasing orders during World War I, the factory expanded and engaged in producing bomb sights and communication instruments. Aichi made good use of its close relationship with the Japanese Navy when it decided to start an aircraft manufacturing branch in 1920. The navy provided comprehensive technical training for Aichi's workmen and engineers at its arsenals, so the company was well prepared to venture into large-scale aircraft production.[19]

In 1923 the navy added two more state-of-the-art floatplanes to its air fleet. The Imperial Maritime Defense Volunteer Association (Teikoku

FIGURE 5.1 This photograph shows two Navy Type Hansa reconnaissance seaplanes, a made-in-Japan version of Heinkel's Hansa Brandenburg W.29. The low-wing monoplane's unusual rudder position provided the gunner with an unobstructed field of fire. These floatplanes were built in large numbers and the beachgoers' scant attention suggests that they have already become a common sight. (Source © Japan Aeronautic Association)

Kaibō Gikai), a foundation for promoting coastal defense, acquired funds to buy two Junkers floatplanes from Germany.[20] These all-metal monoplanes were maritime versions of Junkers's famous F 13, a state-of-the-art passenger aircraft that had set new altitude and endurance records. The association donated the planes to the navy's Yokosuka Naval Air Arsenal for test flights and research on all-metal aircraft design.[21]

The donation of the Junkers float planes received wide press coverage, but two more aircraft from Germany arrived in utmost secrecy. In 1922, when the Washington Naval Treaty limited Japanese naval power to four battle cruisers and six battleships, the navy's strategists put a new emphasis on large submarines.[22] In 1923 Ernst Heinkel, who by then was well known to the Japanese Navy, received a request to build a reconnaissance seaplane that could be carried in a special container inside such a large submarine. To fulfill this almost impossible task, Heinkel referred to the U 1 aircraft he had designed during World War I. In his German workshop, he secretly built a remarkable aircraft. According to his memoirs, a well-trained crew could remove the seaplane from its pressure-proof container and make ready for flight in thirty-one seconds.[23] Heinkel shipped the parts of two dismantled airplanes to Japan, where they arrived at the Yokosuka Arsenal in 1923. After a careful examination of the two submarine aircraft, the arsenal's engineers built their own copy and successfully tested it on one of the navy's submarines.[24]

A New Launching Technology

A major technological breakthrough made it possible for floatplanes to take off from a moving battleship even under rough sea conditions. Naval planners welcomed this development, which promised to overcome the limitations of floatplanes as a "fair-weather" weapon.[25] Once more, the initial impulse came from Great Britain. Frederick Joseph Rutland (1886–1949) was the first pilot to perform a successful takeoff from a gun turret. In June 1917, he took off with his small Sopwith Pup biplane from a six-meter platform installed on the cruiser *Yarmouth*. Rutland demonstrated how a gun turret and platform could be turned into the wind, allowing an aircraft to take off without the ship having to change course. Lieutenant Kuwabara Torao (1887–1975), who stayed in Britain that same year, started similar trials upon his return to Japan. In June 1920, Kuwabara became the first Japanese to perform a takeoff from a small platform mounted on the front part of the seaplane carrier *Wakamiya*. In his memoirs, he vividly describes "how to take off from a battleship at a time when there was no catapult." In a daring trial-and-error procedure, his team successively reduced the initial takeoff distance until Kuwabara nearly lost control of his aircraft:

> We installed a twenty-meter take-off deck on the bow of the Wakamiya with a downward slope of two degrees. . . . With the tail of the aircraft tied to the ship's mast I accelerated the engine to full power. After I gave the signal for release, the airplane picked up speed and took off safely. . . . The next day we reduced the platform first to eighteen meters, then to sixteen

> and to fourteen. When I tried to take off from a twelve-meter-long platform my aircraft first dropped by three meters before gaining enough speed to continue its flight. We decided that a take-off run of fifteen meters would be necessary.[26]

As it turned out, Kuwabara completed his experiments just in time. In 1920 the Japanese Navy commissioned one of the world's largest battleships, the *Nagato*, which—with her eight forty-one-centimeter guns, each weighing more than 100 tons—was the epitome of the big-ship, big-gun doctrine. To combine the advantages of airborne reconnaissance with the firepower of a giant warship, the *Nagato* was to be equipped with a ship-launched aircraft. Once more the navy's officials drew on Heinkel's expertise. To avoid any allegation of violating the Treaty of Versailles, which prohibited the production and export of German military aircraft, the navy employed the Aichi company as mediator for negotiations with Heinkel. In 1925 Japan's Berlin-based naval attaché, Captain Kojima Hideo (1896–1982), encouraged Kaya Masaru (1887–1927), one of Aichi's engineers who had come to Germany, to make personal contact with Heinkel. Referring to the many German engineers already in the employ of other Japanese companies, Kojima told Kaya: "Heinkel is still young, but he seems to be promising. Why don't you meet him? He is the only first-class [aircraft maker] left."[27] It was easy for Kaya to win Heinkel over, as the German engineer welcomed the opportunity of doing more business with Japan. In early 1925, he signed a contract for designing two different seaplanes that could take off by their own power from a battleship's gun turret on a twenty-meter-long rail, which was also to be designed by Heinkel.[28]

During the airplanes' design and construction process, Aichi Tokei's relationship with the Ernst Heinkel Flugzeugwerke company deepened. Both firms contrived a clever scheme to outmaneuver the control commission of the Allied Powers. Aichi's Berlin-based engineers maintained close contact with the Japanese members of the commission. They regularly warned Heinkel about impending inspections, giving the company enough time to hide its military production.[29] This arrangement guaranteed steady progress; already by May 1925 Heinkel had completed two aircraft, the He 25 and He 26, along with the launching device. The business with Japan was of such importance that Heinkel decided to travel to Japan himself to present his work. In autumn 1925, under the watchful eye of its designer, Heinkel's aircraft successfully took off from the *Nagato*'s gun turret (figure 5.2). Aichi was clearly impressed. The next day, the company signed a license agreement for the two aircraft types and the launching device.[30]

Over the following decade, a mutually beneficial relationship developed. Aichi's continuous orders kept the young German company afloat. In exchange, Heinkel developed a large variety of prototypes that he built in Germany and exported to Japan. His designs included several aircraft that could be deployed with catapults, a new technology that the Japanese Navy developed in the late 1920s.[31] The design of these catapult-launched planes was a challenging task because Heinkel had to make sure that they were robust enough to withstand the fast acceleration during takeoff.[32] However, the new technology significantly improved on the older rail-launching method. The power-driven catapults could accelerate an aircraft to takeoff speed within a much shorter distance and made it possible to launch considerably heavier and faster aircraft.

Redefining Naval Airpower: The Early Years of Carrier-Based Aircraft

The remarkable advances in floatplane technology clearly added to the strike power of the Imperial Japanese Navy's battleships. However, a new type of warship was to revolutionize naval battle strategy. Aircraft carriers—basically highly mobile seagoing airfields—greatly extended the range and fighting strength of a battle fleet. The planes that could be deployed from these carriers would no longer be restricted by the floatplanes' cumbersome design, which had to combine seaworthiness with airworthiness. Carrier-based planes could attain a much higher speed, range, and maneuverability.

Shortly after Japanese observers authored their enthusiastic reports about British aircraft carrier technology, a spectacular air strike captured the attention of the military and the public. In the early hours of July 19, 1918, British aircraft staged the world's first carrier-launched air raid. Seven Sopwith Camel biplane fighters took off from the carrier *Furious* and,

FIGURE 5.2 How to take off from the cannon turret of a battleship: two rare photographs from 1925 show the careful takeoff preparations and the successful launch of Heinkel's floatplane. (Source: Mainichi Shinbun, *Nihon Kōkū shi*, 42–43)

after about an hour of flight time, destroyed a German hangar and two airships at Tondern, a small town near the Danish–German border. The operation received wide press coverage; Japanese newspapers took note of the Tondern raid as well.[33]

In the same month, Japanese Navy planners made a bold decision. A Temporary Investigation Committee for Submarines and Aircraft (Rinji sensuikan kōkūki chōsa iinkai) set up the specifications for Japan's first aircraft carrier, the *Hōshō* ("flying phoenix"), which was largely based on the British carrier *Furious*. The Yokosuka Arsenal received the order for construction in October the next year. When the navy commissioned the *Hōshō* in December 1922—only nine months after the United States' first carrier, *Langley*, was put in service—Japan joined the exclusive club of three countries with aircraft carriers in active service.[34]

With the *Hōshō* project, Japan's navy assumed a leading role in the design and construction of aircraft carriers. Rather than confining themselves to merely replicating the British model, the navy used the new carrier for testing and improving advanced equipment, trying out various takeoff and landing techniques, and developing new operational methods and tactics.[35] However, the carrier's integration into Japan's naval force still had to be determined, and many challenging questions had to be addressed. Should the new vessel and its aircraft be used for attacking land targets? Should they support the navy's battleships? Or should the new weapon become an independent part of a long-range strike force? At the same time, operational questions for the carrier design concerning its anti-aircraft armament and flight deck layout had to be answered. Most important, the navy had to develop the specifications and tasks for the different aircraft types that were to be deployed from the carrier. Again, Japan drew on British expertise.

The Arrival of the First British Aeronautical Engineers in Japan

Even though World War I had greatly advanced the rise of British naval aviation, the postwar years led to an equally dramatic decline. With postwar demobilization, only 200 British fighter aircraft remained in service.

The aircraft industry took a nosedive as well. In Britain, as elsewhere, surplus stock and a collapsing demand led to excess capacity. As a result, the number of aircraft makers dwindled from seventy at the end of the war to thirty in 1920. At the same time, the industry's workforce declined from 110,000 to only 3,000.[36]

Chances for survival in an already depressed market further dwindled when the British Treasury introduced the Excess Profit Tax in 1920. This tariff was based on a comparison of a company's prewar profits with its wartime profits, a policy that was especially detrimental for pioneer manufacturers. The year the new tax was introduced, companies like Sopwith Aviation, the manufacturer of many successful World War I aircraft, and Airco (Aircraft Manufacturing Company), still in 1918 considered as the world's largest aircraft maker, went into liquidation.[37] The market for civil aircraft did not offer much relief either. By 1921 all British civil air transport companies had gone out of business. British commercial air traffic resumed only in 1924 when Imperial Airways was newly founded.

With the poor state of Great Britain's aviation industry, many British engineers were looking for an opportunity to offer their skills and knowledge to other countries, including Japan. Such an opportunity arose in 1921, when the Japanese Navy signed a contract with airplane manufacturer Mitsubishi Aircraft for fighters, bombers, and reconnaissance aircraft that were to operate from the new *Hōshō* carrier. The Japanese Navy acted as a mediator for Mitsubishi to find and invite a group of eight British aeronautical engineers to Japan. The team's leader was Herbert Smith (1890–1978), the former chief designer of the now-defunct Sopwith Aviation Company. It seems that the navy's choice was well informed. Having just lost their jobs, the British experts were easily persuaded to accept the Japanese offer. More important, Smith—together with his colleagues—had already designed several aircraft types specifically for carrier operation, most notably the Sopwith Pup and Sopwith Camel.

Smith and his team arrived in Japan in February 1921. They went to work with great zeal and quickly adapted to their new environment, which confronted them with a curious mix of modern and traditional production methods. The memoirs of Ernest Cyril Comfort, one of the group's draftsmen, give a vivid impression of the 1921 Mitsubishi Nagoya factory. He described how modern diesel engines supplied electrical power, bright floodlights illuminated the shop floor, and a railway led right into the factory building. At the same time, the British suffered from soaring temperatures in their poorly ventilated design offices. Comfort was astonished that no cranes were available; heavy items had to be moved by "large gangs of coolies."[38]

Cooperation with the Mitsubishi engineers turned out to be another problem for the British. Smith's contract clearly stated that he was to supply "all the necessary sketches, specifications, calculation sheets of performance and strength." Even more important, Mitsubishi insisted on the British training the company's own engineers. Smith had to agree to "fully illustrate any new design or idea or improvement adequate[ly] enough to enable the principals to prepare the working drawings themselves and to satisfactorily manufacture or produce any aircraft or improvement so designed."[39] But a critical evaluation shows that although Smith and his British colleagues provided the requested blueprints and calculations, they imparted scant know-how about airplane design. Rather than including the Mitsubishi engineers in the design process, the British allowed their Japanese colleagues only to "assist" in making blueprints. Such assistance often involved merely copying the British-made drawings. One Mitsubishi engineer vividly remembered how even this menial task earned him sharp criticism from a British designer. A copy he made that had obviously neglected some minor detail was returned with the written comment: "How to draw a nut. See your school book."[40] This telling anecdote suggests that the Smith team's judgment of Japanese design skills was affected by cultural stereotyping. It also illustrates how the British effectively excluded Mitsubishi's engineers from the design process in an obvious attempt to keep the Japanese from developing an independent design technique.

In spite of these frictions, Smith and his coworkers completed the design and construction of their first airplane just eight months after their arrival. The Navy Type 10 fighter was to write aviation history as the world's first aircraft specifically developed as a carrier-based fighter. The plane was lightweight and had a powerful 300 horsepower engine to be able to take off with ease from the *Hōshō*'s 168-meter-long flight deck. For landing on the carrier, the aircraft had hooks attached to the wheel axle that would connect with arresting wires on the carrier's deck after touchdown. In addition to the Type 10 fighter, the Smith group designed a reconnaissance aircraft, a torpedo bomber, and an attack bomber that were all to be based on aircraft carriers.

The British Aviation Mission to Japan

In April 1921, two months after the Smith team arrived, another group of British experts, known as the British Aviation Mission, came to Japan. To better understand the origins, pursuits, and results of one of the most consequential aviation missions to Japan, we first have to look at some of the tectonic shifts in the military and diplomatic landscape in East Asia.

After the outbreak of World War I, Anglo-Japanese relations—once founded on the seemingly unshakeable bedrock of the 1902 alliance—began to sour. With the 1914 occupation of the German Qingdao colony, Japan had established a strong foothold in China; after the takeover of Micronesia in the same year, the Japanese Navy had seized the strategic control of some of the major shipping lanes in the Pacific. These geopolitical changes directly affected British military strategy in the Far East. In the view of the British admiralty, Japan emerged after World War I as a potential enemy that threatened the British colonies of Malaya and possibly India and Australia. As a result, the admiralty started drafting war plans against Japan in 1919.[41]

In the same year British naval planners began to envision a future armed conflict with Japan, the British government based its armament policy on an entirely opposite assumption. The Ten Years Rule of 1919 assumed that the British military would not be engaged in any major conflict for a decade; the government thus saw no need to direct major funds toward aerial armament. Notably, the rule included relations with Japan. Still, in May 1925 the British Cabinet declared "that in existing circumstances aggressive action against the British Empire on the part of Japan within the next ten years is not a contingency seriously to be apprehended."[42]

Japanese Navy officials were aware of these very contradictory British attitudes. Nonetheless they put great hopes into an aviation mission from Britain. The plan for the mission originated from a group of Japanese naval officers with close access to British aviation technology. During

World War I, Ōseki Takamaru (1883–1925) was an assistant to the Japanese naval attaché in London. Along with Lieutenant Commander Hayashi Sueki and naval engineer Kitajima Kanae, he authored an "Opinion Paper Based upon the Inspection of the Aviation of Western Countries."[43] The paper, submitted to the Ministry of the Navy in June 1919, emphasized the need for Japan's navy to catch up with the naval airpower of Western countries. Therefore, the report went on, Japan must decide about her future naval air strategy and the required aircraft types. The authors argued that with the British Royal Navy being the world's most experienced navy, it would be best "to follow the British model," especially for building a Japanese aircraft carrier fleet. Therefore, the navy should invite British instructors as soon as possible to train Japanese pilots to fly British aircraft. The authors cleverly argued that after the war many discharged British pilots were eagerly looking for employment, and the Japanese Navy should make good use of this opportunity before the most skilled instructors found work somewhere else. In a second report, titled "On Recruiting British Air Force Officers," Ōseki argued that rather than sending a few Japanese trainees to Britain, it would be much more effective to invite a team of British airmen who could instruct a large number of Japanese Navy pilots in all aspects of aerial warfare.[44]

Such a proposal was not as unorthodox as it may seem. Indeed, referring to the "British model" and inviting British instructors was a policy already established during the formative years of the Japanese Navy. In 1867, a team of seventeen British advisers led by Commander Richard E. Tracey (1837–1907) helped establish a training facility that was the predecessor for the Imperial Japanese Naval Academy. A second British naval mission under Lieutenant Commander Archibald Douglas (1842–1913) stayed in Japan from 1873 to 1879. Under the guidance of Douglas and his thirty-three instructors, the Imperial Japanese Navy closely adopted British naval regulations and training policies.[45]

With these prominent antecedents, Ōseki could expect his suggestions to meet with favorable consideration. What's more, his reports successfully conveyed to his superiors a sense of urgency and an opportunity not to be missed. Also in 1919, the Japanese Navy's general staff drafted a paper outlining a new "Policy of Adopting Foreign Aircraft." The memorandum resumed Ōseki's argument that British naval aviation was leading the world and that until now the Japanese Navy "essentially has learned everything from Great Britain." It concluded that making good use of British aircraft would be "highly convenient."[46] The general staff's view received the Ministry of the Navy's full support, especially after Minister Katō Tomosaburō (1861–1923) had to respond to the Diet's challenging questions about the slow buildup of Japan's naval air force. Katō acknowledged the "large gap between Japan's aviation and that of the Western countries" and conceded that it was urgent to procure new aircraft types, train more personnel, and establish more air squadrons and new aircraft factories. Katō assured the Diet members that to achieve these aims the navy would emulate the British navy's air force—and invite British instructors.[47]

The Japanese Navy had no problems allocating sufficient funds for a major aeronautical mission from Britain. According to one source, the navy's reserves for aircraft procurements had been accumulating since 1914, when the outbreak of World War I made it impossible to make large-scale purchases of aeronautical matériel from abroad.[48] With all administrative and financial hurdles overcome, the naval planners drafted a proposal for "Inviting British Air Force Officers" and decided that official negotiations should begin in August 1920.[49] London-based naval attaché Kobayashi Seizō (1877–1962) received the order to initiate informal talks with high-ranking officers of the Royal Air Force. He submitted an official request for aeronautical training and technological assistance in October.

The British response to Japan's proposal was anything but unanimous. The British admiralty strongly advised against any technological support that would lead to a powerful Japanese air arm. As First Sea Lord Admiral David Beatty (1871–1936), put it in 1921: "The naval expansion of Japan might become a considerable menace in the Far East."[50] Yet the British military attaché and the ambassador in Tokyo did not share this view. Both of them strongly supported dispatching a British mission and warned the Foreign Office that any delay would risk increasing French influence and cause Britain to miss the opportunity to make the Japanese fully dependent on the Royal Air Force's assistance.[51] The British aviation enterprises, along with the newly created—yet already powerful—Air Ministry, agreed to keep Japanese aviation in a state of technological dependency.[52] Furthermore, they both welcomed the opportunity for Britain's struggling aircraft industry to receive new purchase orders and get rid of surplus planes that clogged domestic demand. Confident in Britain's huge technological lead over Japan, they were also convinced that the mission would boost aircraft sales and open a lucrative market rather than promote independent Japanese design.

The British Air Ministry eventually reached a compromise and decided to send an "unofficial" aviation mission. Emphasizing commercial over military interests, the principal aim was to reinvigorate the British aviation industry and establish a firm foothold in Japanese naval aviation. Clearly the transfer of know-how in aircraft design and production played a secondary role. In an obvious attempt to dispel the admiralty's concerns about an unchecked rise of Japanese airpower, the Air Ministry strictly prohibited mission members from accessing or passing on any confidential information or technology.[53]

The appointed mission leader embodied a striking mix of aristocratic glamour and dauntless airmanship. Sir William Francis Forbes-Sempill (1893–1965) was recommended by the director general of Civil Aviation "as being a person suitable for the Japanese."[54] A Scottish nobleman educated at Eton, Sempill joined the Royal Flying Corps in 1914. He rose through the ranks quickly and in April 1918 became a colonel in the newly established Royal Air Force. After the war, Sempill served as a test pilot until his retirement from military service in 1919. In January 1921, Sempill reached an agreement with Japanese naval attaché Kobayashi for the "reorganisation, equipment, and training of the Imperial Japanese Naval Air Service."[55] Sempill headed a group of thirty British experts that included six flight instructors, among them Major Herbert George Brackley (1894–1948) and Major F. B. Fowler, known as "two of the oldest and ablest British Pilots."[56] The ground crew consisted of fifteen technicians and four specialists for aerial armament and photography.[57]

From its inception, the character of the British Aviation Mission was entirely different from that of the 1918–20 French Aeronautical Mission that went to Japan with the official approval of the French government. Neither the British Navy Ministry nor the Foreign Ministry nor the Air Ministry took any responsibility for the mission. The mission members were assigned Japanese ranks and provided with a distinct uniform to underscore their unofficial status (figure 5.3). Mission leader Sempill, who had returned to civilian life before coming to Japan, received no official instructions, and he did not have to report to any superior. As we will

FIGURE 5.3 In this picture we see Sir William Francis Forbes-Sempill, the leader of the British Aviation Mission. His unadorned uniform, with his badges of rank barely visible on the lapel, was to emphasize the "unofficial" character of the mission. In the background a student and his flight instructor are getting ready in their Avro 504K, a British biplane that with safe and dependable flying characteristics was ideal for initial flight training. (Source © Japan Aeronautic Association)

see, this apparent lack of official control over the mission's activities in Japan had considerable consequences.

THE MISSION'S AIRCRAFT AND ACTIVITIES

Leaving no room for half measures, the mission members brought more than 100 aircraft, which covered the entire range of British naval aviation (table 5.1). Among these, the airplanes that could be deployed from aircraft carriers were of greatest importance to the Japanese. The Gloucestershire Aircraft Company had designed and built forty-two Sparrowhawk carrier planes according to the Japanese Navy's specification. These aircraft greatly advanced the navy's aerial strategy—ten were specifically designed for being launched from ramps installed on battleships. The

Table 5.1. A list of the extensive aircraft delivery that arrived with the Sempill mission

Trainers	Avro 504K for preliminary flight training land (20) Avro 504L for preliminary flight training sea (10) Gloster Sparrowhawk for carrier training (42)
Torpedo bombers	Sopwith Cuckoo (6), Blackburn Swift (1)
Fighters	Gloster Sparrowhawk carrier-based fighters (10) Martinsyde F.4 (1), Buzzard Fighter (1) S. E.5 A Fighter (1)
Reconnaissance	De Havilland D. H. 9 (1)
Flying boats	Short F.5 (14) Vickers Viking Amphibian (1), Supermarine Channel (1) Supermarine Seal Amphibian (1), Norman Thompson (1)

Notes. The number of each specimen imported appears in parentheses. Adopted from Nagura and Yokoi, *Nichi-Ei heiki sangyōshi*, 390.

Sparrowhawks arrived in components, which allowed the navy engineers to closely examine their design features during assembly. Six aircraft of the Sopwith Cuckoo type, Britain's first carrier-borne torpedo bomber, provided an opportunity for the navy's pilots to become familiar with air-launched torpedo attacks, a new concept of naval warfare. The Sempill mission also brought Britain's latest torpedo attacker, the Blackburn Swift, to Japan. This powerful second-generation carrier-borne torpedo-bomber could carry twice the load of its Sopwith predecessor.

While the British military sought to avoid any association with the mission, Japanese military officials were less reticent and proudly presented the Sempill team to the public and the media. Already on April 15, 1921, the Japanese press celebrated the arrival of the mission's advance party. One article emphasized that British Air Force officers would share their World War I experience with the Japanese Navy pilots and teach them to fly their brand-new machines that compared favorably even with the latest US aircraft.[58] In May 1921, the first delivery of British aircraft arrived just in time for the opening of the navy's new airport at Kasumigaura. The official consecration ceremony for the airport, held in July, was more than a religious formality. The Imperial Japanese Navy orchestrated a spectacular event, attracting more than 30,000 spectators who were thrilled to watch the British officers and their flying machines. The highlight and most dramatic performance of the day was a parachute jump successfully carried out by Major Thomas Orde Lees (1877–1958).

Soon the opportunity for another publicity stunt presented itself. On September 3, Prince Hirohito returned from a visit to Europe aboard the battleship *Katori*. The prince arrived at Yokohama with great fanfare. Gun salutes, wailing sirens, and rousing cheers welcomed the future emperor. The major attraction—apart from the imposing British-built battleship—was, as one reporter called it, "one of the finest exhibitions of aerial work ever seen in Japan."[59] A formation flight of nine airplanes escorted the ship, six of them piloted by British aviators, with three Japanese following. As a special expression of welcome, Major Lees performed another parachute jump, landing right in front of the *Katori*. After this splendid flight show, the mission continued to stage itself to the Japanese public on several occasions. The British pilots dropped thousands of leaflets during the celebration of a railway anniversary in October 1921; they performed a dramatic one-week-long around-Japan flight with three F.5 flying boats that received wide press coverage in April 1922; also that April, they demonstrated another formation flight of thirty-six aircraft on the occasion of a visit of the British warships *Renown* and *Durban* to Japan.

JAPAN'S FIRST CARRIER-DECK LANDING

Amid all the glamour and excitement, the British instructors did not fail to take care of the mission's original purpose. The first training courses started on September 1, 1921, and soon the Central Training Station at Kasumigaura bustled with activity. According to Sempill's account, it was "not uncommon to see forty machines in the air at once."[60] After completing their basic flight instruction, a number of Japanese students specialized in piloting flying boats and seaplanes. Others received advanced training in aerial reconnaissance, fleet cooperation, night flying, and torpedo bombing. There was one ultimate skill that the Japanese Navy was most eager to have its pilots acquire: landing on the deck of an aircraft carrier. Training carrier pilots became an urgent matter: Japan's first aircraft carrier, whose final flight deck layout was completed under British guidance, would soon be ready for launch. In the meantime, the British flight instructors trained their Japanese students on dummy flight decks ashore.

The decision of who was to perform Japan's first carrier deck landing turned into a competition between the British Air Mission and the British design team under Herbert Smith that, it should be recalled, had joined Mitsubishi in 1921. The pursuits of the Smith team and their test pilot, William Jordan (1896–1931), led to increasing discontent among the mission members. Sempill perceived his fellow countrymen as unwelcome competitors. On several occasions, he tried to exert influence on Smith and his colleagues, especially about the question of who was to perform the first deck landing. In a letter to the navy vice minister, Sempill directly asked for permission "to land one of the amphibian machines on the [aircraft carrier] *Hōshō* the same day as Captain Jordan lands the Mitsubishi machine."[61] Herbert Brackley, one of the mission's senior flight instructors, had already flown the Mitsubishi aircraft and considered it "most unsuitable for deck work."[62] Brackley strongly recommended using a British-built plane for the challenging task. The British embassy also seemed to have some misgivings, as can be seen in a rather dismissive report on Smith's latest design, stating that Mitsubishi's new carrier reconnaissance aircraft "is said to be an unpleasant machine to fly; but the Japanese seem well satisfied with it."[63]

The Japanese Navy officials made a clear decision about how to stage the historic event. By insisting that a made-in-Japan aircraft had to perform the first landing on the new carrier, they flatly rejected Sempill's request. On February 5, 1923, two months after the *Hōshō* entered service, Jordan made three successful takeoffs and landings with his Mitsubishi Navy Type 10 Carrier Fighter on the new carrier.[64] He repeated his stunt on February 22 in front of the Yokosuka naval station's commander and about twenty high-ranking navy officers. The event made it into the headlines of the following day's papers, which praised the British pilot's special skills but also strongly emphasized that his aircraft was "made in Japan."[65] Mitsubishi rewarded Jordan with a prize of ¥10,000—roughly ¥10 million in today's currency (or US$95,000).[66]

In the next month, a Japanese pilot was to reenact Jordan's historic feat. On March 5, Captain Kira Shun'ichi (1889–1947), one of Sempill's students, took off with his Type 10 carrier fighter, climbed to a safe altitude, and started his approach toward the *Hōshō*'s flight deck. With many high-ranking navy officials attending, Kira's first attempt resulted in an overrun. In his biplane he plunged fifteen meters down into the sea. Such a failed landing was unfortunate but not uncommon. During earlier attempts, many similar incidents occurred. Some were fatal when pilots crash-landed in the sea and, unable to escape from their cockpit, drowned together with their aircraft.[67] Captain Kira was spared such a grim fate. A rescue team picked him up, and the intrepid pilot immediately took off for a second try. This time he touched down on the aircraft carrier and successfully brought his aircraft to a stop before approaching the end of the landing deck. The attending officers were pleased, and Kira's achievement earned him a gold cup presented by the minister of the navy.[68]

Finally, a member of the British mission received permission to land on the Japanese carrier. On March 13, Brackley landed his Viking amphibious aircraft on the *Hōshō*, impressing the Japanese observers with the airplane's low landing speed and effective wheel brakes, both vital features for a carrier-based aircraft. The British also demonstrated the carrier deployment of their Blackburn Swift torpedo attacker. In December 1923 two more Japanese pilots, Sub-Lieutenant Kamei Yoshio and Sub-Lieutenant Baba Tokuma, made successful deck landings. Along with their comrade-in-arms Kira, they became famous as the "three deck-landing crows"—a playful reference to the Japanese expression *sanbagarasu* (three crows) that designates the three most able people in a given field.[69] These achievements notwithstanding, a problem was emerging. For many years, carrier operation remained a daunting challenge that could be mastered only by a very few skilled pilots. To expand its airpower, the navy had to find a way to train many more airmen to safely take off and land on carrier aircraft.

A MISSION'S EVALUATION AND AFTERMATH

During Sempill's stay, Anglo–Japanese relations soured further. In November 1921, the Washington Naval Conference opened with the aim to forestall an arms race in the Pacific. In February 1922, Japan, the United States, Britain, France, and Italy concluded the Washington Naval Treaty, which effectively limited the size of each country's battleship fleet. The treaty also revealed the growing antagonism between Japan and its former World War I allies. In the wake of a British–American rapprochement, Britain became more willing to give in to US pressure to discontinue its alliance with Japan. With the official ratification of the

Washington Naval Treaty in June 1923, the Anglo-Japanese Alliance came to an end.

Such a geopolitical realignment cast doubt on any prospects for large-scale aviation orders from Japan. Toward the end of the mission, it became obvious that future sales of licenses, aircraft, and aeronautical equipment to Japan would not be easy. An exchange of letters in May 1923 between Brackley and A. Paget, a British business representative in Tokyo, reveals a tense climate that was dominated by fears of license infringements and stiff international competition.[70] Brackley mentioned to Paget that the Japanese were interested in the British Jupiter aero-engine. At the same time, he warned about the "the tendency of the Japanese to copy everything, often regardless of patents or rights." He advised insisting on selling a substantial number of these engines rather than just a specimen that the Japanese would then "take to pieces, make drawings, and copy, afterwards calling it a Japanese engine." Paget expressed his worries about "the very strong French and German competition" and the lack of support from the British government. Brackley shared these concerns. He lamented that "it is such a pity to see the Huns trying to sell or rather almost give away Junker metal machines to the Japanese in their efforts to get the business."[71]

The more optimistic commentators, like Sir Edward Crowe (1877–1960), commercial attaché to the Tokyo embassy, still emphasized the mission's mutual benefits, which allowed Britain to dispose of surplus aircraft while allowing Japan to acquire high-quality hardware at a decent price. In April 1924, Crowe expressed his confidence that the Japanese would increasingly buy British aviation products and thanked Sempill for "affording our manufacturers of aviation material better opportunities for disposing of their goods in a friendly market."[72] It seems that many British experts ignored the fact that the Japanese were not interested in large-scale purchases. Japan's navy and aviation industry instead followed the strategy of acquiring a small number of different aircraft types and learning from them for their own designs.

British hopes for bulk orders from Japan never materialized. According to one source, Sempill contended that he had already procured orders for about £1 million.[73] Some historians give a much lower estimate: between £325,000 and £550,000.[74] Indeed, even though in 1921 Japan bought more than 200 British aircraft, this number declined to under forty the following year and only thirteen during the rest of the decade.[75] Revenue from license fees was also very limited. Apart from the light Avro 504 training aircraft and the all-wooden Short F.5 flying boat, the Japanese Navy did not adopt any British aircraft for licensed production.

Yet another momentous outcome of the British Aviation Mission to Japan deserves attention. The mission reinforced British prejudices about the limited capabilities of Japanese pilots and the Japanese military's inability to develop a strong air arm—a view that many British officials maintained until the beginning of the Pacific War.[76] In March 1922, Brackley complained about his Japanese student pilots that "it has been tiring and trying work with the little fellows."[77] In his view, most of his trainees lacked consistency and sound judgment. Sempill arrived at a similar assessment. Back in Britain, he publicly commented that the Japanese pilots carried out the most dangerous flight maneuvers without hesitation, yet they lacked the "instinctive sense of prompt action" when "thrown into a situation entirely unexpected."[78] Both mission members were skeptical about the future role of aviation in the Japanese Navy's strategy. They criticized the attitude of high-ranking navy officers who still had to be convinced of the practical value of aircraft as useful weapons and were reluctant to discuss questions of air strategy and procurement policies with the British. In a similar way, the British military attaché did not see much hope for the Japanese Navy's air arm. In a 1923 report to the Air Ministry, he ranted, "The Japanese are completely ignorant where the technical problems of naval aircraft are concerned, and the rate of progression depends entirely with the British Mission. . . . The ignorance of senior officers will be an effective brake for some time to come."[79]

The Japanese armed forces, of course, did not share these disparaging views. High-ranking officers like Vice Admiral Wada Hideho (1886–1972) attributed to the mission an enormous impact on the development of Japanese naval aviation. Wada had served as an instructor at the Yokosuka Navy Air Corps and later as the commander of the Kasumigaura Navy Air Corps. He wrote in 1944: "Now we [Japan and Great Britain] are fighting on opposite sides; however, it is no exaggeration to say that the British instructors completely changed every aspect of our naval aviation."[80] According to Itō Masanori, a leading military commentator, Sempill transformed the navy's pilots from mere flyers to "determined" fighters. For Itō, Sempill's legacy was also a change in the pilots' professional attitude. At Kasumigaura these navy pilots studied hard, "like students preparing for an entrance examination." Their intense training covered all aspects of military aviation: carrier plane operation, shooting, formation bombing, torpedo attacks, reconnaissance, and radio communication. Itō concluded that, along with the fast progress in Japan's aviation technology, the navy's fundamental concept of air operations changed dramatically, and for the first time aircraft became an essential element of the fleet's strike force.[81]

NOBLESSE OBLIGE? LORD SEMPILL SPIES FOR JAPAN

As another major—if unintended—result of the Sempill mission, the Japanese were able to recruit a top-level spy with direct access to the British military and aviation industry. This spy was none other than Sempill himself.[82] As laid out already, interministerial quarrels resulted in the mission's unofficial nature, making it relatively easy for the Japanese to exploit this special set of circumstances. From the beginning, Sempill faced a dilemma: with very limited support from a reluctant Air Ministry, he had to fulfill the expectations of the British aviation industry, which wanted lucrative business with the Japanese. On several occasions, Sempill lamented the inadequate cooperation of "those in authority."[83] He was frustrated that despite Britain's superior hardware, flying technique, and design, the Japanese would eventually "seek the aid of other countries more sympathetic to their needs, as has often been the case in the past."[84]

Sempill found his own way to bypass what he perceived as lack of official cooperation. Even before his departure for Japan, a number of officials at the Air Ministry had granted him access to classified information. Throughout the mission in Japan, top-level British officers provided mission members with confidential material. It seems that the Japanese were desperately keen for up-to-date information on British military aviation. In July 1922, shortly before Brackley went for a temporary leave to England, the commander of the Yokosuka Naval Station, Rear Admiral Tajiri Tadatsugu (1874–?), presented Brackley a wish list for "special study in England."[85] At the top of the list were the latest British developments in deck landing, followed by detailed information on another key technology: torpedo bombers. Tajiri wanted to learn everything about the performance and flying technique of these aircraft, torpedo dropping procedures, and the design of torpedo sights. The list finished with the order to "also bring home [to Japan] a collection of representative printed matters on naval air tactics." In November 1922, Brackley could happily report to Sempill that

> many of the [Air Ministry] heads of departments, especially Air Vice-Marshal Salmond and Gen. Bagnold Wild, are very keen to give us all possible help; they wish to keep only one year ahead of other nations with regard to design and quite realize that, for the benefit of the British Air Industry, they must help people who are sent to foreign countries. I got a good deal of information—mostly unofficial.[86]

It is thus easy to conclude that the actual transfer of know-how went well beyond the limits officially set by the Air Ministry.

After Sempill returned to England in October 1922, he maintained his ties to the Japanese Navy. He established close relationships with the Japanese naval attaché in London, Toyoda Teijirō (1885–1961), whom he habitually addressed with "my dear Commander." Sempill's spy activities can be traced back to February 1924, when he passed to Toyoda details and drawings on the development of new bombs for the Royal Air Force. The British authorities soon became suspicious of Sempill's activities. Already in early 1924 MI5, the British counterespionage and defense security department, began to intercept and photograph Sempill's correspondence. Monitoring his phone calls revealed that Sempill used a civilian employee at the Air Ministry for getting confidential technical information. When Sempill arranged a visit for his "Japanese friends" at Westland Aircraft Works to see the new experimental monoplane *Dreadnought*, the Air Ministry intervened. Already aware of Sempill's disloyalty, the ministry instructed the aircraft maker that neither Sempill nor his associates could see the aircraft. Unaware that he was under observation, Sempill kept on passing confidential information to the Japanese, including details about anti-aircraft sound detectors and the latest engine developments. He seemed especially proud to provide extensive documents on the operation of the Royal Air Force's latest carrier aircraft that, in his opinion, "have behind them a wealth of specialized knowledge, of intensive service use on aircraft carriers such as cannot possibly be obtained in any other way or by any other country." Time and again Sempill implored Toyoda with remarks like "please be very careful how you use any information you get. . . . I know exactly how the wind blows, and the need for being super-cautious."[87]

Right under the eyes of MI5, Sempill continued to supply classified information to the Japanese. In 1925, he informed Toyoda about aeronautical equipment, such as aerial cameras, parachutes, and wireless telegraphs. After Sempill procured classified performance data of the new long-range flying boat *Iris*, he publicly talked about the ineffectiveness of the Air Ministry's efforts to keep Britain's latest aircraft developments secret. At this point, the Air Ministry finally denied Sempill access to any military technology. In January 1926, the chief of staff intelligence stated that it was "quite clear that not only is Sempill admittedly furnishing the Japanese with aviation intelligence but that he is being paid for doing so. . . . Steps should be taken to place Sempill on the Blacklist as regards the receipt of information from, and visits to, the Air Ministry or any establishment engaged in the design or manufacture of modern aircraft on behalf of the Air Ministry."[88] In May 1926, Sempill was interrogated and warned at the Air Ministry. He got off lightly. He was spared prosecution for breaching the Official Secrets Act because the British government was loath to call the Japanese naval attaché as a witness—and thus admit that the attaché's correspondence had been under surveillance.[89]

In all its inconclusiveness, Sempill's spy activity seems to have been of little consequence for the advance of Japanese aviation. Some of the "secret" aircraft and engines that appeared in Sempill's reports to Toyoda were later released for sale to the Japanese, who then showed little interest in them.[90] Given the half-hearted attempts of the British intelligence services to stop him, it seems plausible that Sempill—rather than being a risk to national security—was a useful tool for the Air Ministry to fathom the Japanese Navy's plans and interests. Given the increasingly meager results of his espionage, it is also safe to assume that the Japanese might have concluded that Britain had lost its technological edge and that the latest cutting-edge technology had to be found elsewhere.

Conclusion

The planners, strategists, and technocrats of the Japanese Navy cleverly navigated the political, military, and technological disruptions of World War I and effectively seized the opportunities of the immediate postwar era. During World War I Japanese observers could study the latest advances in British naval aviation firsthand, most notably the emergence of aircraft carriers. During the war, the Japanese Navy set up specifications for this new type of warship and in 1921 it launched the *Hōshō*, its first aircraft carrier, which was based on a British design.

The provisions of the Treaty of Versailles helped the Japanese ensure Heinkel's cooperation, and his floatplanes and launching devices greatly improved the striking power of Japanese battleships. The postwar decline of Britain's aviation industry gave the Japanese the chance to call on the expertise of unemployed British engineers to design and build Japan's first carrier aircraft.

One of the Japanese Navy's greatest exploits was the British Aviation Mission to Japan of 1921–23. The navy officials successfully outmaneuvered British interministerial rivalry to ensure the dispatch of the mission. Ignoring the admiralty's strong reservations, the British Air Ministry agreed to a massive transfer of matériel and know-how to Japan. The British mission introduced a wide variety of British aircraft, significantly advancing the skills of the Japanese Navy's pilots, and firing the aviation enthusiasm of the Japanese public. Yet the British soon had to abandon their hopes for extensive exports and dominance in Japanese aviation. To its dismay, the Air Ministry even learned of Sempill's spy activities.

As we will see in the next chapter, Japanese naval aviation continued to advance at a breathtaking pace. Large all-metal flying boats, a second generation of aircraft carriers, and a revolutionary new kind of bomber ultimately enabled the Imperial Japanese Navy to surpass its erstwhile tutor.

CHAPTER 6

Japan's Naval Aviation Taking the Lead

William Jordan's historic deck landing in February 1923 was the apogee of British influence on Japan's naval aviation. A Japanese fighter aircraft, designed by British engineers in the employ of Mitsubishi, landed on Japan's first aircraft carrier, which was modeled on the British carrier *Furious*. This chapter follows the development of naval aviation in Japan after this historic event. I argue that the end of the Anglo-Japanese alliance in 1923 cannot sufficiently explain Britain's loss of dominance in this process. Even after the official end of the treaty, both powers were interested in maintaining good diplomatic relations, and the British government saw "no substantial cause to regard Japan as a potential enemy."[1] The British admiralty took an ambiguous stance to prepare for future conflict but continued to cooperate with its former ally.

Why, then, did Britain fail to keep its role as the driving force behind the rise of Japan's naval airpower? By analyzing the evolution of Japanese large flying boats and advanced carrier aircraft, this chapter reveals a major change in the Imperial Japanese Navy's approach toward aeronautical innovation and technology transfer. While still occasionally referring to British guidance and design, Japanese military planners—along with civilian engineers—cast a much wider net to include German and US cutting-edge technology as well. Equally important, during the 1930s Japan's naval aviation achieved independence from foreign design. Now some of the world's finest flying boats, carrier fighters, and dive bombers forcefully demonstrated that Japan's naval arsenals and civilian aircraft makers had caught up with and, in many aspects, surpassed their former mentors.

Toward an Autonomous Airpower: Large, All-Metal Flying Boats

By the mid-1920s, the catapult launch of floatplanes from battleships had reached a degree of sophistication that satisfied the proponents of big-gun battleships as well as the advocates of naval airpower. However, the emergence of a new type of aircraft that could operate independently (that is, without a mother ship) ushered in a new era of naval aviation. Flying boats had already proved their versatility during the final years of World War I. These aircraft could take off and land even under rough weather conditions, and their spacious fuselages had ample space for fuel, cargo, and weaponry. Unlike landplanes of similar size, they did not need large airfields.

Another consequence of World War I further aroused the interest of Japanese naval strategists in flying boats. In autumn 1914, the Japanese Navy sent two squadrons to occupy German colonial possessions in the south Pacific. Within a month, the Japanese had taken control of the Marshall, Caroline, and Mariana Islands. During the Paris Peace Conference in 1919, the Japanese received British support to assert their claims on the newly won areas, and the Treaty of Versailles granted Japan all formerly German rights in the Pacific territories.[2] For Japan, these islands were of great geostrategic importance because they were situated along the sea route between the US military bases in the Philippines and Guam.

Under these new circumstances, Japanese strategists were especially intrigued by the vast flight range of flying boats, which made them ideal for long-range patrol missions in the Pacific. The navy planners began to envision a chain of seaplane stations in Pacific lagoons, from which they could deploy these aircraft not only for transport, search, and rescue but also for reconnaissance and bombing missions. Flying boats were the ideal means for establishing a military presence within the former German protectorates in the Pacific, especially since Japan had signed the covenant of the League of Nations, whose Article 22 did not allow building any military bases on these islands.

In the early 1920s, deliveries from Britain helped the Imperial Japanese Navy launch its own flying boat production. The arrangements of Sempill's mission included importing eight Short F.5 flying boats along with the components of six more aircraft. The all-wooden Short F.5 biplane was an obsolete World War I design, and its low cruising speed and poor takeoff performance were well known.[3] Nevertheless, the Japanese Navy was eager to master the construction of this aircraft type and opened an entirely new workshop for manufacturing flying boats in October 1921. The Short Brothers company sent twenty-one engineers to Japan for a nine-month instruction course on producing the F.5. The Hiro Arsenal, together with aircraft maker Aichi, built forty of these aircraft until 1929.

THE MARVELS OF A NEW MATERIAL: DURALUMIN

A path-breaking invention from Germany revolutionized the design and construction of large flying boats. In 1903 Dürener Metallwerke invented an alloy that was nearly as strong as steel but weighed only a third as much. Duralumin, as the company aptly called its new alloy, was of great interest to German aircraft designers during World War I. The lightweight construction material was a perfect replacement for wood and allowed more precise stress calculations. Aeronautical engineers now could refer to the methods of structural analysis developed for designing steel bridges and apply these techniques to the construction of airframes that combined minimized weight with proven structural strength (figure 6.1). The new material provided one more step in a process where quantitative science replaced the more intuitive "art" of aircraft building. Equally important, with its uniform strength and malleability, duralumin lent itself to the mass production of standardized parts by unskilled workers.

The Japanese gained ready access to this key technology with the postwar peace settlement. German aircraft designer Adolf Rohrbach became the key person for this important technology transfer. In Germany Rohrbach was known as the "purist of all-duralumin design."[4] For the aviation experts in the Japanese military, he was a household name as well. He was associated with the design of their most spectacular war trophy,

FIGURE 6.1 This workman apparently easily holds the central duralumin spar that connected the wings of Adolf Rohrbach's giant 1920 passenger airplane with its fuselage. The picture, which appeared in one of Rohrbach's sales brochures, aims to demonstrate how the new material allows the design of large lightweight structures. It also illustrates how the all-metal design borrowed the use of rivets and braces from steel bridge construction. (Source © Bundesarchiv-Abt. Militärarchiv, BArch RH 8/3606 fol.91v)

the Zeppelin-Staaken R.XV giant bomber, examined in detail in chapter 3. During the early 1920s, the German designer had established friendly relations with a member of the Inter-Allied Commission of Control, Captain Takada Yoshimitsu, who made numerous inspection trips and thus emerged as the Japanese Navy's expert on German aviation.[5] For Takada, Rohrbach offered to build any flying boat according to the Japanese Navy's request. When Takada submitted a favorable report on Rohrbach with the German engineer's proposal, the Japanese Navy's Technical Department became well disposed toward a future collaboration.[6]

Clearly the more conservative navy technocrats favored Rohrbach's design method over that of his competitors because of its affinity with shipbuilding. Wada Misao (1889–1981), a future vice admiral and chief of the Naval Air Headquarters, was one of the outspoken advocates of an alliance with Rohrbach. Initially trained as a shipbuilding engineer, Wada previously gained experience in the design of seaplanes during his work with the Short engineers in 1921. When in the following year the navy sent Wada to Germany, he soon became aware of Rohrbach's unique design principles, which seemed familiar to him and at the same entirely different from anything he had seen before: "As compared to Dornier, Rohrbach's airplanes had a completely new and original structure. . . . Because I have studied the structural strength of ship hulls, I thought that Rohrbach's use of duralumin was outstanding, and I immersed myself in the study of Rohrbach's method."[7] Immediately after the war, Rohrbach took over the management of Zeppelin's former Staaken factory and started the construction of a large all-duralumin passenger aircraft, the Zeppelin-Staaken E.4/20. Rohrbach's plane impressed during its successful 1920 test flight with its high cruising speed and long flight range—but the Allies ordered it to be destroyed, which meant the end of the Staaken factory.[8] Having lost his aircraft and livelihood, Rohrbach was nonetheless a much sought-after engineer. Aircraft makers Fokker and Junkers and even the Soviet government showed their interest in his skill and expertise.[9]

Yet Rohrbach received the most attractive proposal from Japan. The Japanese Navy planned to develop an entirely new generation of large seaplanes with a design similar to that of Rohrbach's Staaken monoplane. In 1922 Rohrbach signed a contract with representatives of the Japanese Navy for the construction of four long-distance reconnaissance airplanes, with two more to be built in Japan. The navy put strong emphasis on including the training of Japanese engineers and workmen in the agreement. In addition, the contract provided for the Japanese Navy to receive blueprints, calculations, and the right to produce an unlimited number of aircraft. Obviously, the project was hugely important for Rohrbach as well. A central statement in his account of the "establishment and economic growth of the Rohrbach Metall-Flugzeugbau company" testifies to the vital nature of his cooperation with Japan:

> The contract reveals the complete trust [of the Japanese] as at that time I did not have any factory, engineers, or workmen. Only by means of this contract was I able to establish the Rohrbach-Metall-Flugzeug GmbH in Berlin in August 1922 and at the same time in Copenhagen the Rohrbach Metal Aeroplan Co. Aktieselskab.[10]

MITSUBISHI TIES UP WITH ROHRBACH

The navy decided to include the Mitsubishi Company in the production of the all-metal flying boat. Mitsubishi's aviation department eagerly seized the opportunity to learn more about Rohrbach's advanced technology. As early as autumn 1922, the company sent its engineers to Germany to receive advanced training at Rohrbach's factories.[11] For several of these designers, the trip was the start of a distinguished career. Ōtsuka Keisuke, the first Mitsubishi engineer to arrive at Rohrbach, became the head of Mitsubishi's Ki-20 "Superbomber" project. The following year, Ōtsuka was joined by Hattori Jōji, the future head of Mitsubishi's airframe design section.[12] A third engineer, Shōda Taizō, who replaced Ōtsuka in 1924, became a staunch supporter of German aviation technology—he was convinced that "because of its theory and logic, the German design method was leading the world."[13] Shōda was appointed vice president of Mitsubishi Heavy Industries after World War II. By early 1924, the Japanese workers and engineers finished their training and returned to Mitsubishi. Rohrbach praised them as "diligent and skillful, having reached the level of our best German staff."[14]

The minutes of a Mitsubishi board meeting held in April 1923 reflect the high hopes invested in the affiliation with Rohrbach. Board members emphasized the "promising future" of metal aircraft and left no doubt that the Japanese Army and Navy would both adopt these airplanes in the near future. They considered it an "extremely advantageous plan" to establish a tie-up with Rohrbach, who was the "leading expert in Germany." The board unanimously agreed to sign a contract with Rohrbach for the production of metal aircraft.[15]

Apart from the prospect of gaining access to an entirely new technology, cooperation with Rohrbach held more appeal. To elude the Allies' ban on constructing military aircraft, Rohrbach outsourced his production to Denmark, to a newly established branch factory, Rohrbach Metal Aeroplan Co., near Copenhagen. This clever scheme made it easier for his Japanese customers to maintain the appearance of respecting the Treaty of Versailles even after the official arrangement had begun.

Rohrbach's deliberate separation of design from production had another important consequence. Because of the geographical distance between his creative and productive branches, Rohrbach adopted the "American system" by using specialized machines to produce interchangeable parts. He incorporated elements of Henry Ford's factory management and Frederick Taylor's time studies to further rationalize his production. The Berlin headquarters devised all designs, calculations, and drawings. It came up with an elaborate system of blueprints and parts lists that enabled the Copenhagen plant to manufacture aircraft with a largely unskilled workforce. According to Rohrbach, "the shop floor can only work at low-cost if there is nothing left to think about."[16] After this organizational division was tried and tested, it became easy to establish another production line abroad—say, in Japan.

After Rohrbach received the Japanese Navy's initial order, he started the development of the Ro II flying boat. He chose a rectangular cross section for the hull to increase its freight capacity, and a relatively small wing with a high load bearing capacity to allow a high cruising speed. According to Rohrbach, such a wing required a sophisticated design, which was based on the company's unique method of scientific strength calculations.[17] The Ro II made its first successful test flight in November 1923; the next year, the aircraft set world records in speed and flight range, clearly demonstrating its suitability as a long-range reconnaissance aircraft or bomber.[18] In 1925 the Japanese Navy invited Rohrbach to Japan.[19] Under the personal supervision of Rohrbach and his head engineer, Paul Ludwig, the experts of the Yokosuka Naval Arsenal assembled Rohrbach's airplane and duly named it the Experimental R-1 flying boat.

In hopes of receiving additional orders, Rohrbach took the opportunity to promote his design principles to the Japanese military. In a lecture to officers of the Imperial Navy, he tried to convince his audience of the benefits of hydroplanes over carrier-based aircraft. Clearly the navy had not disclosed its recent advances in carrier technology to him. Doubting the feasibility of aircraft carriers, Rohrbach estimated that it would take fifteen to twenty minutes after each landing to clear the carrier for the touchdown of the next aircraft.[20] He therefore envisioned future naval aviation to rely mostly on flying boats that either were launched from catapults or took off from water.

During his stay in Japan, Rohrbach began negotiations with Mitsubishi that resulted in the establishment of Mitsubishi-Rohrbach Ltd. in June 1925. Mitsubishi held a majority stake of 60 percent in the new enterprise, which was to oversee the licensed production of Rohrbach's aircraft at Mitsubishi and at the navy's arsenals.[21] Whereas the first flying boat had been assembled at the Yokosuka Naval Arsenal, the second was to be built at the newly founded company. For this purpose, Mitsubishi established a new plant, which, according to one employee, became the "cradle of Mitsubishi's light metal aircraft production technology."[22] As part of the thorough preparations, the company compiled and printed a German–Japanese dictionary that included translations of all technical terms on Rohrbach's blueprints.[23] Although Mitsubishi still had to import the major parts of the flying boat, made-in-Japan components were used as well. Most notably, the aircraft was powered by two 450 horsepower engines that were produced by Mitsubishi and provided a 25 percent increase in power compared to the German original (figure 6.2).[24]

It was important for the Japanese Navy to have their arsenals adopt the new technology as well. Navy officials therefore decided to build the third flying boat at the Hiro Naval Arsenal, the navy's first airplane factory, established near Hiroshima in 1920. Wada Misao, who had been trained at Rohrbach, supervised the construction. While referring to Rohrbach's basic design, the arsenal's engineers refined the shape of the aircraft's wing and floats and installed their own Hiro Lorraine engines on specially designed struts. The result was an aircraft with improved takeoff performance and a remarkable flight endurance of twelve hours.[25]

In spite of their impressive payload, speed, and endurance, Rohrbach's flying boats could not meet Japanese expectations. As it turned out, these

FIGURE 6.2 This picture shows a Mitsubishi Type R flying boat with its crew of six. The short wings incorporated Rohrbach's contested design principle of a relatively short wing bearing a high weight. (Source © Japan Aeronautic Association)

airplanes performed poorly during takeoff and touchdown in rough sea conditions. Ironically, this problem was a direct result of Rohrbach's proclaimed design philosophy, which favored a strong all-metal structure with a high wing load in which a relatively small wing carried a comparatively high weight. In several of his lectures, Rohrbach promoted his original idea, arguing that such aircraft would have superior maneuverability. The drawback of such an advanced design was the need for sufficient engine power to reach the required takeoff speed. Even though Rohrbach's made-in-Japan flying boats were equipped with significantly stronger Japanese engines, their takeoff characteristics were still unsatisfactory, and the navy did not adopt the aircraft. Mitsubishi discontinued production and dissolved the Mitsubishi-Rohrbach aircraft company in 1926.

ROHRBACH'S LEGACY AND THE "MOST OUTSTANDING WATER-BASED COMBAT AIRCRAFT"

Even after its sobering termination, Rohrbach's cooperation with Mitsubishi had a major influence on the development of Japan's aviation sector. As discussed already, three years before the delivery of Rohrbach's first aircraft to Japan, engineers at Mitsubishi and the navy arsenals received extensive training at Rohrbach's factory. They became proficient in the design of all-metal aircraft and mastered the various processing methods of duralumin. According to one source, this group of German-trained Japanese experts played a central role in the development of the metal aircraft industry in Japan.[26] This claim can be further substantiated when we consider that during the early 1920s, training for aeronautical engineers even at Japan's foremost universities was still very rudimentary. Sanuki Matao (1908–97) graduated from the Department of Aeronautics at Tokyo Imperial University in 1931. According to his memoirs, all that his professors taught him during the design-related lectures was the specialized terminology of aircraft design. At the same time, Japanese students learned to admire German scientists and engineers who—as their teachers told them—had laid down the fundamentals of aerodynamics and aircraft design. Sanuki concluded that the transfer of advanced German know-how "was the ultimate reason behind the purchase of Rohrbach's flying boats."[27]

Certain features of Rohrbach's original design ushered in the era of what can be called a second generation of all-metal aircraft in Japan. One of the breakthrough innovations that he introduced to the Japanese was the stressed-skin design principle. German engineer Herbert Wagner (1900–1982), who joined Rohrbach's company in 1924, provided the crucial invention. He explained it in the following way:

> Rohrbach covered all outer-wall frame structures which were subjected to higher loads with thin metal, similar to the fabric-covered frame structures. I suggested omitting the diagonal rods within the frame and mounting the vertical rods closer together, so that under load the obliquely folding sheet metal skin takes up the lateral (vertical) forces. Comparative tests showed, for the same weight of the wall, a nearly doubling of the allowable lateral force; the workshops welcomed the simplification.[28]

Wagner's pioneering innovation allowed the design of a highly resilient airplane structure that could carry a considerably larger payload. The new technology eliminated the need for struts inside the fuselage, thus minimizing the amount of required material. At the same time, it provided more space for passengers and cargo—or bombs. It is important to note that Wagner also developed an appropriate theory that allowed designers to perform stress calculations, which eliminated the need for expensive and time-consuming breaking-load experiments.

According to Mitsubishi's chairman of the board of directors, Funakoshi Kajishirō (1870–1962), even though the Rohrbach cooperation ended early, it "laid the foundation for the production of metal aircraft and the successful adoption of this technology."[29] This emphasis on the transfer of knowledge rather than machinery helps explain the rather short existence of the Mitsubishi-Rohrbach venture. Mitsubishi ended the formal affiliation after only nine months in May 1926, even with Rohrbach's aircraft still under construction at Mitsubishi's factory.[30] The records of Mitsubishi's board meetings do not reveal the details leading up to the decision. But it can safely be assumed that, with the training of engineers, the handover of blueprints, and the arrival of one sample aircraft, Mitsubishi had achieved sufficient know-how and material to make further cooperation unnecessary.

The navy's decision to assemble Rohrbach's aircraft at three different places—near Tokyo, Nagoya, and Hiroshima—led to a wide diffusion of the technology in Japan. The Hiro Naval Arsenal in particular started a series of large flying boats that were made from duralumin and featured the Wagner thin-board tension structure. In 1931 the arsenal's Navy Type 90-1 flying boat made its maiden flight. Rohrbach-trained navy engineer Wada Misao had been in charge of the project. The impressive three-engine aircraft went down in aviation history as the first large all-metal plane entirely designed and built in Japan. With a takeoff weight of nearly twelve tons and maximum speed of 230 kilometers per hour, the flying boat was twice as heavy and more than 40 percent faster than its Mitsubishi-Rohrbach predecessor.

To further spread the new technology, the navy encouraged aircraft maker Kawanishi to build flying boats as well. The company had manufactured its first aircraft in 1921 and pioneered in regular mail and passenger flights with its Nippon Kōkū airline, which was founded in 1923 (figure 6.3). In 1928 Kawanishi completed Japan's first commercial wind tunnel under the guidance of Theodore von Kármán (1881–1963), director of the Aeronautical Institute at Aachen University, and his assistant, Erich Kayser.[31] Initially it seemed that Kawanishi was a most unlikely candidate to rise to Japan's foremost manufacturer of flying boats: it had

FIGURE 6.3 Two elegant young women, one dressed in traditional clothes, one in Western garb, are about to embark on this Kawanishi K-7A seaplane. The Nippon Kōkū airline started to use the single-engine transport plane for its Osaka–Fukuoka route in 1925. Presumably this picture was to entice more hesitant air travelers to board the narrow passenger cabin with its triangular windows. As it

no experience in designing and building large flying boats. Furthermore, Kawanishi was suffering from a series of recent setbacks. In 1928 the Aviation Bureau stopped the aircraft maker's ambitious trans-Pacific flight project and ousted Nippon Kōkū from the market with the new Japan Air Transport company (Nippon Kōkū Yusō KK), which started its service with heavy state subsidies.

In need of profitable military contracts, Kawanishi welcomed the navy's 1929 request for a large long-range flying boat based on Short Brothers' latest technology. In February 1928 Short had attracted worldwide attention with the Calcutta, a three-engine flying boat. The British aircraft maker built the biplane to specifications of British air transport company Imperial Airways for its Far Eastern route; it had a flight range of approximately 1,000 kilometers. The Japanese Navy dispatched its chief engineer, Hashiguchi Yoshio, to Britain to participate in the conversion of the civilian airliner Calcutta into a heavily armed military flying boat, the H3K1. The redesigned aircraft was shipped to Japan in 1930. Even before the navy had formally accepted the flying boat, Kawanishi bought the production rights for ¥1 million and built a new factory for the construction of the H3K1 at Naruo, halfway between Osaka and Kobe.

In spite of its outdated biplane structure, the H3K1 satisfied the navy officials with its high load capacity of seven tons and flight range of 1,500 kilometers. The aircraft was accepted as the Navy Type 90-2 flying boat in 1932, and Kawanishi began the licensed production of what was then the largest flying boat in the Pacific region. Short sent ten engineers to Japan to supervise the production for one and a half years. However, with its conservative (if not outdated) design, the flying boat did not contribute much to advance Kawanishi's expertise. For instance, when Kawanishi engineer Kikuhara Shizuo wanted to learn from the British experts more about strength calculations and how the aircraft's maximum load factor was determined, he received only the laconic remark, "we do this based on our experience." This approach was a far cry from Rohrbach's scientific methods. It led Kikuhara to the conclusion that "after the departure of the British engineers we still have to do a lot of research and development on our own."[32]

Ironically one of the aircraft's few original design features led to the early termination of its production. The plane's unorthodox mix of material—duralumin above the hull's waterline and stainless steel below—resulted in galvanic corrosion between the components. As a result, on several occasions the flying boat's bottom fell off during flight. Even though Kawanishi's engineers eventually fixed this fatal flaw by putting a special coating between metal boards, Kawanishi stopped production of the Type 90-2 flying boat after the fourth specimen in 1933.

Even after getting stuck in such a technological deadlock, Kawanishi did not abandon its efforts to improve its flying boats. The company could still count on the support of the Naval Arsenal, which kept on developing advanced flying boat designs. When the arsenal began the production of its Type 91 flying boat in 1932, it had Kawanishi set up a parallel production line. With support from navy engineers, Kawanishi could draw on the arsenal's expertise—above all, Rohrbach's all-metal technology—and benefit from the navy's ongoing experiments and design upgrades.

At the same time, Kawanishi significantly expanded its research facilities. A new wind tunnel and improved water tanks and strength-test stations allowed the engineers to overcome British empiricism and follow the German scientific approach. Making good use of the upgraded facilities, Kawanishi developed the H6K, a new flying boat that far outranged its US and British competitors with a flight range of more than 6,700 kilometers.[33] In 1937 the navy adopted the H6K as its Type 97 flying boat and ordered Kawanishi to build more than 200 of these airplanes. The success boosted the Kawanishi engineers' confidence and made the company decide to completely dispense with any foreign assistance. As one engineer declared: "If there is money available to buy [the license for] a new foreign aircraft it makes much more sense to do our own research and development and to build our own prototypes."[34]

Kawanishi's long series of outstanding flying boats culminated in the H8K, dubbed "Emily" by the Allied code-name system. The aircraft was adopted in 1942. Its heavy armor and protected fuel tanks made it an extremely difficult target for enemy fighters. With a top speed of more than 460 kilometers per hour and a load capacity of up to fourteen tons, the H8K became a hallmark of Japanese aeronautical proficiency. In the view of some commentators, the H8K was the "most outstanding water-based combat aircraft of the Second World War."[35]

The Next Generation of Japanese Aircraft Carriers

Along with its efforts to build up a fleet of advanced flying boats, the Japanese Navy launched an equally ambitious project to expand its aircraft carrier force. Ironically, the Washington Naval Treaty's restrictions on naval armament provided a major stimulus for Japan's carrier program and aerial armament. In an attempt to compensate for the cap on battleships, the Japanese Navy decided to direct its funds toward an expansion of its airpower. As one newspaper article quite openly stated, "following the Washington Naval Conference the navy had to suspend building capital ships and now regards the manufacturing . . . of aircraft as extremely important for national defense."[36]

The treaty regulations indeed brought the construction of battle cruiser *Akagi* and battleship *Kaga* to a sudden halt. Yet instead of scrapping the half-finished hulls, the navy decided to convert these ships into aircraft carriers. The Japanese referred to Article VII of the Washington Naval Treaty, which set a maximum of 81,000 tons to Japan's total aircraft carrier tonnage. This was a generous regulation considering that the displacement of Japan's only carrier, the *Hōshō*, was less than 7,500 tons.[37] Rather than bringing naval expansion plans to a complete halt, the Washington treaty actually encouraged the Japanese Navy to dramatically enlarge its carrier fleet.

Remodeling of the warships began in 1923. In this critical stage of aircraft carrier development, the Japanese Navy continued to rely on espionage as an important if precarious avenue for technology transfer. Soon after Sempill's departure from Japan in 1922, Frederick Joseph Rutland, the British aviation pioneer, whom we already met in the previous chapter, offered his services to the Japanese. Born in 1886 into a working-class family, Rutland entered the Royal Navy at the age of sixteen. In 1914, he became one of the first pilots to join the newly established Royal Naval Air Service. Rutland turned out to be an intrepid airman with an innovative attitude. During the Battle of Jutland, fought between the Royal Navy's Grand Fleet and the German Hochseeflotte in 1916, he excelled in a daring reconnaissance mission. His endeavors earned him the Distinguished Service Cross along with the sobriquet "Rutland of Jutland."

In 1917 Rutland took the initiative to design and build a platform for a trial takeoff from a seaplane carrier. His successful launch caught the admiralty's interest and gave the decisive impulse for the installation of flight decks on British cruisers and—even more important—for converting the cruiser *Furious* and the battle cruisers *Glorious* and *Courageous* into aircraft carriers.

With the Sempill mission coming to an end and diplomatic relations with Great Britain in decline, the Japanese were still looking for British advice for the buildup of their carrier force. Rutland's experience and expertise made him a perfect candidate for such a job. He seemed amenable to the idea. His age and background offered few prospects for a further career in British naval aviation and led him to think of resignation and pursuing new opportunities. Obviously, the British government was well informed about his intentions. In 1922, the Government Code and Cipher School at Bletchley Park had broken the code of the Japanese naval attaché and intercepted messages where Rutland proposed to the Japanese Navy to impart his knowledge and skills. From then on, Rutland was under observation. Afraid of compromising its "very delicate" source of information—and in a move foreshadowing Sempill's treatment—MI5 took no disciplinary action.[38]

While the British were still reluctant to keep Rutland on a tight leash, the Japanese seized the opportunity. Captain Kobayashi Seizō, the naval attaché who had been instrumental in setting up the Sempill mission, and Lieutenant Commander Hara Gorō (1886–1940), the future commander of Japan's first aircraft carrier *Hōshō*, arranged a two-year contract between Rutland and Mitsubishi's trading company, Mitsubishi Shōji. The British airman handed in his resignation and left the Royal Air Force in October 1923. Rutland took up residence in the outskirts of Paris while staying in contact with the Japanese. At that stage, British Military Intelligence already concluded that "this individual has been heavily bribed to betray his secret knowledge to the Foreign Power whose service he is now entering."[39] In July 1924 British misgivings materialized when Rutland left Marseilles for Japan.

Officially Rutland stayed in Japan for designing airframes for Mitsubishi. Questioned by the British military attaché in Tokyo, Rutland maintained that he visited Mitsubishi's offices in Tokyo several times a week "to design aeroplane chassis." Obviously, he engaged in designing oleo shock absorbers for undercarriages that were successfully tested at Kasumigaura. Efficient shock absorbers play a vital role in the development of carrier aircraft—they help prevent bounced landings, which can lead to dangerous overruns. However, Rutland's real usefulness to the Japanese Navy lay elsewhere.

During the crucial period between launching (1925) and commissioning (1927–28) the carriers *Akagi* and *Kaga*, the Japanese drew regularly on Rutland's expertise. Historian of Japanese naval aviation Ikari Yoshirō emphasizes the importance of Rutland's advice for developing the carriers. To cover up Rutland's direct involvement, plainclothes navy officials visited him in his home every week and asked him to answer detailed questionnaires.[40] In spite of the navy's precautions, the British military attaché was well aware of Rutland's close relations with Japanese Navy officials. When the attaché visited the 1925 Air Pageant at the Yoyogi Parade Ground, he reported his astonishment to see Rutland "ensconced in a place of honor, closely attended by Lieutenant Commander Kuwabara, . . . the Chief Flying Officer of the I.J.N. [Imperial Japanese Navy], and the man more directly connected with deck flying in the Japanese Navy, than any other."[41]

The Kure and Yokosuka Naval Arsenal launched the new carriers in 1925. Fitting out took another two years for the *Akagi* and three for the *Kaga*. As a result, the navy could add two aircraft carriers to its fleet that were a quantum leap forward in carrier design. With a displacement of more than 37,000 tons and space for about ninety aircraft, the ships' size and carrier capacity exceed that of their predecessor *Hōshō* by a factor of five. Both carriers' extended flight decks provided sufficient runway length for deploying faster and heavier aircraft that required longer takeoff and landing distances. In addition, the two carriers featured a unique layout with three flight decks on different levels, allowing for simultaneous takeoff and landings.

Rutland left Japan in October 1927, a month after the *Akagi* had been commissioned. In 1933, he went to the United States to continue his covert activities for Japan under the guise of being a businessman. A few days after the 1941 attack on Pearl Harbor, he returned to Britain, where he was imprisoned for two years without formal charges. After having vanished into obscurity he committed suicide by gas poisoning in 1949.

A Second Generation of Carrier Planes

Although the Japanese Navy had some of the world's most advanced carriers under its command, it was facing a major problem. The development of the carrier-based aircraft did not keep pace with the fast advances in aircraft carrier technology. Even after the introduction of the *Akagi* and *Kaga*, the design and use of carrier aircraft were still experimental. These planes could not yet serve as a feasible alternative to the navy's flying boats. Until the early 1930s, the navy used its carriers mainly for training pilots, testing equipment, and developing operational methods and tactics. During this trial period, landing on a pitching and rolling aircraft carrier was still notoriously hazardous. Flying the navy's first generation of carrier airplanes required special skills for which, according to one estimate, only the top 2 or 3 percent of the navy's pilots qualified—even after an extensive training period. Under these conditions, navy leadership worried that reliance on such a small group of elite pilots would limit the expansion of the carrier fleet.[42] Clearly a new type of carrier plane was needed to allow safe operation by enough pilots who could be trained within a reasonable time.

The navy's first official competition for a new carrier-based prototype turned into a contest among Japanese, German, and British design. In 1926 aircraft makers Mitsubishi, Aichi Tokei, and Nakajima received orders to develop a new carrier fighter.[43] The new aircraft would replace the obsolete Mitsubishi Type 10, designed in 1921 by Herbert Smith's team. The navy's main requirement—for the aircraft to be able to float after an emergency water landing—clearly reflects persistent problems with engine reliability, insufficient flight range, and navigational errors. At the same time, the navy emphasized the need for high speed and a long flight range that would expand the plane's dogfight capability and operational radius.

These conflicting demands made the design and construction of a new aircraft a challenging task. Mitsubishi wanted to defend its position as the navy's sole supplier of fighter planes and put Hattori Jōji, one of their most experienced designers, in charge of the project. Hattori had been trained at Rohrbach in Germany and had worked under Smith. He and his team attached great importance to the floating capability of the aircraft by making the fuselage and wing sections waterproof. To facilitate an emergency touchdown on water, the aircraft was equipped with a fuel-dumping device and a droppable undercarriage. The Taka ("falcon") fighter, as the new plane was called, was also equipped with wing flaps. By providing extra lift during takeoff and landing, these devices significantly reduced takeoff and landing distance, an important safety feature for carrier operation.

Aichi also went to great lengths to win another navy contract. In spring 1926 the company's managing director, Masumoto Toshisaburō (1881–1951), traveled to Germany to pass the navy's request on to Ernst Heinkel. The German designer closely followed the navy's specifications for high performance and also added several advanced safety features: Heinkel's aircraft could take off within a distance of only ten meters and, to stay afloat after a forced landing, its fuselage and lower wing were watertight. Heinkel added an elaborate mechanism that, in case of engine failure, kept the propeller in a horizontal position, thus preventing damage during touchdown.[44] The German company built two prototypes of the Heinkel HD 23 and sent them to Japan in 1927.

Nakajima outsourced its prototype development as well. It ordered the British Gloucestershire Aircraft Company to modify one of its all-wooden fighters for carrier operation and ship it to Japan. By deliberately ignoring the navy's request for flotation, Nakajima's engineers could keep the plane much lighter than those of their competitors. During the final evaluation to decide which aircraft would be adopted for mass production, the navy was impressed with the maneuverability of Nakajima's lightweight aircraft and accepted it as its official Navy Type 3 Carrier Fighter in 1929. The decision met with harsh criticism. Heinkel argued that conflicting requirements like high speed and short takeoff distance would always result in an unsatisfactory trade-off.[45] One Japanese commentator stated that this "obscure judgment" that preferred Nakajima's British design over Mitsubishi's original work valued copying foreign designs over a "spirit of building made-in-Japan aircraft" and thus delayed the development of Japanese aviation technology for several years.[46] Indeed, the next generation of carrier fighters that replaced the Type 3 Carrier Fighter in 1932 was the Nakajima Type 90, which still relied heavily on US and British designs.[47] It took four more years until, in 1936, Nakajima's A4N and Mitsubishi's A5M clearly demonstrated Japanese independence from foreign design, with the A5M being the world's first monoplane carrier fighter and the predecessor of the iconic Zero Sen.

A New Role for Carrier Aircraft: Preemptive Air Strikes

Even with these apparent inconsistencies in the navy's procurement policy, the expansion of Japan's naval airpower increased in pace. Paradoxically, further efforts for international disarmament accelerated this process. In 1930 the London Naval Conference attempted to close several loopholes of the Washington Naval Treaty. The London Naval Treaty, signed by Japan, the United States, the United Kingdom, France, and Italy in April 1930, included submarines and light cruisers but still did not impose any restrictions on naval aircraft. The treaty was highly controversial among Japan's military, and Prime Minister Hamaguchi Osachi (1870–1931) decided to win the naval general staff's support with increased appropriations for naval aviation. During the first half of the 1930s, the Ministry of the Navy therefore could launch a large-scale expansion of the Imperial Japanese Navy's air-strike capability.[48]

Japanese Navy planners began to put unprecedented emphasis on carriers and their planes as an attack force. This fundamental shift incorporated a new, aggressive doctrine of preemptive air strikes (*sensei kūshū*) to overcome the limited range of naval artillery. Such air attacks on a distant enemy's aircraft carriers would disable the opponent's strike power and establish Japanese air superiority. Only then would Japanese battleships and cruisers fully engage in what the naval strategists called the main fleet's decisive battle (*shuryoku kantai no kessen*).[49] Such a high-risk strategy relied on two important prerequisites. First, Japanese aircraft had to be the first to locate the enemy's ships and thus needed a superior flight range that allowed them to outrange the enemy's aircraft. Second, to achieve the required hit ratio, the notoriously poor accuracy of carrier-based bombers had to be improved significantly. The proponents of the "first strike capability" doctrine envisioned that once the technology for such precision bombing became available, it could lead to victory even against a numerically superior enemy.

During the late 1920s, a new bombing technique nourished the hopes of the navy's airpower faction and seriously challenged the supporters of the "big-ship, big-gun" doctrine.[50] Dive bombing promised to be a very effective way to attack fast-moving ships that were heavily armed and armored. A dive bomber pilot would typically approach from a high altitude, then begin a high-speed dive over the target and release his bombs at a height of around 500 meters. This technique was to ensure a high hit ratio while minimizing the bomber's exposure to anti-aircraft fire. However, dive bombing put great demands on airmen and their aircraft, which needed outstanding structural strength to withstand high acceleration forces while carrying bomb loads with sufficient explosive force.

The Japanese Navy learned about the latest advances in dive bombing in 1930 when Nagahata Jun'ichirō, an engineer of the Navy Technical Research Institute (Kaigun Gijutsu Kenkyūjo), visited several US aircraft manufacturers.[51] In 1926 the US Navy had carried out its first successful dive bombing experiments with bombers diving at angles of over seventy degrees from an altitude of 12,000 feet.[52] In 1929 the Curtiss Aeroplane and Motor Company presented a prototype of the company's F8C, aptly named "Helldiver," to the US Navy and Marine Corps.[53] Nagahata had an opportunity to watch a bomb-dropping demonstration by the new Helldiver. Clearly impressed by the performance, he convinced his superiors to have the Japanese Navy start its own dive bombing experiments. Aware of the dive bomber's potential to radically transform existing battle doctrines, the navy declared the project top secret. It even avoided the phrase "dive bomber" (*kyūkōka bakugekiki*) and used the designation "special bomber" (*tokushu bakugekiki*) instead.[54]

In 1931 the navy turned to Nakajima, then one of the country's largest aircraft makers, for a joint development of Japan's first carrier-based dive bomber. The navy provided the basic design based on Nagahata's US experience, and Nakajima developed two prototypes of the Experimental Kūshō 6-Shi Special Bomber. However, in late 1932, the first test flight ended in a disaster. Nakajima's chief test pilot, Fujimaki Tsuneo, climbed to a sufficiently high altitude and started his steep descent. As the aircraft accelerated, he could not pull the bomber out of the dive. The plane crashed, burying itself two meters deep in the ground and killing its pilot.[55] In spite of this devastating failure, the navy's Yokosuka Arsenal and Nakajima continued their efforts and built several prototypes according to new specifications to ensure a safe recovery from a high-speed dive.

While the naval arsenal and Nakajima were still trying to improve their designs, the navy ordered Aichi to present its own prototype for a competitive trial scheduled for 1934. As before, Aichi turned to Ernst Heinkel. After Heinkel received the navy's specifications, he began with consecutive improvement of two prototypes and then built the Heinkel He 66. Two Aichi engineers, Miki Tetsuo and Ozawa Yasushirō, oversaw the acceptance flight in Germany. Miki was especially fascinated by how the He 66 went into a vertical dive and dropped a mock-up bomb exactly over its target.[56] After Heinkel's aircraft arrived in Japan, Aichi's engineers further improved the German design. They equipped the plane with a special bomb-release mechanism, a redesigned undercarriage to withstand rough carrier landings, and a stronger Japanese 560 horsepower engine with a propeller made from metal instead of wood. These features convincingly proved the high degree of expertise the Aichi team had acquired. During the competition, the modified dive bomber impressed the navy's officials with its structural strength and combat performance. With its prototype clearly surpassing that of its Nakajima competitor, Aichi won the production contract.

Aichi's success secured a large order for the company. Between 1934 and 1940, it produced nearly 600 D1A1 carrier dive bombers, including the improved D1A2 version that featured a more powerful 730 horsepower engine. Even more important, after having outdone its competitors, Aichi became the exclusive large-scale manufacturer of the navy's dive bombers. The navy deployed Aichi's dive bombers in large numbers in the Second Sino-Japanese War. In 1937, the D1A2 gained international notoriety when one of these dive bombers attacked and sunk the US gunboat *Panay* on the Yangtze River.

Heinkel's designs continued to shape Japanese dive bomber design. His inventions included speed brakes that could effectively control a bomber's speed during its dive and an elaborate mechanism for foldable wings that increased the number of aircraft that could be stowed on a single aircraft carrier. In 1935, a year before the conclusion of the German-Japanese Anti-Comintern Pact, Aichi imported Heinkel's He 70 Blitz ("lightning"). As it turned out, the only feature of the He 70 that impressed the Aichi engineers was the high-speed elliptical wing design, which they adopted during the development of the D3A1 carrier dive bomber. Another remarkable feature for which the Aichi engineers alone earned the credit was the D3A1's outstanding radius of operation. Based on Japanese intelligence information on the performance of US naval aircraft, the new dive bomber was designed to have a flight range of nearly 1,500 kilometers, outranging the enemy's aircraft by a factor of three.[57] Mass production of the D3A1 began in late 1939, and it became the navy's main strike force in the Pearl Harbor attack.[58]

Aichi's cooperation with Heinkel lasted from 1925, when the first contract for seaplanes launched from cannon turrets was signed, to the last import of a Heinkel aircraft in 1935. During this decade, the Japanese company managed to reduce and finally eliminate reliance on Heinkel's designs. This process is all the more remarkable because it thwarted Heinkel's declared strategy to keep his customers in a semi-dependent position. Until the mid-1930s, Heinkel ran his enterprise largely as a design company with a large research and development department that had only a small production section attached.[59] The firm developed ten or more new designs each year, built one or two prototypes of each type, and sold them for mass production to foreign companies. Heinkel never exported technology for aircraft production to Japan, nor did he concern himself with training Japanese engineers. Under these conditions, Heinkel could focus on developing state-of-the-art aircraft in the hopes that his Japanese customers would not catch up with or overtake him with their own designs. This strategy worked well only until the mid-1930s, when the expertise of Aichi's engineers had reached a level that allowed them to successfully modify and improve Heinkel's latest designs.

Britain's Waning Influence and a Fateful Legacy

As demonstrated, Japanese advances in technological independence undermined Heinkel's business strategy. They also compromised British efforts to regain a foothold in Japanese naval aviation. Imports of British aircraft to Japan, and the exchange of engineers, continued throughout the 1920s—even though the results were often less than satisfactory. Only two imports from Britain resulted in a noteworthy Japanese production:

the Gloster carrier fighter and a Blackburn-designed carrier bomber. Their contribution to the advancement of Japanese design capabilities was negligible. In both cases the complete design and prototype construction were carried out in Britain without the participation of Japanese engineers.

Even after the discovery of Sempill's and Rutland's spy activities, the exchange of officers between the Imperial Japanese Navy's air arm and the British Royal Air Force continued. Obviously both sides were interested in maintaining a pragmatic relationship. In 1927 the Japanese Navy dispatched one of its ace pilots to Britain. Lieutenant Kamei Yoshio had gained fame as one of the first Japanese pilots to perform a deck landing in 1923, and he joined a fighter squadron of the Royal Air Force's Fleet Air Arm for flight training.[60] Based on this experience he drafted the "Teaching Methods for Air Battle" (*Kūchūsentō kyōhan*) in early 1929. Kamei's manual provided the first systematic guideline for the navy's flight training and crew evaluation.[61] In November 1929, Kamei became the leader of the Yokosuka Flying Corps. In this new position, he emphasized the importance of carrier-based fighter aircraft to navy officials.[62] Kamei continued to be closely involved in developing Japan's carrier air force, especially in adopting the technique of deck landings at night, a notoriously dangerous task. He served as the commander of the *Kaga* and *Ryūhō* air squadrons before rising to the position of a rear admiral.

In the early 1930s, a second British aviation mission came to Japan to instruct the Imperial Japanese Navy's pilots—but on a very modest scale. From November 1930 to the end of February 1931, the Japanese Navy employed two Royal Air Force instructors, R. W. Chappel and E. E. Wingate, as "provisional instructors of air battle" (*rinji kūchūsentō kyōkan*).[63] Squadron Leader Chappel already had experience with Japan during his time as a language officer at the British embassy in Tokyo from 1926 to 1927. His companion, Flight Lieutenant Wingate, was an experienced Royal Air Force shooting instructor. The Japanese, it seems, welcomed the British advisers with open arms. At a time when the Japanese military became increasingly secretive, Chappel was granted free access to the Yokosuka airfield, the Tateyama air corps in Chiba prefecture, and even Nakajima's Ogikubo aircraft factory.[64]

The Japanese Navy sent eight of their top pilots to be trained by the British instructors. Among this hand-picked group was Genda Minoru (1904–89), an ambitious fighter pilot. In his 1961 memoirs, Genda recalls that "there was not so much to learn about [British] flying and shooting technique. . . . When we received the lectures and training from the British officers, we won the air battles against them, and during target practice we got better marks than they did."[65] Chappel soon realized the proficiency of his supposedly untrained students. On several occasions, he reported the skill of Japanese pilots, especially in torpedo bombing, to the Air Ministry.[66] Even though the British mission did not contribute much to the advancement of Japanese flying skills, Chappel's lectures on air tactics made a strong impact on Genda. The British squadron leader's insistence on assigning a key role to fighter aircraft and starting a battle with a surprise air attack convinced Genda of the advantages of an aggressive first-strike air attack and became the accepted doctrine of Japanese naval aviation from the early 1930s onward.[67]

With the fast rise of its naval airpower, the Japanese Navy became increasingly confident about the skill of its pilots and the performance of its aircraft. British hopes for the Imperial Japanese Navy's continuing dependence on the Royal Air Force further faded. By the early 1930s, Japan no longer needed imports and instruction from Britain, and the strategic landscape began to change toward a more confrontational trial of strength in the Pacific. In September 1934 Japan decided to abrogate the Washington Naval Treaty. The revised 1936 National Defense Policy (*Teikoku kokubō hōshin*) added Great Britain as a potential enemy. Accordingly, the navy's planning changed toward a more aggressive strategy against British bases in East Asia. By 1939, to even the most casual observer, the ascendancy of Japan's naval aviation was without question: the Japanese Navy had eight aircraft carriers and five seaplane carriers under its command, while the British China Fleet had only one carrier at its disposal.

Japanese officials who were dispatched to Britain consolidated this view. Genda Minoru served as a naval attaché in London from December 1938 to September 1940.[68] He witnessed the Battle of Britain, which led him to compare the training, aircraft types, and proficiency of the British fighter units with their Japanese and German counterparts. He concluded that the Royal Air Force and the Luftwaffe were "considerably inferior" to the air arm of the Japanese Navy. He was especially convinced of the superiority of Japanese naval airpower in air battles over the open sea.[69] As it turned out, Genda's London experience together with his former British instructor's advocacy of surprise air attacks had profound implications. In

1941 Genda became an operations officer of the First Air Fleet, where he carried out the complete tactical planning for carrier allocation and large-scale aircraft deployment for the Pearl Harbor attack.

It is difficult to fathom how Britain could turn a blind eye to the eventual consequences of an increasingly powerful and aggressive Japanese naval buildup. Inconsistencies in British policy making might offer one explanation. The warnings of the admiralty and MI5 about the emergence of a dangerous potential enemy could be easily ignored in a political environment that emphasized doing business with Japan. Furthermore, many British decision makers seemed to have found comfort in a self-administered Ten Year Rule that eliminated any confrontation with Japan by definition and spared the country from worries about a major military conflict.

Finally, deep-rooted prejudices seem to have played an important role in underestimating the rapid rise of Japanese naval airpower.[70] The Smith team's patronizing of Mitsubishi's designers, the British flight instructors of the Sempill mission complaining about the incompetence of the "little fellows," and the British military attaché's self-satisfied reports about the ignorance of Japanese militaries appear to exemplify an attitude of many British officials who refused to deal with their Japanese counterparts on an equal footing. Even in 1939, the Royal Air Force's "Handbook on the Air Services of Japan" still reiterated the familiar themes of Japan's dependence on foreign assistance and lack of original research. The handbook assured its readers:

> It is remembered that as yet the Japanese have shown no national inventive genius in the whole wide realm of mechanics. Their undoubtedly efficient machines derive entirely from Western models, copied or modified. It is not to be supposed therefore that in the highly specialised and ever advancing technique of aircraft design they will do more than follow western ideas for a considerable period. Thus in this particular, their efficiency vis à vis Western Power is limited.[71]

Conclusion

By 1937, on the eve of the Second Sino-Japanese War, the Imperial Japanese Navy's airpower had made considerable progress, and Japan's naval aviation had reached technological independence. Even though this process was remarkably successful, it did not follow a clear-cut line. Initially, the transfer of British technology and know-how was instrumental for the early progress of Japan's naval air force. In the early 1920s, licensed production of the F.5 flying boat allowed the Japanese to gain insight into the design and construction of these wooden aircraft. However, importing German all-metal aircraft in 1925 made the F.5's design outdated, and the Japanese Navy resourcefully joined forces with the domestic aircraft industry for an ambitious program that ultimately led to some of the world's finest flying boats.

With the help of the British navy, British engineers, and British spies, Japan succeeded in building its first aircraft carrier and gained experience in the operation and deployment of carrier aircraft. These airplanes—imported from England or designed by British engineers in Japan—played key roles in pilot training and establishing carrier landing techniques. With the *Akagi* and *Kaga*, a second generation of large carriers allowed the takeoff and landing of faster and heavier airplanes. In a highly contested move, the Japanese Navy once more turned toward British imports for its new carrier-fighter designs while relying on German dive bomber technology.

After 1930, apart from small cargo planes, Japan no longer imported military aircraft from Britain.[72] The only British aircraft that could be seen in the skies over Japan were a number of light civilian aircraft owned by Japanese newspapers, which used them for news delivery. Japanese domestic design was making such quick progress that British aviation technology had lost its competitive edge. Within a decade the aeronautical engineers of the navy's arsenals and the Aichi Aircraft Company had achieved technological self-sufficiency. Aichi bought its last aircraft from Heinkel in 1935. Three years later, the engineers of the Yokosuka Arsenal started the development of the Suisei ("comet") under the leadership of Yamana Masao (1905–76).[73] With its sophisticated, streamlined design, the Suisei was the fastest carrier-based dive bomber of World War II, clearly demonstrating that Japanese designs could surpass those of their former mentors.

Rapid and comprehensive technology transfer not only provided the Japanese Navy with the essential hardware but also led to fundamental doctrinal change. Initially, floatplanes effectively supported the navy's battleships with their reconnaissance and artillery-spotting missions but left the big-ship, big-gun policy largely unquestioned. Then the new technology of large, all-duralumin aircraft fired the imagination of military planners. It demonstrated the feasibility of huge military flying boats that could autonomously carry out long-range patrol and bombing missions in the Pacific. Finally, new dive bomber technology turned the aircraft carrier into a powerful weapon that ultimately challenged the raison d'être of battleships. These technological and doctrinal advances in naval airpower vastly increased the Imperial Japanese Navy's strike force and emboldened its planners to envision an aggressive first-strike strategy.

Once more the sharp difference between the two branches of Japan's armed forces should be emphasized. The army continued to depend fully on the country's civil aircraft makers for developing its new aircraft. In contrast, the navy played an active role in a remarkable process that began with the acquisition, indigenization, and diffusion of imported technology and ultimately led to independence from foreign design. Unlike the army's ill-fated Air Technical Research Institute, the Yokosuka Naval Air Arsenal evolved into one of the nation's foremost centers of aeronautical innovation. Its comprehensive array of activities included theoretical aerodynamics, the design and testing of prototypes, and aircraft manufacturing. As we have seen, the Naval Air Arsenal also cooperated closely with civil companies, such as Mitsubishi, Kawanishi, Aichi, and Nakajima, for the transfer and diffusion of advanced aviation technology and the setup of mass production. Furthermore, by organizing regular gatherings across institutional boundaries, the Naval Air Arsenal succeeded in forging links between academic researchers and civil manufacturers. According to Tani Ichirō (1907–90), a noted aerodynamics expert at Tokyo Imperial University, these meetings soon gained a reputation for their insightful high-level discussions in a frank and open atmosphere that allowed for interpersonal communication and an extensive interchange.[74]

By 1933, then, Japan's naval aviation clearly surpassed that of Great Britain and Germany.[75] Ironically, the two countries whose aviation specialists had decisively contributed to the development of Japanese naval airpower neglected the modernization and growth of their own maritime air forces. Interministerial rivalries, a conservative naval doctrine, and the emphasis of the Royal Air Force on heavy bombers seriously hampered the expansion of British naval airpower.[76] The single most crucial factor was a budget cut that forced the admiralty to repeatedly postpone its ambitious carrier program. Only in 1935 was Britain's first post–World War I carrier, the *Ark Royal*, laid down. Until then, the Royal Navy had to settle for a small and increasingly obsolete carrier fleet that, because of its limited carrying capacity, also delayed the procurement of new carrier aircraft.

The history of Japan's and Germany's naval aviation illuminates the strong interaction between military doctrine and technological development—in opposite ways. After 1935 the German Luftwaffe under Aviation Minister Hermann Göring (1893–1946) insisted on its monopoly of "all things flying."[77] Lacking any clear doctrine for developing naval airpower, the Luftwaffe showed no interest in providing dive bombers or effective torpedo aircraft for the Kriegsmarine. Until 1945 the Japanese Navy had been building thirty-two aircraft carriers, but the two aircraft carriers being constructed by Germany were abandoned at early stages in 1939 and 1943. Lacking effective naval aviation, the Luftwaffe could not cooperate with the navy's U-boats to interdict ship transport to Britain, a failure that had decisive consequences for the outcome of World War II.

PART IV

Toward Pearl Harbor and Beyond, 1937–45

CHAPTER 7

US Know-How for Japanese Aircraft Makers

On the morning of December 7, 1941, more than 300 Japanese carrier-based airplanes attacked the military base at Pearl Harbor in Hawaii. Exactly two hours later and more than 10,000 kilometers away, Japanese long-range bombers started their first air raid on Singapore.[1] On December 10, Japanese bombers attacked and destroyed the British warships HMS *Prince of Wales* and HMS *Repulse* off the Malayan coast. For the first time in naval history, capital ships were hit and sunk solely by aircraft on the open sea. Japan's aggressive assaults met with international indignation. There was also widespread astonishment that their allegedly backward aviation industry could provide the military hardware for such seemingly impossible air strikes.

By the late 1930s, Japanese aeronautical design had reached an advanced stage that allowed the country's aviation industry to build airplanes like Nakajima's Type 97 carrier attack bomber or Mitsubishi's Zero fighter. These aircraft combined outstanding performance with enormous flight range. One year after the Pearl Harbor attack, the US Intelligence Service still cautioned US pilots that the Japanese Zero fighter had "superior maneuverability to all our present service type aircraft" and issued a warning: "Never attempt to dog fight the Zero."[2] Japan's aircraft makers not only designed these state-of-the-art airplanes but also built them in increasing numbers. Modern production methods and machinery allowed the mass production of airframes, engines, and propellers in unprecedented quantities. Between 1935 and 1941, Japan's aviation industry had increased its output tenfold.

This chapter analyzes the influence of the United States in the rapid development of aviation in Japan. It focuses on the decade between the 1931 invasion of Manchuria and the Pearl Harbor attack, an era of escalating political, economic, and military tensions that culminated in the US embargo against Japan. It addresses the question of how, under these conditions, Japan could maintain and even intensify the import of US aviation and production technology. The chapter also examines how importing US aeronautical material and know-how not only promoted technological advance in the design and production of Japanese aircraft but also shaped the mutual perception of the two countries before they went to war. As we will see, continuing US stereotypes of Japanese backwardness and derivative engineering led US intelligence to underestimate the fast progress of Japanese aviation.

At the same time, the Japanese armed services had to deal with an impending US embargo on aeronautical material and production tools as well as disturbing reports on the increasingly superior US industrial power. Forced to compete against the clock, the Japanese military put their trust in advanced aviation technology while urging their pilots to keep up a superior fighting spirit even when outnumbered by the enemy.

Late Japanese Interest in US Aviation

Despite the Wright Brothers' pioneering flight in 1903, Japanese interest in US aviation had a late start. A look at the early years of aviation in the United States might help explain this phenomenon. In 1914 the annual production of US aircraft makers had not even reached fifty planes.[3] When the United States declared war on Germany in April 1917, US airpower rested on fewer than 300 airplanes. Yet with a rapid increase in appropriations, the US aviation industry expanded and quickly adopted the country's already highly developed methods of mass production. In May 1917, the US War Production Board ordered the development of a new aero-engine consisting only of tried-and-tested interchangeable parts. The mass production of the iconic Liberty engine began just six months later, with more than 13,000 exemplars built before the end of the war.[4] At the same time, US airframe manufacturers boosted their annual output to an impressive rate of nearly 14,000 war planes.[5]

By 1918, three defining features of the US aviation industry were already taking shape. All of them incorporated the characteristics of US-style mass production: a strong focus on standardization and interchangeability, a preference for a conservative design that concentrated on efficient production and reliable operation, and the capability of boosting production numbers within a short time.[6]

In spite of these achievements, the US military aircraft industry sunk into a postwar slump. Budget cuts and a conservative naval doctrine of "battleship first" took their toll on the advancement of military aviation. In 1920 Congress reduced expenses for military aviation by over 60 percent, and within four years the Air Service's aircraft fleet dwindled from over 3,000 to fewer than 800 airplanes. The continuing reluctance of the US military to consider the potential of aerial warfare became front-page news in 1925 when Colonel William Mitchell, an ardent advocate of airpower expansion, was court-martialed and discharged from service.

Even in the early 1930s, US military aviation development lagged behind that of Japan and several European powers. The US Air Corps still had to rely on aircraft that differed little from their World War I predecessors. Fabric-covered biplane bombers like the 1931 Keystone B-6A with a cruise speed of about 160 kilometers per hour and an effective radius of action of less than 700 kilometers clearly had limited striking power. This was in marked contrast with Japan, where companies like Kawasaki and Mitsubishi supplied the military with advanced all-metal monoplanes that excelled in speed and operating range.

The Japanese military was well aware of its technological lead and did not hold US military aviation in high regard. In 1928 the Aviation Headquarters of the Imperial Japanese Army carried out an investigation on "American Warplanes and Engines" and concluded, "when compared with current aircraft of other foreign countries, the performance of US military aircraft is not especially prominent."[7]

What really caught the interest of the Japanese experts was the US civil aviation sector, which had catapulted itself into world leadership within less than a decade. Several factors contributed to this remarkable process. The 1925 Kelly Act provided funds to establish a nationwide air mail service, which encouraged commercial aviation and the manufacture of civil aircraft. One year later, the multimillion-dollar Daniel Guggenheim Fund for the Promotion of Aeronautics stimulated the development of civil aviation, aeronautical science, and research. The single most important event that led to a boom in civil aviation, of course, was Charles Lindbergh's much-celebrated solo Atlantic crossing in May 1927. In the wake of the "Lindbergh phenomenon,"[8] the US aircraft industry more than tripled its sales within two years. In a grim twist of fate, the 1929 stock market crash forced many small companies into bankruptcy.

To stay in business, many US aircraft makers invested great effort into putting the latest results of aeronautical research into practice. Driven by a commercial market in which a lighter plane with higher speed directly translated into greater profits, US engineers devised a new generation of passenger aircraft on their drawing boards.[9] Streamlined fuselages, enclosed cockpits, and sophisticated engine cowlings reduced air resistance and allowed for a major speed increase; advances in all-metal aircraft design resulted in a significant weight reduction; efficient engines and propellers combined high power with low fuel consumption; and slotted wings lowered the aircraft's minimum speed to ensure safe takeoffs and landings.[10]

As it turned out, these design features were of great interest to the Japanese. As the admiral and former naval engineer Okamura Jun put it: "When, around 1933, American aviation all of a sudden made big progress, the Japanese aviation world became interested in it." Okamura emphasized that the ensuing importation of US civilian aircraft to Japan "wielded enormous influence on Japanese technology and spurred the advance of Japanese design."[11]

US Aviation Technology Comes to Japan

Before the mid-1930s, the export volume of US aviation technology to Japan was still moderate. During the first half of the 1930s, the annual sale of US aeronautical material to Japan never exceeded US$500,000 and only in 1936 went up to close to US$1 million.[12] Yet it should be emphasized that these figures alone tell us little about the direct US impact on Japanese aviation. Instead, they reflect a tried-and-tested purchasing strategy of Japanese aircraft manufacturers who used imports primarily for improving domestic production. Therefore, rather than passing wholesale orders for many identical airplanes and engines, the Japanese attached much importance to getting a variety of aeronautical material. In many cases they selectively shopped for advanced aeronautical parts and components like engines, propellers, and instruments rather than purchasing whole planes.

It is important to note that the United States did not reveal cutting-edge aviation technology indiscriminately. A case in point was research on a revolutionary wing design that became known as the laminar flow airfoil. The curved surface of such a wing yielded considerably more lift and induced much less air friction. This new wing type would dramatically increase an aircraft's speed and flight range. In 1939 the National Advisory Committee for Aeronautics (NACA) publicly announced the successful design of a laminar flow airfoil. But due to the classified nature of the research project, NACA made sure not to disclose any design details. Around the same time, a team of Japanese researchers led by Tani Ichirō was working on a similar project at the Aeronautical Research Institute in Tokyo. Aware of the recent US breakthrough but lacking any specific information, the Japanese independently developed a laminar flow airfoil from their own detailed calculations and experiments.[13]

AIRCRAFT ENGINES

Clearly the US aviation industry was less hesitant about exporting advanced technology to Japan. America's two largest aero-engine makers, Curtiss-Wright and Pratt & Whitney, established close business ties with their Japanese counterparts, Nakajima and Mitsubishi. The licensing transactions reveal how Japanese companies successfully acquired detailed knowledge about the design and production of advanced US aircraft engines. Japanese trading companies like Mitsui Bussan and Mitsubishi Shōji played important roles in this process.

In 1934 Nakajima, by then Japan's largest aircraft manufacturer, acquired a production license for Curtiss-Wright's new 400 horsepower Cyclone R-1820 engine. The contract included training five Nakajima engineers in the United States.[14] Dispatching these Japanese specialists was an important move: they not only received instruction on Curtiss-Wright's engine design but also gained firsthand information on US production methods. Sakuma Ichirō, the head of Nakajima's design section, stayed for seven months at the Curtiss-Wright factory in Patterson, New Jersey. He received systematic training on each step of engine production: processing material, assembling engine parts, and testing the finished engine. Sakuma familiarized himself with the principles of Taylor's scientific management and Ford's assembly-line production with the aim of introducing both to Nakajima's factories.[15] Kiyooka Shōichi was another Japanese expert who visited the US engine maker. The army engineer, who came to Curtiss-Wright in the guise of a Nakajima employee, was overwhelmed by the sheer size of the plant: it was ten times larger than Nakajima's factory. He became aware how, at a time when Nakajima produced two to three engines each month, Wright's advanced production methods allowed an output of two to three engines *each day*. Kiyooka also admired the uniform quality of Wright's engines. For him Japanese products were still "no match" (*oyobu tokoro dewa nai*) for their US counterparts.[16]

The two engineers returned to Japan with several pieces of US production machinery, which they combined with their experiences to revamp Nakajima's production process. They also brought the Cyclone engine's blueprints and specifications and three specimen engines. Rather than mounting these motors on planes and putting them into use for flight, the Nakajima engineers disassembled them—as a US intelligence report put it—for "research purposes."[17] Nakajima built an engine prototype based on the Cyclone that was approved by the Japanese Army and Navy in 1934.[18] The engine was put into series production and became the highly successful Hikari ("light") engine that powered many fighters, dive bombers, and reconnaissance aircraft.[19]

Mitsubishi Aircraft, Japan's second-largest aircraft maker, also relied on US engine technology. In 1934 the Japanese company bought the production license for Pratt & Whitney's R-1689 Hornet engine. Like its competitor Nakajima, Mitsubishi modified the US design and presented it to the navy, which adopted the engine under name Kinsei ("Venus") Type 4 in 1937.[20] Mitsubishi built more than 12,000 of these engines. The Kinsei underwent several modifications and was used before and during the Pacific War aboard a wide range of aircraft types, most prominently on Mitsubishi's twin-engine G3M bomber and Aichi's D3A dive bomber.

US AIRLINERS FOR THE JAPANESE MILITARY

In the early 1930s, a new generation of US commercial aircraft made its appearance in civil aviation. These "airliners" featured advanced technology, such as autopilots for long-range flights, retractable landing gear for higher cruising speeds, and deicing equipment for flying in severe weather. Within the US aviation industry, the Douglas Aircraft Corporation had a reputation for innovative design that incorporated the latest aeronautical research. The company's DC-1 airliner received credit for being the "first scientifically designed American airplane."[21] Nakajima was one of the first Japanese manufacturers to show interest in these new passenger aircraft. In the early 1930s, Nakajima sent several engineers to Douglas. When Douglas's DC-2 passenger aircraft made its first flight in 1934, Nakajima promptly bought the production license, blueprints, engines, and aircraft parts for ¥1.3 million (approximately US$400,000 at the time).

Such an expensive license purchase was not without controversy. In June 1935, *Asahi* published the regretful comment of a Japanese aircraft engineer: "we can make such an aircraft by ourselves if we were provided with such an enormous amount of money."[22] Such criticism obviously ignored the fact that Nakajima's cooperation with the US aircraft maker granted the company full access to the most recent advancements in US technology. In addition to the license, Nakajima could buy a wide range of the latest machine tools to "improve its workshop practice" and state-of-the-art aeronautical material: cockpit instruments, several Pratt & Whitney engines, and the latest hydraulic systems for wing flap operation and landing gear retraction.[23] Nakajima incorporated many of the latest US design features in its AT-2 transport plane that the company delivered to the army, Manchurian Airlines, and Japan Air Transport (figure 7.1).

Even though Douglas had designed its aircraft for commercial air transport, most of its technology could be used for military purposes as well. The US military was fully aware of the passenger aircraft's dual-use character. In 1935 the US military attaché commented on Nakajima's procurement of the DC-2 that a "reliable source states that the new Nakajima bomber will be adapted from the Douglas."[24] In the following year a US report on an "Experimental Bombing Plane" stated that "Nakajima consistently denied that the Douglas would be copied as a bomber . . . although an officer of the company at the last inspection of the plant said regarding the Douglas 'this should be the future bombing plane of the Air Corps.'"[25]

US misgivings were well founded. The Experimental LB-2 Long-Range Attack-Bomber was Nakajima's first all-metal twin-engine monoplane. The bomber's wing design and hydraulically operated landing gear were strikingly similar to those of the DC-2. The aircraft made its successful maiden flight in March 1936. Based on this experience, Nakajima's engineers skillfully incorporated the DC-2's advanced concepts in virtually all of their future designs. As a result, retractable landing gear, wing flaps, and variable-pitch propellers became standard features that dramatically improved the performance of Nakajima's fighters, bombers, and reconnaissance aircraft.

The Japanese soon turned their attention to one of the era's most iconic airplanes: the Douglas DC-3, which made its first flight in 1935. With a maximum speed of 370 kilometers per hour, it was the fastest commercial aircraft of its time. With its reliability and low operating cost, the DC-3 revolutionized the airline industry by finally turning passenger transport into a profitable business. In 1937 the Mitsui trading company placed an order for twenty Douglas DC-3s, which were all to be delivered to Dai Nippon Kōkū KK (Greater Japan Airlines Co., Ltd.) at a unit price of US$104,000.[26] Several of these planes were equipped with the latest high-tech accessories, which included Wright Cyclone engines, Hamilton constant-speed propellers, and Sperry automatic pilots.[27]

The Japanese Navy became interested in the new US airliner as well and decided to acquire a DC-3 for flight tests. Following the navy's secret instructions, Mitsui bought the production license for the DC-3 in

FIGURE 7.1 What appears to be a fleet of bombers on their way to conquer the world are actually civilian passenger aircraft of Japan Air Transport's "Super Express in the Sky." The stylized airplanes (in all likelihood Nakajima AT-2 transports) stand for the airline's promise to fly its passengers "within the same day" to Manchuria, north China, or Taiwan. (Source: National Air and Space Museum Poster Collection)

February 1938 and had the aircraft delivered to the civil airport at Haneda. A US intelligence report commented on this move: "This may have been a blind to delude or influence the American State Department."[28] Mitsui sublicensed Nakajima and the Shōwa Aeroplane Company to build the airplane as a personnel and cargo transporter for the navy. The two Japanese companies produced nearly 500 of these aircraft, clear proof that they had mastered not only the construction but also the mass production of the US airplane.

In its eagerness to acquire the latest US aviation technology, the Japanese Navy ordered Mitsui to purchase the only DC-4E airliner that Douglas ever built. The four-engine prototype was twice as big as its DC-3 predecessor but much less successful. Because of its complex design and cost overruns, the DC-4E never went into production.[29] Nevertheless, in March 1938, just three months before the aircraft's maiden flight, Mitsui signed a contract to buy the transport plane and a manufacturing license for the considerable amount of US$725,000.[30] The trading company ostensibly purchased the DC-4E for Dai Nippon Kōkū (figure 7.2). However, immediately after its arrival, the aircraft was handed over to Nakajima, which used it as a model for its first four-engine heavy bomber, the Shinzan ("mountain recess"). Nakajima's new bomber made its first flight in December 1939. Its speed and bomb load were similar to those of America's latest heavy bomber, the Boeing B-17 Flying Fortress, which was introduced to the US Army Air Corps in 1938. Yet with a flight range of more than 4,200 kilometers, the Nakajima bomber could outrange its US competitor by nearly 1,000 kilometers. However, due to the sheer complexity of the aircraft, the Shinzan was beset with similar design problems as was the DC-4E. Problems of excessive weight and poor engine reliability added to these troubles; only five exemplars of the heavy bomber were ever built.

THE IMPORTANCE OF THE VARIABLE-PITCH PROPELLER

The Japanese also acquired advanced US propeller design, which led to a leap in aircraft performance. Propellers play a simple but crucial role: they convert engine power into forward thrust. An ordinary propeller with fixed blades is designed for a predetermined air speed. It delivers optimal thrust for either takeoff, climb, or cruise flight. A variable-pitch propeller adjusts its blades according to the aircraft's speed. It uses engine power

DC-4 Off For Japan

THE crane above is lifting the prototype Douglas DC-4 off American soil for the last time. When this picture was taken, the giant four-engined airliner was being loaded aboard a freighter bound for Japan where it is to be delivered to Japan Air Transport, its purchasers. The ship was sold to the Japanese for a price said to be in the vicinity of $700,000. It was reported that the cost of shipping the DC-4 to Japan was approximately $50,000.

The Japanese airline bought the huge transport following exhaustive and extensive tests by Douglas and airline engineers. Just prior to its sale it had returned from an elaborate transcontinental tour, mostly under the auspices of United Air Lines. Incidentally, United recently placed an order for six DC-4's and delivery is expected to start some time in the spring of 1941. The DC-4 is powered with four 1,500 h.p. engines and carries 42 passengers.

FIGURE 7.2 "DC-4 Off for Japan": This short report appeared in the December 1939 edition of the US magazine *Popular Aviation*. The author notes with a tinge of regret that this aircraft, a Douglas DC-4E (Experimental), is leaving "American soil" for good. The article fails to mention that Douglas had shelved further development of the DC-4E because of its inadequate performance and prohibitive maintenance costs. (Source: *Popular Aviation*, December 1939, 52) much more efficiently, allowing steeper climbs, higher cruising speeds, and shorter takeoff runs—an essential feature, especially for carrier-based aircraft. In the words of John David Anderson, the variable-pitch propeller was "one of the most important technical developments in the era of the mature propeller-driven airplane."[31]

Japanese aircraft makers fully depended on foreign licenses to produce this important high-tech device. In 1930 US propeller manufacturer Hamilton Standard introduced the first workable controllable-pitch propeller that could be set into two positions for takeoff and cruise. In April 1934, Mitsui Bussan bought the license for such a propeller for US$70,000. From June to October 1934, the propeller division of Sumitomo Metal Industries sent several of its engineers to Hamilton's factory near Hartford, Connecticut. Sumitomo delivered its first two-position variable-pitch propeller to the Japanese Navy's air arm in October 1935. The following year, Sumitomo reached an annual production capability of 600 variable-pitch propellers, all based on the Hamilton design.[32]

In 1935 Hamilton presented a breakthrough invention: the variable-pitch constant-speed propeller. The device, which could be best described as an automatic gear shift for the aero-engine, automatically adjusted a propeller's blades according to the aircraft's speed, allowing the engine to run always at maximum efficiency. The same year, Sumitomo urged Hamilton to sell a production license for the constant-speed propeller, but, following a policy not to export military technology immediately after its first production, the US government intervened and prohibited the sale of the Hamilton patents. According to a US intelligence report, Sumitomo "offered the Hamilton concern fantastic sums to allow one of their experts to come to Japan to show them how to manufacture propellers according to the Hamilton license but so far, the Hamilton company has refused to permit their men to accept this offer."[33] In 1937, the US government lifted the export ban and allowed Hamilton to sell a production license. Within two years, Sumitomo fully mastered the production process, an achievement that led to a major advance in Japanese aviation. In April 1939, Japan's first constant-speed propeller was installed on the navy's experimental carrier fighter A6M1, a prototype of the famous Zero fighter. The new propeller type dramatically improved the fighter's performance. As a Japanese aviation expert put it: "the decisive factors of the Zero fighter's success were its outstanding airframe design . . . and the Hamilton constant-speed propeller."[34]

A Craving for US Machine Tools

Along with its advances in the design of aircraft and aero-engines, the US aviation industry also made significant progress in its production methods. Well-planned factory organization and the efficient use of machine tools became the keys to survival in a competitive market that demanded both an ever-increasing output and the economical use of material and manpower. The US aircraft makers greatly benefited from the tried-and-tested practices of the US automobile industry. T. P. Wright, vice president of Curtiss-Wright, said in 1939: "we were all surprised at the tremendous number of automotive production methods applicable to aircraft work and helpful in reducing production costs, but of which we had not previously been aware."[35] Mechanization became the key issue. For each stage of the production process—cutting, forming, fastening, and assembling—the US manufacturers used a wide array of advanced machinery. What became to be known as "American methods of aircraft production" included the use of giant hydraulic presses, batteries of drop hammers, precision lathes, drills, and grinders, as well as automatic riveting machines.[36]

After the outbreak of the Sino-Japanese War in July 1937, the Japanese aviation industry began to attach increased importance to efficient aircraft production as well. As a response to the military's demand for more warplanes, all major aircraft makers launched an important shift to large-scale industrial production. In the wake of this development, the character of technology transfer from the United States changed significantly. Instead of importing aircraft, engines, and aeronautical equipment, Japanese manufacturers increasingly sought access to US production machinery. The results were astonishing: between 1937 and 1939, machine tool imports from the United States grew by more than 500 percent.[37]

Mitsubishi offers a good example to illustrate the heavy impact of US production technology on Japan's aircraft makers that transcended the mere import of machine tools. In 1937 the head of Mitsubishi's aircraft engine production department, Fukao Junji (1889–1977), visited the Pratt & Whitney plant in East Hartford, Connecticut. Renowned industrial architect Albert Kahn (1869–1942) had designed the factory for efficient mass production that guaranteed a continuous workflow from raw materials to finished engines.[38] Fukao became convinced that this facility could

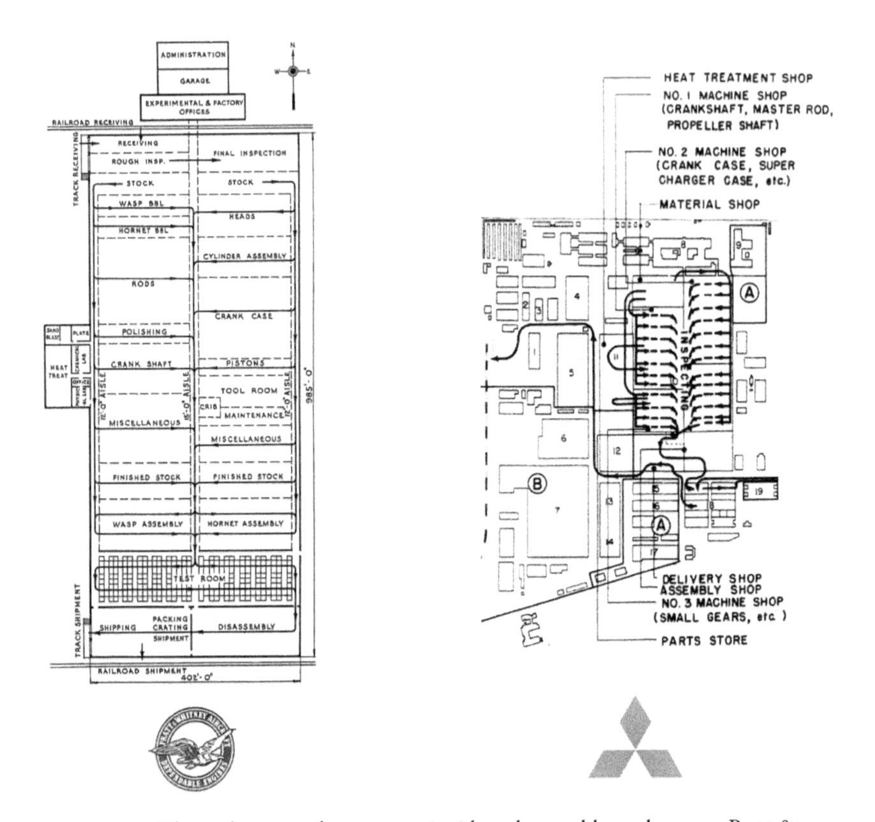

FIGURE 7.3 These plans reveal a not so coincidental resemblance between Pratt & Whitney's engine factory in East Hartford, Connecticut (left), and Mitsubishi's No. 4 Engine Works in Nagoya. Note especially the close similarity between the US factory's optimized layout and material flow and its Japanese counterparts (upper right quadrant). (Sources: Nelson, *Industrial Architecture*, 169; United States Strategic Bombing Survey, *Mitsubishi*, 146)

be the perfect model for Mitsubishi's new engine factory. While still in the United States, Fukao enthusiastically reported the US factory's innovative architecture, ventilation, and illumination. He recommended a careful study of the plant's layout to build the "ideal factory" in Japan.[39]

Only one year later, Mitsubishi opened its new Daikō plant in Nagoya. According to Yamazaki Eiji, a Mitsubishi employee in charge of engine production, the Daikō factory was closely modeled after Pratt & Whitney's plant (figure 7.3). With its high degree of mechanization and systematic assembly process, it ushered in a new era of manufacturing in Japan and could have served as a model factory for other Japanese enterprises.[40] In a similar vein, Mitsubishi manager Fujimoto Akio expressed his excitement when he visited this "US-style mass production factory" for the first time. Fujimoto was impressed by the 50,000 square meter building that even allowed free passage for heavy trucks. For him, the well-organized, step-by-step production process was clearly a "one-to-one adaptation of Pratt & Whitney's production methods."[41]

Even with the remarkable developments in Mitsubishi's mass production, many Japanese observers still doubted whether their country's aviation industry would ever catch up with the manufacturing power of the United States. In 1938 a civilian inspection trip set out with the goal of comparing the state of Japan's aviation industry with that of its US counterpart. Three Kawanishi engineers and three employees of Manshū Hikōki Seizō KK (Manchurian Airplane Manufacturing Company) spent one and a half months in the United States. Their visits to twenty-six aviation companies—including all major makers of aircraft, engines, propellers, and aeronautical instruments—made them confident in concluding: "We inspected most of the US aircraft-related factories so that we can accurately judge the state of the US aircraft industry."[42] The group members were less confident about the state of their country's aviation industry. Their US experience made them especially concerned with the low productivity of Japan's aircraft makers. They expressed their worries in a four-volume "Inspection Report about the US and European Aviation Industry" that was published immediately after their return.[43]

Kawanishi engineer Kōno Hiroshi authored the summary section of the report. He noted that even though the performance of Japanese aircraft and engines was not inferior to that of any other country, Japanese production methods and industrial output were far behind. In Kōno's view, Japanese manufacturers were suffering badly from inferior material and a lack of modern machinery, whereas their US counterparts made good use of machine tools like hydraulic presses, drop hammers, and riveting machines. The high degree of mechanization in US factories resulted in lower labor costs and higher productivity, with an output per worker that was two to three times higher than in Japan. For Kōno, the Wright Company—with a production capacity reaching 500 engines a month—was a case in point. Referring to a recent US plan to increase the annual production to 10,000 aircraft, Kōno warned that America's rich resources and powerful industry would indeed allow the country to fully expand its air force by an "amazingly large amount" of military aircraft.[44] He admonished his readers: "We really envy the US companies for their use of a large number of superior machinery and we should learn from their efficient work."[45]

THE ROAD TO THE US EMBARGOES AND A SHOPPING FRENZY'S ABRUPT END

The Kōno report spurred Japanese aircraft manufacturers to upgrade and expand their production with the help of US know-how and machinery. Yet after the launch of the China campaign in July 1937, they became aware that access to US aeronautical material and production tools would become increasingly difficult. By August 1937, Japan had started bombing Chinese cities. These large-scale air raids included indiscriminate bombing of civilian targets. The first air attacks on Nanjing in September 1937 led to worldwide protests. In an October 5 address, US President Franklin Roosevelt condemned the Japanese acts with these words: "without a declaration of war and without warning or justification of any kind, civilians, including women and children, are being ruthlessly murdered with bombs from the air."[46] In December 1937, US–Japanese relations received a further blow when Japanese aircraft attacked and sunk the US gunboat *Panay* while it was evacuating US citizens near Nanjing. Even though the *Panay* incident caused a public outcry, the US government was still loath to impose sanctions against Japan. US diplomats argued—somewhat casuistically—that the 1935 Neutrality Act, which forbade the sale of arms to any country at war, could not be invoked because neither China nor Japan had formally declared war.

In the meantime, Japanese bombing intensified. In February 1938, the air services of the Japanese Army and Navy began a long series of attacks on the provisional capital of Chongqing. These assaults, along with air raids against Canton in May and June, killed thousands of civilians.[47] The American public now increased its pressure on the US government. One of the most vocal groups was the American Committee for Non-Participation in Japanese Aggression. In a 1938 booklet the committee distributed widely, one contributor expressed deep concern with the words "Japan furnishes the pilot—America furnishes the airplane, gasoline, oil and bombs for the ravaging of undefended Chinese cities."[48]

On July 1, 1938, Roosevelt took a first hesitant step to curb Japan's airpower. He imposed a nonbinding "moral embargo" as a compromise to stop the exports of US aircraft to Japan while still allowing the delivery of armament to China. Without mentioning Japan by name, the US government advised all US aircraft manufacturers and Japanese trading companies that it strongly denounced the bombing of civilians and therefore opposed the "the sale of airplanes or aeronautical equipment, which would materially aid or encourage that practice in any countries in any part of the world." Therefore, the Department of State "would with *great regret* issue any licenses authorizing exportation, direct or indirect, of any aircraft, aircraft armament, aircraft engines, aircraft parts, aircraft accessories, aerial bombs or torpedoes to countries, the armed forces of which are making use of airplanes for attacks upon civilian populations."[49]

Amid escalating US–Japanese tensions, US machine tools had gained such a strategic importance for Japan's aviation industry that the Japanese Army and the Navy sent their own purchasing teams to the United States. These "aeronautical missions" played an important role in the massive transfer of US aviation and production technology to Japan. Equally important, the missions allowed the Japanese military to deftly combine purchasing aviation material with gathering intelligence. The dispatches of these groups illustrate the close cooperation among the Japanese military, trading companies, and aircraft makers for procuring US machine tools.[50] It is remarkable that the US consul in Tokyo was well aware of this practice and—obviously without consequences—cautioned the Department of State against it.[51]

The first major Japanese aeronautical mission came to the United States in 1937.[52] Its leader, Colonel Okada Jūichirō, belonged to the Japanese Army's Air Technical Research Laboratory and was uniquely qualified for obtaining firsthand information on the state of the US aviation industry. Already during two earlier years-long stays in the United States, he had studied the country's military carefully. He also was an accomplished engineer who participated in the design of the long-range *Kōkenki* research airplane (which I examine in further detail later in this chapter).[53] Okada had received instructions to examine US production methods and, above all, capabilities to expand aircraft production, especially under wartime conditions.[54]

The mission's official purpose was to purchase aeronautical material and machine tools for the sizable sum of about US$1 million. The US manufacturers on Okada's visiting list had good reason to expect lucrative contracts with the Japanese. At the same time, US military intelligence was aware that the Japanese, rather than putting the US aircraft and engines into practical use, would purchase them "for study and adaptation."[55] It is astonishing how the group's spending power granted its members access to all the major aviation companies, military installations, and research centers.

These activities were coordinated with the US military and were deemed "agreeable to the War Department" and the Navy Department, which "has been consulted and interposes no objection."[56] As a result, Okada and his group obtained a comprehensive picture of US military aviation and the civil aviation industry. Moreover, the Japanese mission purchased a Chance Vought V-143 fighter and a DC-3 transporter, along with a wide range of aeronautical material that included engines, autopilots, propellers, and machine tools. For the designers at Nakajima and Mitsubishi, the US fighter's advanced design proved to be especially useful. They adopted features like the V-143's retractable landing gear for the Hayabusa fighter, which became the Japanese Army's mainstay during the Burma and Malay campaigns as well as for the B5N torpedo bomber and the Zero-sen that made up the bulk of Japanese airpower during the Pearl Harbor attack.[57]

In 1938 another mission, led by Colonel Kanda Minoru, traveled to the United States to purchase equipment that then would be loaned to the civilian manufacturers.[58] Clearly the mission members were driven by a sense of urgency, as they were aware of the continuing low standard of the Japanese machine tools industry and deteriorating relations with the United States. They focused on selected machinery that could significantly expand the Japanese aircraft makers' production capability. Kanda and his team chose manufacturers that guaranteed prompt delivery. To make the group's stay as efficient as possible, Mitsui's New York branch had arranged, in advance, visits to the aircraft manufacturers Douglas, Wright, and Northrop as well as to several producers of aeronautical instruments and accessories. The trading company had also gathered all necessary information about US machine tools, especially for mass producing metal propellers and aero-engines.

With the help of Mitsui and Japan's other three leading trading companies, Ataka, Mitsubishi, and Ōkura, the mission acquired more than 300 machine tools for the staggering sum of nearly US$2.5 million. Their purchase included riveting machines, drop hammers, a 2,000-ton press, and a crankshaft-hardening machine that were to be delivered to Kawasaki, Mitsubishi, Nakajima, Kawanishi, and Hitachi. All parties involved in these transactions cooperated so well that the mission leader could gleefully report to the Ministry of the Army that "we could obtain excellent results—much more than we expected."[59]

In September 1939, Colonel Kanda came to the United States for a second time and placed another massive order for more than 600 machine tools at a selling price of more than US$2.4 million.[60] The mission's detailed order and shipment lists reveal the continuing dependence of the Japanese aero-engine industry on precision machinery from the United States. At the same time, they give an impression of the strategic importance of importing these machines. For instance, a large portion of the purchased machine tools was to be used for producing hardened steel parts like crankshafts, cams, rods, and valves—all vital parts for aircraft engines that not only were crucial for an engine's performance but also determined its reliability.

Kanda made his purchases just in time. In December 1939, the Department of State tightened its boycott to aim directly at Japan's aviation industry. An enhanced moral embargo now included any material for plane production. The US government announced that it would revoke the 1911 Treaty of Commerce in January 1940, opening the way for a full embargo. In July 1940, the Export Control Act authorized the US president to ban the export of aircraft, aluminum, aviation fuel, and—equally important—machine tools to Japan. After Japan invaded the southern part of French Indochina in July 1941, the US Cabinet froze Japanese assets in the United States, a move that effectively resulted in a total trade embargo against Japan.

Know Your Enemy: US Assessments of Japanese Airpower

Why was it possible for the Japanese to get large-scale access to US aviation and production technology until late 1939? This section examines the factors that led the US military into a gross underestimation of Japan's advances in aircraft design and production. Such misapprehension is all the more astonishing, since by the 1920s and 1930s professional US intelligence on Japan's military capabilities was already neatly organized. The Tokyo-based US military attachés systematically collected data on the training, equipment, organization, and battlefield doctrines of Japan's armed forces. Language officers were in charge of providing firsthand information on all military matters. They regularly joined units of the Japanese Army as observers. Based on their reports, the assistant attaché for air compiled the annual Aviation Statistics, to which he added his evaluation of Japan's military aviation, flight schools, and aircraft industry. The number and scope of US Military Intelligence Reports clearly show that until the outbreak of the Sino-Japanese War in 1937, the Ministry of the Army and Navy quite open-handedly granted the US military attaché and his staff access to Japanese arsenals, aircraft factories, and air units.

Notions of Japanese backwardness and an image of the Japanese as mere imitators were common threads running through most of these intelligence reports. A 1927 "General Summary [of the] Japanese Air Service" contended that, even fifteen years after Japan's first flight, the country's aviation still was underdeveloped. The survey estimated that with the Japanese being "great copyists and poor originators," it would be another twenty years before their engineers could compete with Western designers. The report surmised that the Japanese military's obsolete equipment, rudimentary tactics and strategy, and inferior battle strength might result from the army considering aviation "to be something of a toy" and that the "the terrific power of bombing and attack supported by pursuit is probably undreamed of in the Japanese Army."[61]

The image of the Japanese as unimaginative copiers was remarkably persistent. Even in 1937, ten years after the foregoing report, a "Summary of Aeronautical Inspections" asserted that "the Japanese are apparently to date producing no materiel except that copied from other countries." The survey argued that Japan would never catch up with Western nations. The authors made the simple calculation that due to the unavoidable time span between the start of production of a new aircraft type in a foreign country and the copying and production of the same aircraft in Japan, the country would "continue to be approximately three years behind," and thus the "Japanese Air Corps cannot be a menace to any first class power at present."[62]

After the war with China started, it became increasingly difficult for foreign observers to properly evaluate the state of Japan's aviation. The revision of Japan's Law for the Protection of Military Secrets (Gunkihogohō) in August 1937 made it illegal to produce any photographs, sketches, or records about military aircraft, airfields, or aviation-related factories, a crime that was punishable by imprisonment of up to seven years.[63] The tightened secrecy measures also affected the work of the US military attaché. In December 1939 he expressed his discontent that since June 1937, his office had been granted only one inspection, and the July 1940 Aviation Statistics simply noted: "No inspections are permitted."[64] As a result, the assistant military attaché had to rely on press reports, aviation magazines, and discussions with Japanese officers and "reliable and competent foreign observers." The attaché was nonetheless confident in corroborating earlier assessments that Japan's aircraft industry was unable to produce original designs and was still depending on foreign technology and patents.[65]

Most of the annual aviation reports included estimates for Japanese aircraft production. The 1939 assessment called into question any further production increase in Japan's aviation industry. According to the authors, the "present stringent control of imports"—along with Japan's inadequate supply of machines, lack of skilled labor, and dearth of raw materials—made the implementation of around-the-clock production with a three-shift system highly "impracticable." Therefore, Japan's aircraft production was reaching its limit of fewer than 1,900 aircraft a year.[66] The report also pointed at Japan's lack of strategic materials, such as nickel and chrome, which had to be replaced with inferior substitutes, and concluded that the "performance of Japanese aircraft might be expected to fall off in consequence." As it turned out, with actual aircraft production growing by an annual average of 45 percent, the military attaché's production estimates after 1937 were way off the mark (table 7.1).

Even though in 1937 Japanese aircraft factories became off-limits to the US military attaché and his staff, a number of US specialists could gain direct insight into the production capabilities of Japan's aircraft makers. In 1939, two years after the tightened version of the Law for the Protection of Military Secrets had been passed, the Army Aviation Headquarters

Table 7.1. Japan's annual aircraft production: Speculations and facts

	1935	1936	1937	1938	1939	1940
US estimates	735	662	1,015	1,460	1,825	1,730
Actual	476	557	1,150	2,227	3,489	3,365

Notes. US estimates: Annual production statistics submitted by the military attaché to the War Department. *Source:* Lester (ed.), *U.S. Military Intelligence Reports*, reel 30, 497–540; Military Attaché: Production Statistics. Actual numbers: Nihon Kōkū Kyōkai, *Nihon kōkūshi Shōwa zenki hen*, 872.

invited Curtiss-Wright engineers K. E. Sutton and E. B. Parke to Japan. The US experts were to lecture on the mass production of aero-engines and introduce the "aircraft factory for the future."[67] In a nine-day lecture course, Sutton and Parke disclosed details about Wright's organization, factory layout, and quality control. They advised the Japanese about the critical transition from small-scale experimental manufacturing to large-scale mass production.[68]

In return, the US engineers were granted access to Nakajima's and Tachikawa's production plants, a privilege that enabled them to arrive at a fairly accurate estimate of production capabilities. Having witnessed the enormous expansion of Japanese engine factories, Sutton reckoned that Japan would double its output of military aircraft. Their memorandum concluded that the Japanese could produce 3,000 aircraft between March and December 1939 and more than 4,000 airplanes in 1940.[69] Indeed, Sutton's estimates were much more accurate than those of the US military attaché. Even though his report is mentioned in a confidential note to the US chief of staff, it received no attention and was not taken into account in the US military attaché's production estimates.

ADVANCES IN JAPANESE AVIATION ADVERTISED AND IGNORED

During a critical period from 1937 to 1941, stereotypes and overconfidence in US technological superiority seemed to have shaped America's image of the Japanese aviation industry. Most military and civilian observers were convinced that Japan could produce only derivative aircraft in limited numbers.[70] However, in spite of Japan's increasingly tightened military secrecy, several milestones of Japanese aviation should have attracted the attention of US aviation experts and prompted them to reassess their judgment.

In 1937 two highly publicized record flights showcased the rapid development of made-in-Japan aircraft. Mitsubishi's new reconnaissance aircraft, the Ki-15, made its first flight in May 1936. Soon the newspaper *Asahi Shinbun* showed interest in Mitsubishi's aircraft. *Asahi* planned to set up a new flight time world record from Japan to Britain on the occasion of the coronation of King George VI in May 1937. The Army Aviation Headquarters readily offered its cooperation, expecting that *Asahi*—as with its successful 1925 "visit Europe flight"—would again promote the advancement of Japan's aviation. In March 1937, the newspaper bought Mitsubishi's second Ki-15 prototype, a converted version of the reconnaissance aircraft with all armament and cameras removed. The aircraft's name was selected from over 12,000 proposals that readers submitted to *Asahi*: *Kamikaze*, or "divine wind." (Only after Japan's first suicide air attacks in autumn 1944 was this name associated with spiritual fanaticism.) On April 1, 1937, *Asahi* held an elaborate naming ceremony at Haneda Airport with more than 10,000 invited guests. The event featured military music, throngs of singing schoolchildren, formation flights, and the mass release of pigeons and balloons. It attracted attention from all over Japan and was broadcast live by the national broadcasting station.

The *Kamikaze* departed for its record flight from Japan on April 6 and arrived less than four days later at Croydon Airport in south London, where its crew was welcomed with great fanfare. Immediately after the plane's successful landing in faraway London, the *Asahi* head office in Tokyo flushed with excitement: even at just after midnight Japan time, thousands of people surrounded the *Asahi* office, relentlessly shouting "banzai" with paper strips flying all over the place. The aircraft also left its imprint on Japanese culture: the *Kamikaze*'s distinct shape appeared on children's kimonos and on a whole series of new year's cards under the motto "Japan's Great Leap Forward." Japanese composer Ōsawa Hisato (1906–53) even dedicated Piano Concerto No. 3, "Kamikaze" (1938), to the record flight (figure 7.4).

FIGURE 7.4 This photograph, taken in July 1937, shows a large group of schoolchildren gazing with obvious admiration at Asahi's record-breaking aircraft. The airplane's fuselage tells the whole story. From right to left we see the aircraft's name "Kamikaze," followed by the newspaper's logo and "from Tokyo to London in ninety-four hours" in small lettering on the tail fin. (Source © Japan Aeronautic Association)

The World Air Sports Federation (FAI) approved the new flight record of the *Kamikaze* crew, who had covered the 15,360 kilometers between Tokyo and London in an actual flying time of fifty-one hours, seventeen minutes, resulting in an average speed of 300 kilometers per hour (the overall travel time was ninety-four hours). Tanakadate Aikitsu (1856–1952), a physicist who had become one of Japan's most eminent aeronautical scientists, commented: "For a long time the Japanese aviation world had no world flight record, but now finally we can be proud before the world."[71] Mitsubishi's engineers also took pride in their achievement. The *Kamikaze* was the first aircraft entirely designed and built in Japan to fly to Europe. Western observers were impressed—in a somewhat different way. They remarked that the *Kamikaze*'s designers had skillfully incorporated advanced features like flush rivets ("some of the finest we have ever seen"), an engine cowling based on advanced US design, and a made-in-Japan version of the US Hamilton propeller, whose elaborate center part they considered "an excellent example of Oriental attention to detail."[72] Obviously such ambiguous praise focused more on Japanese craft skills and a talent to replicate than on original technological prowess.

In the month following the *Kamikaze*'s record flight, another remarkable Japanese aircraft made its first flight. The futuristic-looking *Kōkenki* (short for *kōkūkenkyūsho shisaku chōkyoriki*, or aeronautical laboratory experimental research airplane) was designed and built to break another world record and secure its place as a symbol of Japan's cutting-edge technology (figure 7.5).[73] In May 1938, the *Kōkenki* became the first airplane to cover a distance of more than 11,000 kilometers, clear proof that such an aircraft's flight range would allow for a nonstop flight from Tokyo to New York. Its record-breaking pilot, Fujita Yūzō (1898–1939), who had skillfully steered the aircraft in a cramped cockpit for sixty-two hours, was declared a national hero.[74] One week later, a US intelligence report duly noted the record. However, rather than evaluating the record flight's connection with the development of Japan's airpower, the author contented himself with describing the aircraft's low fuel consumption and the aerodynamic efficiency of its airframe and propeller.[75]

The following year, the American public experienced Japan's advanced aviation technology firsthand. In August 1939, a converted Mitsubishi G3M long-range bomber took off from Tokyo for a spectacular air journey around the globe. In the words of its sponsor, the *Mainichi Shinbun*, the goodwill flight was to "promote international understanding and amity."[76] But the driving force behind *Mainichi*'s project was the success of its competitor *Asahi*. The president of *Mainichi Shinbun*, Okumura Shintarō, declared that "when the whole of Japan was seething with excitement over the success of [*Asahi*'s] Kamikaze, I was deeply filled with admiration but also with envy."[77]

When *Mainichi* approached the Japanese Navy for a suitable aircraft, the naval general staff initially hesitated to send one of its long-range bombers abroad. In late 1938, when navy officials became confident that a new generation of attack bombers would soon be at hand, they made one of their Mitsubishi G3M attack bombers available to *Mainichi*.[78] The Mitsubishi engineers remodeled the G3M into a long-distance civil aircraft. They removed the heavy armament and installed an additional window for

FIGURE 7.5 High-tech made in Japan: the long-range research airplane *Kōkenki*. The aircraft's thin fuselage and stretched wings with a perfectly smooth surface incorporate maximum aerodynamic efficiency. A snow-covered Mt. Fuji in the background adds to the appeal of this rare photograph from 1938. (Source © Japan Aeronautic Association)

a navigator in the airplane's nose section. Additional fuel tanks increased the aircraft's range to 6,000 kilometers. Advanced navigation devices—including instruments for celestial navigation, a radio direction finder, and a drift indicator—would ensure accurate long-range navigation.

On August 26, 1939, *Mainichi* staged a grand send-off ceremony with 1,000 invited guests and more than 30,000 students and pupils attending. At 10:27 AM the aircraft, whose name, *Nippon* ("Japan"), was selected from more than eleven million entries, took off from Haneda Airport toward Sapporo. From there the crew set out on the journey's most daunting passage: a 4,340-kilometer nonstop flight across the northwestern Pacific (figure 7.6). After a nearly sixteen-hour-long struggle battling dismal weather conditions and a lack of oxygen, the *Nippon* crew landed safely in Nome, Alaska. From there, the Japanese embassy had already procured the US War Department's approval for flying over the United States with stopovers in Seattle; Los Angeles; Chicago; New York; Washington, DC; and Miami. The embassy had also received the US secretary of war's assurance to "extend every practicable assistance."[79] To counter US worries about spy activities, especially taking aerial photographs of the US coastline, the Japanese crew agreed to seal all cameras on arrival in Alaska.

US commentators who saw the *Nippon* expressed their disbelief in Japanese original design. For them the plane seemed "to be a cross between a Douglas DC-2 and a Lockheed Electra," and some surmised that its engine must have been made in Germany.[80] Such disparaging evaluations further evince the occluded view of the fast progress of Japan's aviation technology. On its way to North America, the *Nippon* had already demonstrated that it easily could outrange not only America's most notable airliner, the Douglas DC-3, but also Boeing's latest bomber, the B-17 Flying Fortress (these aircraft had a maximum range of 2,400 and 3,200 kilometers, respectively).

LATE REVELATIONS OF THE JAPANESE FILES RESEARCH PROJECT

Only after the outbreak of the Pacific War did US intelligence began to tap a unique source for crucial firsthand information on the Japanese aeronautical industry. In 1942 a small group of thirty US researchers launched

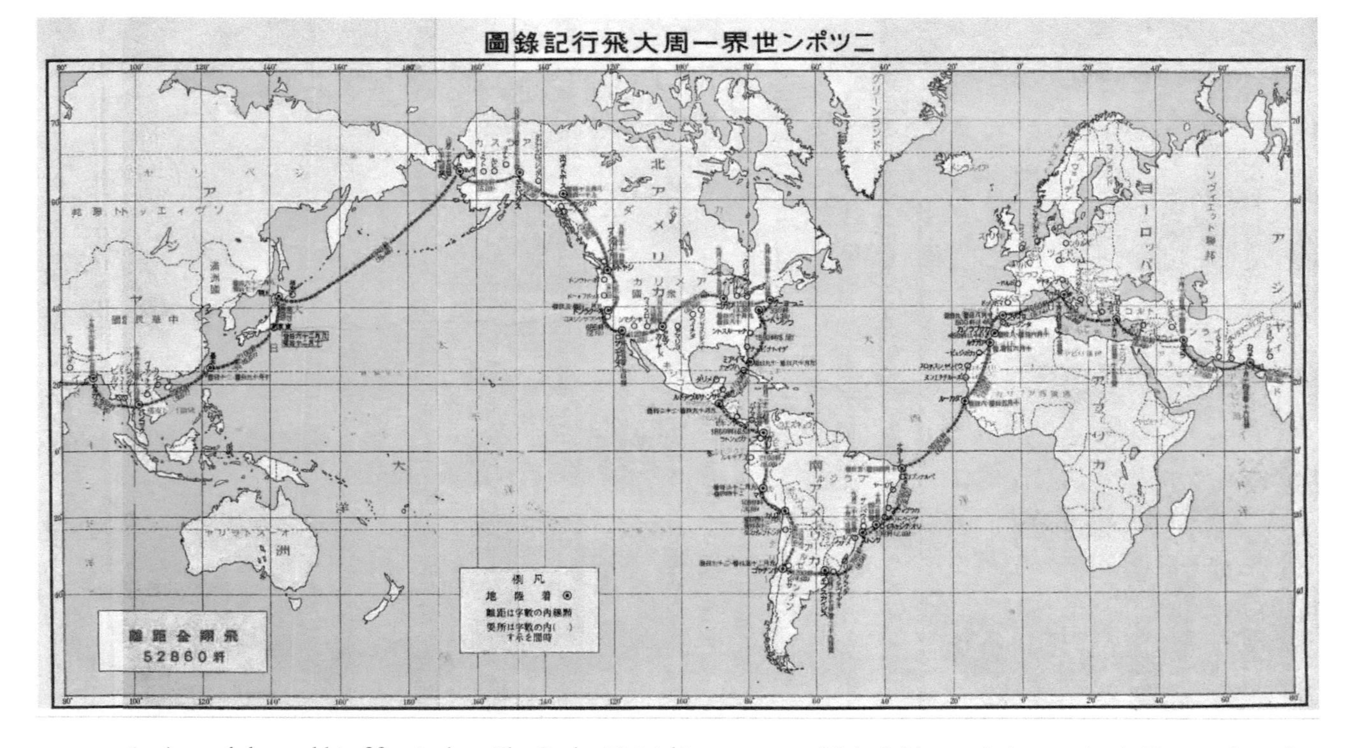

FIGURE 7.6 Around the world in fifty-six days: The Osaka *Mainichi* newspaper published this map in its 1940 book *Nippon sekai isshū daihikō* [The great around-the-world flight of the *Nippon*]. Its thin red line shows the 52,860-kilometer trip of the *Mainichi's* long-range aircraft. (Source: *Nippon sekai isshū daihikō*, 6–7)

an unparalleled investigation under the aegis of the War Division in the Department of Justice. The Japanese Files Research Project aimed to secure economic intelligence based on original files confiscated from the US branches of all major Japanese trading companies (figure 7.7). By tracking the destinations of exported US production machinery and raw materials, the group gathered detailed information about the geographical location of Japanese industrial plants and their production capabilities. This information was important. It was the only reliable guideline for selecting bombing targets before US forces recaptured the Mariana Islands in summer 1944. Only then could reconnaissance flights take off from air bases in Saipan and Tinian to gain more information through aerial photography. When the file researchers presented their 4,500-page report in 1944, they categorized it as the "only primary authoritative source material on Japanese industry in the United States that is available today."[81]

The report devoted large sections to Japan's major aircraft makers, including the army and navy arsenals, Nakajima, Mitsubishi, Kawasaki, and Kawanishi. Most of these accounts followed a step-by-step analysis that began with data on each maker's factory size and production capacity. The researchers used blueprints and order lists to infer the types of engines and airframes produced. A list of acquired US manufacturing licenses, inquiries, orders, and shipments concluded the investigation.

Although the reports served their purpose of retrieving invaluable information about Japan's wartime industry, we can read them as a disclosure of a significant intelligence debacle and a criticism of US interwar trade policy. Based on its prewar sources, the research project revealed, in unprecedented detail, the industrial production capacity, expansion capability, and conversion to wartime production of Japan's industry before the Pearl Harbor attack. The analysis can be also understood as an implicit condemnation of US intelligence that failed to gather and use these data much earlier. The researchers commented on the files of Mitsui Bussan and Mitsubishi Shōji, the two companies that handled about half of all US exports to Japan, that they "reflect a fair cross-section, not only of Japan's demand-requirements for vital war materials, but, also, her long-range planning in securing most of them, and her frantic last-minute efforts to secure some of them."[82] Such a statement suggests taking a much more critical view of the role of prewar US–Japanese trade in Japan's war capabilities. Finally, the researchers' evaluations of the Japanese

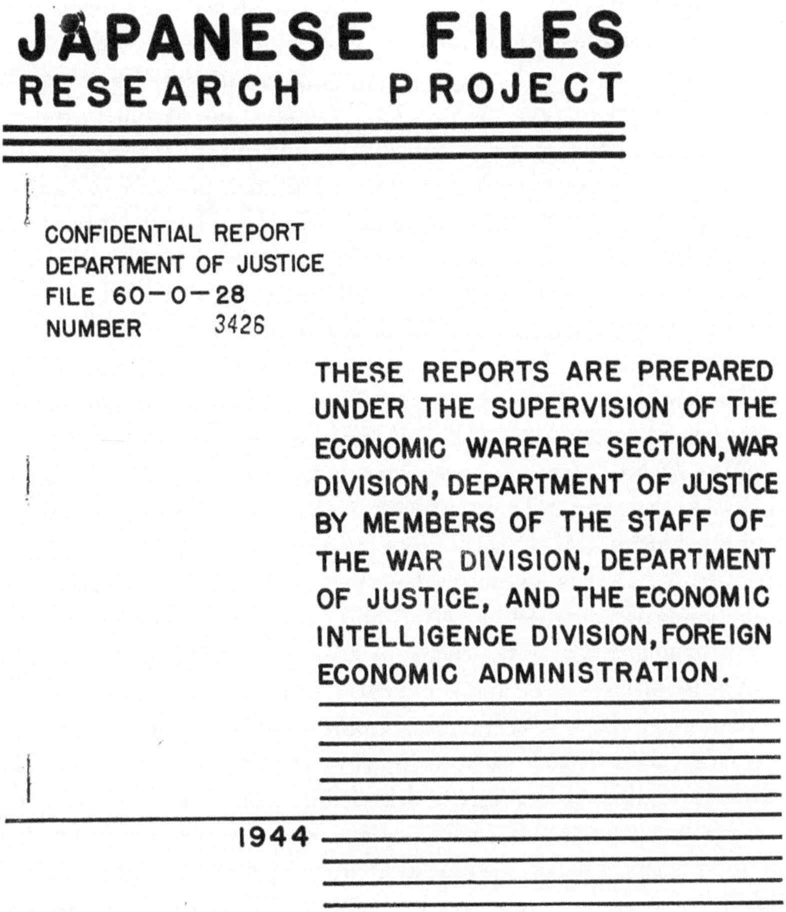

JAPANESE FILES
RESEARCH PROJECT

CONFIDENTIAL REPORT
DEPARTMENT OF JUSTICE
FILE 60-0-28
NUMBER 3426

THESE REPORTS ARE PREPARED UNDER THE SUPERVISION OF THE ECONOMIC WARFARE SECTION, WAR DIVISION, DEPARTMENT OF JUSTICE BY MEMBERS OF THE STAFF OF THE WAR DIVISION, DEPARTMENT OF JUSTICE, AND THE ECONOMIC INTELLIGENCE DIVISION, FOREIGN ECONOMIC ADMINISTRATION.

1944

FIGURE 7.7 An unspectacular cover of a spectacular report: the front page of one of the Japanese Files Research Project reports of 1944.

missions to the United States in 1937 and 1938, such as "Colonel Kanda and his party came close on the heels of the Okada Mission to continue the purchasing and the espionage,"[83] disparaged not only Japanese duplicity but also the policy of US industrialists who sacrificed secrecy for business.

Japanese Perceptions of the US Aviation Industry

Historians have noted that Japanese intelligence was remarkably effective in collecting information on a tactical level but failed to do so for a wider strategic view.[84] Immediately after World War II, US analysts concluded that Japanese short-term estimates of US airpower two months before the Pearl Harbor attack were astonishingly "timely, adequate in coverage, substantially accurate and in considerable detail."[85] Their study also shows that in December 1941 Japanese war planners were in possession of surprisingly accurate estimates of US aircraft production. The figures not only correctly predicted the monthly production to double within the next six months but also projected, with an error of less than 1 percent, a fourfold increase in annual production: a staggering number of 85,900 airplanes.[86]

Japanese newspapers presented their own estimates about the expansion of the US aviation industry. In late October 1941, a few weeks before the Pearl Harbor attack, *Asahi Shinbun* launched a series of articles titled "America on the opposite shore" (*Kaigai no America*) that tried to ease its readers' worries about present American industrial power while warning about future developments. The paper declared, "No matter if we fight with America as an enemy or we keep company with it as a friend, we have to pay attention to America."[87] *Asahi* offered to its readers an "expert analysis" of US armaments and US resources. Referring to Roosevelt's May 1940 announcement to aim for an annual production of 50,000 aircraft, the paper assured its readers that the president's declaration shocked the US aviation industry, which was apparently unable to meet such a high figure. Even when the order was downgraded to 21,000 planes, the US aircraft makers, who were struggling to keep up a monthly production rate of 1,200 aircraft, could not comply.[88] According to one of the articles, the United States would need more time for converting its automobile manufacturers into aircraft makers. Only after mobilizing US resources and production machinery could full-scale mass production start. As one Japanese expert put it: "At this moment we do not have to be scared of America's [economic] power to wage a war." Continuing in a more foreboding tone, he predicted: "But if we give them one more year . . . the US economy will be fully prepared for war."[89]

Even though it is difficult to judge the effects of such a ticking-clock scenario, we now know that Japanese intelligence reports had little influence on the military's long-term strategic planning. Japan's war planners paid little heed to the precise and detailed information available to them. They ignored warnings about the effects of an impending embargo and brushed aside an August 1941 report that stated that US strategic material would outnumber Japan tenfold.[90] Japanese command and staff officers produced their own long-term estimates. However, their forecast of US Air Service expansion was far off the mark, predicting 15,000 US aircraft by the end of 1943, whereas the actual number turned out to be more than 26,000.[91] Once the war began, Japanese strategists also underestimated the US capability to recover from the Pearl Harbor attack. At the same time, they based their decisions on a fantastic overestimation of Japanese fighting capability, which made them confident about achieving air supremacy even at an overwhelming ratio of 1:10 to US airpower.[92]

Conclusion

During the interwar period, the US aircraft industry played an increasingly important role in the advancement of Japan's aviation. Japan resourcefully profited from the advances in US civil aviation. A selective import of dual-use aeronautical matériel allowed the Japanese to further advance the already high standard of their military aviation. The 1937 start of the war with China became a turning point that trapped the Japanese in a dilemma. The further growth of Japan's airpower increasingly depended on the country's access to US production technology. At the same time, the use of airpower over China led to escalating US sanctions that threatened to halt this vital supply. With Japan's heavy dependence on

US production machinery, America's impending machine tool embargo created a running-clock scenario that led to a close cooperation among Japan's military, trading companies, and aircraft makers in procuring US technology. During the short period between 1937 and 1941, Japan deftly combined large-scale purchasing of aeronautical matériel and production technology with widespread espionage and intelligence gathering. These efforts met with little resistance and came to an end only with an effective US embargo in summer 1941.

During the same period US intelligence was clouded by stereotypes of Japanese slavishly copying and continuing their full dependence on Western technology. After the Japanese shut off access to military bases and production plants, the US intelligence community showed little initiative to tap alternative intelligence sources. Even highly publicized record flights did not bring about any reevaluation of Japan's airpower. As a result, US intelligence assessments grossly underestimated the performance of Japan's latest fighters and bombers as well as the dramatic production increase of the aircraft industry before the Pearl Harbor attack.

Japanese intelligence was more efficient in providing accurate estimates of US production capability. Yet Japan's military planners fatefully paid little attention to these reports. It seems that civilian experts were much more aware of the consequences of fighting a war against the industrial strength of the United States. Upon receiving the news about the Pearl Harbor attack, Mitsubishi's star designer, Horikoshi Jirō, expressed his stunned disbelief by comparing Japan to a "country bumpkin challenging a champion sumo wrestler."[93] Upon hearing about Japan going to war with the United States, aircraft manufacturer Nakajima Chikuhei purportedly burst out in an even more desperate mood: "This is hopeless for Japan" (*kore de Nihon wa dame da*).[94]

CHAPTER 8

Jet and Rocket Technology for Japan's Decisive Battle

The start of the Pacific War in December 1941 catapulted Japanese aviation into a new phase. Military officials pressed aviation manufacturers for new aircraft types with higher performance. The more the tide of war turned against Japan, the more this pressure increased. The traditionalists' dogma of a superior Japanese fighting spirit found its counterpart in technocrats' belief in a superior technology that would compensate for Japan's limited resources and industrial capacity. Such technological fanaticism accounts for the breathtaking pace of Japanese aeronautical development—a process that accelerated right up to the last days before Japan's surrender in August 1945. This chapter looks at two revolutionary propulsion technologies that epitomize this quest for radically new designs. Jet and rocket engines ignited the imagination of Japan's militaries to a point that made them gather the elite of the country's aeronautical engineers and send these specialists into the fray in a race against the clock.

By the late 1930s, piston engine technology was approaching its limits. To meet the demand for ever greater engine power, engineers designed piston motors that were increasingly complex and heavy. As an escape from this predicament, turbojets and rocket motors offered an alternative means of propulsion and began attracting attention from aeronautical engineers in both Japan and the West. The potential of these engines was enormous, and once success could be achieved, aviation would be revolutionized. To master these technologies, researchers had to overcome entirely new challenges. High-speed aerodynamics, materials science, and applied chemistry became the key disciplines. Aeronautical engineers had to design airframes that could handle the enormous thrust and acceleration. Scientists started searching for alloys that were able to withstand temperatures exceeding 1,000°C. Chemists began developing high-energy propellants that could react powerfully enough to accelerate a rocket-driven aircraft to its designed speed.

From a technical point of view, turbojets and rockets offer considerable advantages. The power-to-weight ratio of these propulsion systems surpasses that of piston engines. They also show superior performance at high speeds and high altitudes. Turbojets consist of a compressor that forces air into a combustion chamber, where it burns with fuel to produce jet gases. These gases operate a turbine, which, in turn, drives the compressor. A rocket motor is typically fed by two chemical components that react in a combustion chamber, resulting in a hot gas stream leaving the nozzle. Because of their relatively simple structure and high exhaust velocity, rocket engines are lightweight and powerful.

For the Japanese military, these advanced technologies offered an escape from the country's most pressing predicaments. By mid-1944, the need for a fast and high-flying rocket interceptor to protect the homeland became disturbingly clear. Japanese military leaders, who had already received news about the disastrous air raids on Germany, anticipated a similar fate for Japan. Their fears materialized in June 1944 when the first US bombers took off from Chinese air bases to attack Japanese cities. The situation was further aggravated with the loss of Saipan in early July 1944, bringing Tokyo within flight range of large formations of Boeing B-29 bombers. A race against time began. The military faced a stark choice of rapidly building up a fleet of high-performance interceptors or risking the annihilation of Japan's aircraft industry.

Jet engine technology also promised a means of coping with the increasingly critical situation of Japan's wartime economy. Intensifying Allied attacks on supply lines led to the collapse of Japan's sea transport, resulting in a severe shortage of aviation fuel.[1] Furthermore, since early 1944, the production and reliability of Japanese piston aero-engines had rapidly declined. On the home front, aggressively orchestrated campaigns to save material and accelerate production resulted in a chaos in engine production. On the battlefield, pilots had to cope with reduced maintenance standards and inferior fuel, which led to frequent engine breakdowns. Jet engines offered an innovative way to compensate for these shortcomings. Compared with a traditional piston engine, a jet engine's design was relatively straightforward and held the prospect of easy mass production. Furthermore, because jet engines do not require high-grade fuel, they would not be afflicted by the acute shortage of high-octane aviation fuel.

This chapter once more challenges the view of Japanese engineers as unimaginative imitators. Western-language accounts often treat Japan's first rocket-powered interceptor and jet attack bomber as mere copies of their German counterparts.[2] However, as we will see, it was precisely the scarcity of German matériel that spurred the creativity of the Japanese engineers to come up with original design solutions. In addition to unprecedented technological challenges, these engineers successfully coped with grueling time pressure, dwindling resources, and the constant threat of all-out air raids. Literally a few weeks before Japan's final surrender, they presented two working prototypes that catapulted their nation into the rocket and jet age.

Early Japanese Experiments

The development of rockets in Japan had a slow start and was not well coordinated. In 1935 the army started experimenting with liquid oxygen and alcohol as a propellant, launching its first liquid-fuel rocket in 1937.[3] Just two years later the project was shelved. Only in 1942 did Mitsubishi's ordnance factory at Nagasaki resume research on rocket propulsion and begin working on a rocket-driven torpedo. After a series of experiments, the company's engineers arrived at the conclusion that a combination of hydrogen peroxide (H_2O_2) and hydrazine (N_2H_4) was the ideal rocket propellant. As it turned out, this was the same mixture German engineers were using for their rocket-driven Messerschmitt Me 163.[4]

Japanese jet engine research was a joint effort that included the Naval Air Arsenal, university research laboratories, and civilian companies. In fall 1938, Captain Tanegashima Tokiyasu (1902–87) returned from a study trip to Europe and brought several French and Swiss turbosuperchargers to the Naval Air Arsenal. Even though these devices were designed to boost the power of piston engines, Tanegashima envisioned a new engine type based on the turbo-supercharger principle. He convinced the Naval Air Arsenal at Yokosuka to cooperate with civilian steel manufacturers Sumitomo, Daidō, Nippon Steel, and Hitachi in developing heat-resistant alloys. Furthermore, he procured the collaboration of the Imperial Universities of Nagoya and Tōhoku at Sendai.[5] Tanegashima also enlisted the cooperation of three leading experts, Hibi Yoshitarō from Mitsubishi Nagoya and two lieutenants, Katō Shigeo and Tamaru Nario.[6] The group began designing jet engine components and carrying out experiments with high-speed bearings and heat-resistant materials.

Among this group of four, Katō made the decisive step that brought the navy's arsenal very close to developing an operational jet engine.[7] In 1942, he designed a high-performance gas turbine that was to increase the power of a piston engine. The device was driven by the exhaust gas of the piston engine for which it provided compressed air, allowing the engine to operate at high altitudes. However, the turbine performed poorly. In a stroke of genius, Katō decided to radically change the arrangement (figure 8.1). He replaced the piston engine with combustion chambers that would drive the turbine. The turbine was connected to a shaft that turned the compressor. The compressor, in turn, would feed high-pressure air directly

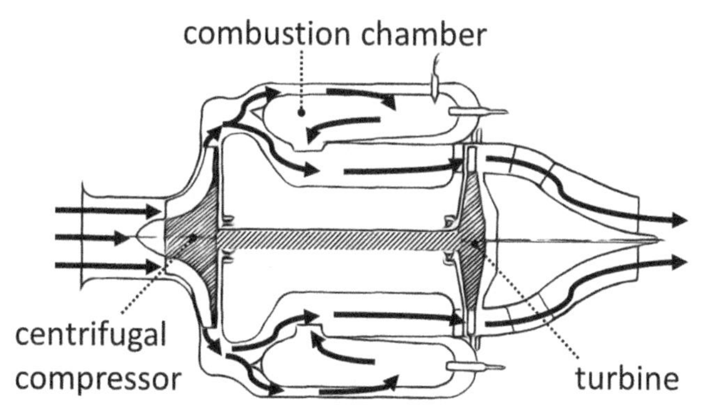

FIGURE 8.1 A functional diagram of the TR-10, Japan's first turbojet engine. Note the strong deflection of the engine's internal airflow on its way from the engine inlet to the combustion chambers. (Source: Hayashi, "Aru enjin sekkeigijutsusha," 82, modified by the author)

into the combustion chambers. Katō started a series of experiments with his new TR 10 jet engine in June 1943.

Even though Katō could prove the viability of the new design, he had to cope with frequent breakdowns of turbine blades, bearings, and combustion chambers.[8] Yet he and the rest of Tanegashima's team continued tackling fundamental problems like gas turbulence in the combustion chamber, distortion of the fuel nozzle because of thermal strain, and overheating of the turbine bearings. Gradual development led to the Ne-10 engine and its improved version, the Ne-12.[9] Engine reliability was still an issue. Cracking turbine blades posed an especially obnoxious problem that occurred, with persistent regularity, within less than thirty minutes of a full-power test run.

German Technology to Japan

Around the same time the Naval Air Arsenal began jet engine research, two German aircraft makers pioneered the development of rocket- and jet-driven aircraft.[10] On June 20, 1939, Ernst Heinkel's He 176, the world's first aircraft powered by a liquid-fueled rocket, made its maiden flight. On August 27, 1939, the first jet-driven airplane, the Heinkel He 178, successfully passed its test flight, proving the feasibility of the new propulsion technology. In the same year, Willy Messerschmitt (1898–1978) began developing a fast-climbing rocket interceptor, the Me 163. Two years later, on October 2, 1941, test pilot Heini Dittmar (1911–60) took off with his Me 163 for a record flight that reached a speed of more than 1,000 kilometers per hour, nearly 250 kilometers per hour faster than the world's fastest piston engine aircraft.[11]

News about these revolutionary new technologies soon arrived in Japan. In spring 1941, after four years of service at the Berlin embassy, Commander Kumazawa Toshikazu returned to his homeland and informed the engineers at the Naval Air Technical Arsenal about German research on gas turbines and the development of a radically new type of aircraft engine.[12] But before such technological marvels could find their way to Japan, several hurdles had to be cleared.

Geographical distance, of course, had always been an obstacle for technology transfer. Through the 1930s, German aeronautical matériel could be sent via Siberia. After the German attack on the Soviet Union in June 1941, overland transport was no longer an option. The Axis Powers soon abandoned their initial plans to establish a regular long-range flight connection between Germany and Japan, leaving blockade runners as the only possibility for the transport of personnel, documents, and matériel.[13]

German efforts to shroud their aviation technology in secrecy posed yet another problem. In 1943 Foreign Minister Joachim von Ribbentrop (1893–1946) and Japanese ambassador to Germany Ōshima Hiroshi (1886–1975) signed an official Agreement on Economic Cooperation.[14] Nevertheless, the Germans were unwilling to inform their ally about their latest weapon development, and within a month after concluding the treaty, Hitler prohibited the export of advanced aviation technology to Japan.[15] It took until March 2, 1944, when the German Foreign Ministry's economic division and the economic attaché at the Japanese embassy, Matsushima Shikao (1888–1968), signed another contract that granted Japan free manufacturing rights for military equipment in exchange for delivering raw material to Germany.[16]

The increasingly effective US blockade against Japan substantially changed the framework of German exports to Japan. After Allied submarines and aircraft began to intercept or destroy surface blockade runners, submarines played a central role as means of transport between Germany and Japan. But with their limited freight capacity, these vessels were ill-suited for the transport of bulky war matériel and machinery. Instead, German know-how in the form of blueprints, descriptions of production methods, and small prototypes became the preferred cargo.

Under these new circumstances, the Japanese, still eager to learn more about Germany's secret rocket and jet fighter projects, submitted an official request for further information on these "miracle weapons" (*Wunderwaffen*) in early 1944. After rushed negotiations among Field Marshal Erhard Milch (1892–1972), Reichsminister of Aviation Hermann Göring, and naval attaché Abe Katsuo (1891–1948), the Germans finally agreed to release basic outlines, drawings, and descriptions of two new aircraft types.[17]

Naval engineer Iwaya Eiichi (1903–59) became the central figure in transferring German rocket and jet technology to Japan. Iwaya arrived in Germany via submarine in late 1943. The Ministry of Aviation granted him access to blueprints for the rocket interceptor Me 163 on March 27, 1944.[18] The next day, Iwaya visited BMW's Berlin factory, where engineers showed him the operation of a BMW 003 jet engine and informed him about the progress of its series production.[19] Iwaya had the chance to see the new planes himself. On April 6, 1944, he arrived at the Messerschmitt plant at Augsburg. He took a close look at the Me 163 rocket aircraft, which impressed him with its exotic tailless design. He also witnessed a test run of the interceptor's rocket engine, which impressed him with its deafening sound and powerful shockwaves.

It seems that Iwaya was even more amazed by a flight demonstration of the jet fighter Me 262:

> After almost gliding along the runway this aircraft that had no propeller took off. Soon it approached us as fast as an arrow for a low-level flyby. The aircraft's speed surprised me: Its engine sound followed only after the aircraft had already passed by. "Excellent" is the only [appropriate] word. Probably I was the first Japanese to see the flight of a jet aircraft. . . . I could not help but show my respect to the inventiveness of German technology.[20]

On April 13, the German Ministry of Aviation handed several coveted plans and blueprints to Iwaya. Their disorderly state gave the impression that they had been prepared in a hurry. Iwaya traveled to the German submarine base at Lorient, France, and from there, he left for Japan. It took the Japanese I-29 submarine three months to complete its 28,000-kilometer journey. When the vessel arrived in Japanese-occupied Singapore, Iwaya decided to shorten his trip by air travel, carrying the German documents with him. As it turned out, he was lucky with his choice of transport. During its underwater voyage from Singapore to Japan, the I-29 was attacked and sunk by a US submarine, with all its cargo lost and only one crewmember surviving. Spared the grim fate of his fellow travelers, Iwaya arrived in Tokyo on July 19 and presented his documents to excited navy officials.

Japan's First Rocket Aircraft

Obviously, a spirit of nearly desperate urgency was the driving force behind Japan's rocket fighter project. When Iwaya arrived at the navy's Aviation Bureau, the first question was, "Did you bring material about the Me 163?" Only then did it become clear to him how impatiently the navy had been waiting for these documents.[21] On July 20, 1944, just a day after Iwaya's arrival, the navy decided to build a new rocket interceptor with the name Shūsui ("polished sword"), with a design based on the Me 163.[22] To expedite the project, army and navy officials even settled their long-standing rivalry and agreed on a division of labor. The navy's Air Arsenal would oversee the development of the Shūsui's airframe, and the army would be responsible for designing the rocket engine. Yet as we will see, what might have become a historic moment of interservice cooperation soon came to an end with the navy monopolizing the whole project.

Rapid progress, it seems, was of ultimate importance. Under normal circumstances, it would have taken three years from the initial design phase to the final acceptance flight. The navy's tight schedule allotted only one year for the concurrent development of the Shūsui's airframe and engine and for setting up rocket fuel production facilities. Not everyone agreed with such a hurried method. Lieutenant Nakaguchi Hiroshi, who was in charge of the airframe's design, argued that the simultaneous experimental production of the airframe and an as-yet-unknown engine was an extremely unorthodox and risky attempt that could result in a waste of material and efforts. The navy turned down his suggestion to start experimental production of the airframe only after the motor had taken shape.[23]

THE DESIGN AND PRODUCTION OF AN EXOTIC AIRFRAME

Mitsubishi Aircraft became a principal if reluctant partner in the rocket project. In early August 1944, the Japanese Navy ordered the company to start preparations for the experimental production of the Shūsui airframe. Mitsubishi initially declined, arguing that its engineers had no experience in designing such an exotic tailless aircraft. However, the navy insisted and promised technical support through the Yokosuka Naval Air Technical Arsenal. As it turned out, Mitsubishi's misgivings were not unfounded. At an early stage, the company's airframe designers had to cope with several problems. They were only offered a twenty-page leaflet with a vague design description of the German Me 163's fuselage and coordinates of the wing cross-sections.[24] They could also refer to the Me163's operation manual, which, according to one engineer, was "very basic, more like a collection of photographs." To the engineers' further bewilderment, the navy did not provide any performance data; the new aircraft just had to be "like the Me 163."[25]

In a meeting on August 10, 1944, the Mitsubishi team set up performance parameters that closely followed the specifications of the German model: the aircraft should be able to climb in an amazingly short time of only three and a half minutes to an altitude of 12,000 meters, where it was to reach a speed of 900 kilometers per hour in level flight. The airframe was to be larger than the German version to provide enough space for a different fuel tank arrangement and a modified hydraulic system.

The Japanese engineers had to find a way to make up for the incomplete information in the German documents. They decided to carry out preliminary aerodynamic studies with unpowered gliders with the same shape as the future Shūsui. A series of test flights would examine the stability and maneuverability of the airframe with a focus on the longitudinal balance, a known weakness of tailless aircraft.[26] These gliders were to be flown by a group of sixteen handpicked fighter pilots that would belong to the future Shūsui flying corps.[27] To make them familiar with the extraordinary operation of an aircraft that could climb to an altitude of twelve kilometers within a few minutes, these airmen had to undergo altitude training in a low-pressure tank. Pilots had to get used to wearing a rubber flight suit, which would protect them from hazardous rocket fuel. This outfit, along with high-altitude masks and oxygen hoses, made them look more like space travelers than World War II fighter pilots.[28]

Experiments with the unpowered Shūsui gliders started at Hyakurigahara, a small naval air base about ninety kilometers northeast of Tokyo. On December 26, 1944, a propeller aircraft towed the first glider with Lieutenant Inuzuka Toyohiko at its controls to its release altitude. Inuzuka safely returned with the glider to the airfield, proving the viability of the airframe design. Less than two weeks later, a "heavy glider" with added ballast to compensate for the missing engine and armament successfully underwent the same procedure.[29] To the engineers' great satisfaction, both glider types showed stable flight characteristics and even proved to be much more maneuverable than a conventional aircraft.[30]

The construction of the rocket aircraft's airframe made fast progress. By early December, Mitsubishi had completed its first Shūsui airframe. Before any tests could be carried out, unforeseen disaster struck twice. In the early hours of December 7, a major earthquake hit the Nagoya metropolitan area, causing more than 1,000 casualties. The Mitsubishi Ōe Factory, where the Shūsui had been built, suffered major damage, bringing production to a halt. On December 18, the US Army Air Forces launched the first of more than thirty air raids on the Mitsubishi airframe works. A large formation of Boeing B-29 Superfortresses, the very aircraft the Shūsui was designed to intercept, destroyed the few facilities that survived the earthquake. In great haste, the Mitsubishi workers transferred the Shūsui prototype from Nagoya to the navy arsenal at Yokosuka.[31] Continuing air attacks rendered the Ōe factory completely unusable, and Mitsubishi decided to disperse its production. The Shūsui's design group relocated to Matsumoto in Nagano prefecture; construction was transferred to Komatsu in Ishikawa prefecture and Fujinomiya in Shizuoka prefecture.[32]

Under the double pressure of increasingly disastrous air raids and a dispersal program that endangered the whole project, the navy decided to start series production of the airframe already in early January 1945. This was a bold decision, because the tests of the unpowered Shūsui were still on their way, and no working prototype of the rocket motor existed yet. Mitsubishi and three more aircraft makers—Nippon Hikōki, Fuji Hikōki, and Nissan—received an order for a grueling production plan that aimed at over 1,200 Shūsui by September 1945. Clearly ignoring the limited production capacity of these manufacturers, the navy decreed an overall number of 3,600 rocket interceptors to be delivered by March 1946.[33]

The navy's extreme demands might have been an attempt to place the blame for the military's debacle on civil industry.[34] They certainly put a heavy burden on the aircraft manufacturers. Consider Nippon Hikōki, which mainly produced training aircraft for the navy. As soon as the company received the necessary construction documents from Mitsubishi, it converted its production to airframes for the Shūsui. The government-imposed plan aimed for Nippon Hikōki to produce 235 aircraft by September 1945. However, due to material shortage, lack of engines for test flights, and confusion from the dispersal of facilities, Nippon Hikōki produced only one single airframe by August 1945.[35] Fuji Hikōki and Nissan fared even worse: these companies were still preparing for production when the war ended.

THE "SPECIAL ROCKET ENGINE"

While designing the tailless Shūsui airframe already tested the limits of Japanese engineering, developing the rocket engine was an even more daunting technological challenge. As mentioned already, the military had high hopes for an aero-engine driven by chemical rocket fuel and thus not dependent on the dwindling supply of aviation fuel. Furthermore, they envisioned a small but powerful engine of barely 200 kilograms that, compared with conventional aircraft engines, would require much less material and man hours for production. Ignoring the previous agreement to entrust the army with the design of the rocket motor, the navy put the Naval Air Technical Arsenal along with Mitsubishi in charge of building the Toku-Ro (*tokubetsu rokketo*), the "special rocket engine."

After initial study of the materials Iwaya brought from Germany, the engineers of the Naval Air Technical Arsenal were in high spirits. They were convinced that a rocket-powered interceptor could be a spectacularly powerful weapon. They were especially impressed with the ingenious structure of the engine's combustion chamber and jet pipe and were inspired by the idea to use a high-pressure turbine pump instead of compressed air for delivering propellants into the burner.[36]

Mitsubishi's engineers took a more dispassionate view. They were aware of the radical departure from a previous design doctrine that emphasized engines for long-range operation. Now they had to design and build a new engine that provided thrust for a seven-minute high-speed flight. On July 29, 1944, the chief engineer of Mitsubishi's Nagoya engine plant, Narita Toyoji, ordered a stop to the development of a new high-power diesel engine that was intended to be used for a long-range attack aircraft. He assigned Mochida Yūkichi as the head of a team that was to immediately start designing a rocket engine for an interceptor. Within two days, the diesel engine group changed to rocket technology. Mochida was well aware of the technological challenge and the enormous time pressure: "The military gave us only six months to complete a liquid-fuel rocket monster, which none of us had ever seen or heard of before."[37]

As if the situation was not bad enough, the engineers had to carry out their exacting project under appalling working conditions. Their daily life was beset by food shortages, a constant fear of exploding rocket fuel, and intensifying air raids. Oguri Masaya, a research engineer at the Aeronautical Research Institute at Tokyo Imperial University, joined Mitsubishi in November 1944. He gives us an idea of what it was like to be a Japanese wartime rocket engineer:

> To fight our hunger, we went to the beach, where we gathered and ate some black shells, the name of which we did not even know. I thought that we were developing the one and only aircraft that could intercept the B-29 bombers. But the only kind of support we received were just some empty words of favor. Nevertheless, we continued our experiments amidst air attacks and frequent accidents from explosions. The [rocket fuel] hydrogen peroxide completely destroys everything. Even a tiny quantity will burn a hole into your clothes, eat into your shoes, and destroy your skin.[38]

Oguri also noticed the strange contrast between the rocket scientists and their helpers, who were wearing traditional craftsman clothes. Most of the assistants were elderly drafted workers who had worked as carpenters or wooden clog makers. During the test runs, only a concrete wall separated them from the running engine. Especially during the early experiments, the combustion chambers often exploded. We know of at least one incident when an explosion killed a staff member; during another experiment, many were injured.[39]

Detonations inside the factory were not the only danger. On December 13, 1944, Mitsubishi's Daikō engine plant became the target of a full-scale air raid, with seventy-five B-29 bombers dropping their deadly payload on the factory.[40] The rocket laboratory detonated, inflicting heavy casualties. The navy decided to carry out all further research activities at its Oppama facility near Yokosuka. On December 18, Mitsubishi's rocket team departed for their new workplace with eight trucks carrying the experimental rocket engine, laboratory equipment, and drawings.

In addition to these extreme conditions, the Japanese engineers had to overcome major technological obstacles. Again, they could only refer to the incomplete information Iwaya provided. The small booklet gave only a rough outline of the German Walter-type rocket engine. Its diagrams contained no details or numbers; neither did they explain the structure and function of the engine's components. The drawing for the turbine drive, the central part of the engine, was reduced to a size of one fifth of a postcard, with the crucial pump unit missing. In the words of Mitsubishi engineer Makino Ikuo: "These were not design data, just design hints" (*sekkei no hinto*).[41]

Consider the rocket engine's pump unit. The task of this small, crucial engine component is to constantly feed the combustion chamber with the precise amount of fuel necessary. However, there was no Japanese or foreign textbook or any special literature available that could provide a description of such a pump.[42] Professor Kasai Taijirō, an expert on fluid dynamics from Kyūshū Imperial University, provided crucial guidance for a series of experiments. The engineers solved a number of persistent problems, such as the reflux of the rocket fuel and bubbles forming inside the pump. In March 1945, they presented a pump and an improved quantity and pressure controller that met the performance requirements.

Among all engine components, the combustion chamber posed the most daunting challenge. A high combustion temperature was essential for the rocket motor's efficiency. The engineers determined that the combustion chamber had to withstand temperatures of up to 1,800°C. Successive refinement of the chamber's casing, pipes, and nozzles led to considerable improvements.[43]

In January 1945, engine testing started. As a remarkable testimony to the Toku-Ro team's perseverance, skill, and ingenuity, the rocket engine successfully passed the final extensive test run by the end of March. The engineers were satisfied that their machine delivered a maximum thrust of 1,500 kilograms, showing exactly the same performance as the German Walter rocket engine.[44]

FUEL FOR THE ROCKET

While the design of the new aircraft was under way, the problem of how to secure a sufficient supply of fuel still had to be solved. During its short flight time, the rocket interceptor was to burn about two tons of propellant. For any meaningful deployment of the fighter, the large-scale production of rocket fuel had to be secured. Japanese industry had already mastered producing limited amounts of hydrogen peroxide and hydrazine. Nevertheless, the Japanese once more turned to Germany for detailed descriptions for industrial-scale production of the rocket chemicals.[45] However, the German submarine U-864, which was to bring the crucial documents to Japan, was sunk by a British torpedo on December 5, 1944. Now all hope rested on the navy's Munitions Bureau, which was to launch a national effort for the large-scale production of rocket fuel.

The Japanese planners did not shy away from unorthodox methods to boost the output of rocket fuel. Platinum electrodes played a crucial role in hydrogen peroxide production. Soon a nationwide campaign for the enforced purchase of platinum from civilians began. In November 1944, the Ministry of Munitions launched "platinum alerts" in the nation's newspapers, threatening those who failed to deliver their platinum jewelry with imprisonment of up to ten years.[46] Japanese traditional craftsmen were also enrolled in the high-tech project. Two of the "Six Ancient Kilns" of Japan received orders to produce special pipes and containers made from white porcelain. These corrosion-resistant utensils were necessary for safely conveying and storing the highly reactive hydrogen peroxide.

Among all these efforts, Japan's chemical industry became the key factor. Edogawa Industries, Japan's largest hydrogen peroxide producer, had already increased its monthly output to 120 tons of hydrogen peroxide in 1944. The Munitions Bureau aimed at a monthly production of 3,000 tons of high-concentrate hydrogen peroxide. It therefore set up new plants at the First Naval Fuel Depot near Yokosuka and at the Chōsen Chisso Company in Korea.[47] The large-scale production of hydrazine, the second rocket fuel component, was an even bigger challenge. By 1944 the highly explosive and extremely corrosive chemical could be synthesized only in a few laboratories. Responding to the military's demand, a wide range of chemical factories—among them Nippon Synthetic Chemical Industry, Mitsui Dyestuff, Chōsen Chisso, Tōagōsei, and Dai Nippon Chemical—successfully started the industrial production of hydrazine.

One More Miracle Weapon: Jet Airplanes

Within less than a year after Iwaya had brought his documents from Germany, the Japanese experts mastered the production of rocket airframes, engine, and chemicals. During the same time, Japan's jet engine program advanced rapidly. Before Iwaya's return to Japan in July 1944, the Naval Air Technical Arsenal had built up considerable expertise in jet engine design, but it needed a new idea to break an apparent technological deadlock. But it seemed that Iwaya could not offer much to the engineers. All he had was one blueprint that showed a longitudinal cross-section of a BMW 003 jet engine on a scale of 1:20 (figure 8.2).

To Iwaya's astonishment, this single piece of paper sparked a wave of euphoria among the Japanese engineers. Nagano Osamu, an engine designer at the Naval Air Technical Arsenal, remembered: "Even though this diagram did not reveal any design or production details, we welcomed it with wild joy and felt as if a light suddenly illuminated us on a moonless night."[48] Tanegashima, the head of the arsenal's jet engine team, was even more exuberant. His reaction placed him somewhere between a spiritually enlightened monk and a no-nonsense technician:

> This one single picture that Commander Iwaya showed me was enough. When I saw it, in a split second I understood it all. The principle was completely the same as ours. But instead of a centrifugal compressor, an axial-flow compressor was used. Moreover, the speed of rotation was low, and the turbine was designed with [masterful] ease. The combustion chamber allowed an unrestrained air flow. By just looking at it I thought: Well done![49]

With the new material in their hands, the army and navy officials agreed on two important decisions. First, they ordered a stop to the development of any other jet propulsion engine. Second, the navy—in collaboration with Ishikawajima Heavy Industries—would be in charge of developing a new turbine jet engine.[50]

Recent developments at the war front gave further momentum to the jet engine project. In October 1944, after the disastrous loss of nearly thirty warships in the Battle of Leyte Gulf, the Japanese Navy felt driven

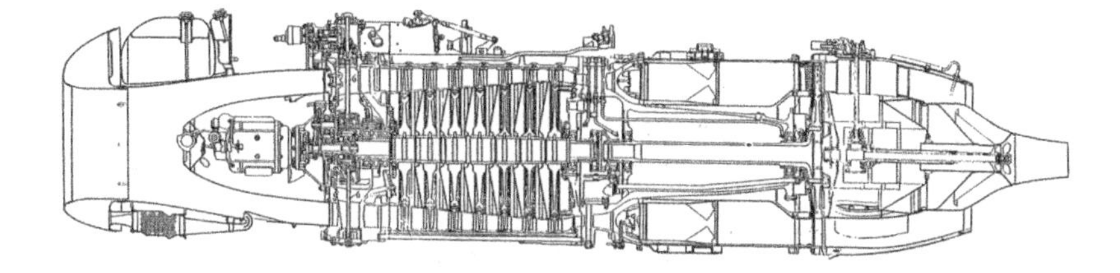

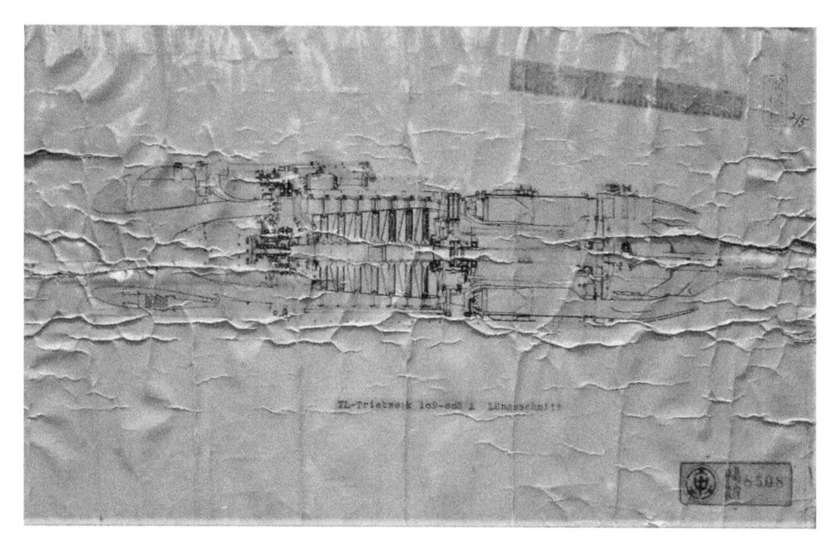

FIGURE 8.2A & B This technical drawing of the BMW 003 jet engine made Japanese aviation history. Below is the only still extant 1944 copy. The imprint *Geheime Kommandosache*, barely visible in the upper right corner, classifies the document as top secret. Note the absence of any measurements and descriptions. (Sources: BMW Group Archiv, Kanazawa Institute of Technology, Library Center)

into a corner. Naval officials decided that rather than rebuilding the fleet, all efforts should be directed toward strengthening airpower. They ordered civilian aircraft manufacturers to maximize their output while the naval arsenals were to focus their efforts on mass producing the new jet engine.[51]

Design and construction of the new jet engine prototype, named Ne-20, made fast progress. Japanese engineers solved a considerable number of problems to which the German blueprint offered no answers. They found suitable high-grade steel that was not only heat resistant but easily workable to allow mass production. They designed special ball bearings

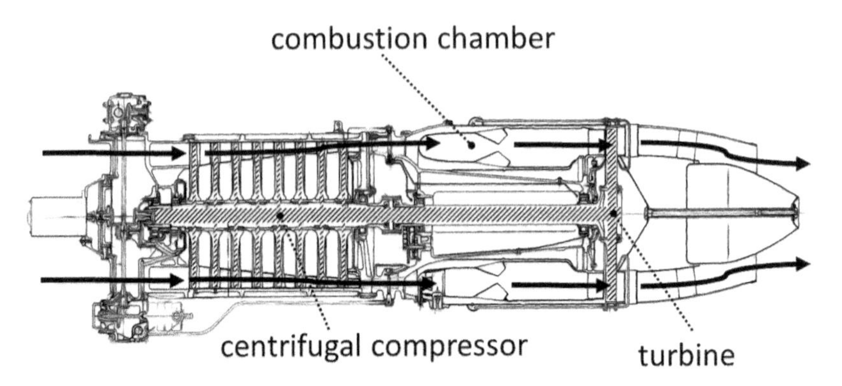

FIGURE 8.3 This functional diagram shows the Ne-20, the engine that powered Japan's first jet fighter. Note the relatively small engine diameter and smooth internal airflow compared with the TR-10 engine (fig. 8.1). (Source: Hayashi, "Aru enjin sekkeigijutsusha," 82, modified by the author)

for turbine discs that could withstand a load of two tons while rotating at a speed of 11,000 revolutions per minute. Engine specialists worked out the optimal shape of the engine's combustion chambers and the most efficient method of fuel injection (figure 8.3).[52] In February 1945 the design of the Ne-20 was completed, and within a month the first engine prototype was built.[53] Engineers started test runs during which time pressure, the unpredictability of a new technology, and the constant threat of air raids taxed their technical and psychological capacities. The first full-power test run was disappointing. The axial-flow compressor did not even reach two-thirds of the expected pressure.

Ignoring this shortcoming, the increasingly impatient navy officials decided to start series production for the jet engine even though it was still in an experimental stage. It is easy to imagine that such an unusual decision was a production engineer's nightmare. It meant that new test results had to be incorporated into an ongoing production process. For instance, when experiments showed that it was necessary to change the angles of the compressor blades, the navy immediately ordered Ishikawajima and the Naval Arsenal to incorporate this improved design into their production lines.

In the meantime, the Ne-20 design team kept improving the engine's performance and reliability. The engineers solved problems with backfiring, turbine blades cracking, and damage to bearings. They smoothed the internal airflow, changed the shape and size of the blades, and further improved the bearings' design.[54] The last of a total of six Ne-20 prototypes successfully passed its five-and-a-half-hour test run on June 22, 1945.[55] The first in-flight test took place five days later with the Ne-20 installed as an additional engine on a torpedo bomber. The engineers of the Naval Air Technical Arsenal could proudly announce that their jet turbine was ready to power Japan's first jet fighter.

BUILDING THE KIKKA AIRFRAME

In one of aviation history's ironies, Japan's most technologically advanced engine was to power an aircraft that had to be produced as cheaply as possible. In September 1944, the navy set up specifications for Japan's first jet fighter. The emphasis was on low-cost mass production, and the manufacturing process was to be cut down to a third of what was required for a standard Zero fighter. In November 1944, Nakajima received the navy's order for trial production of the new aircraft, named Kikka, or "orange blossom."[56] To meet the navy's requests, Nakajima designers simplified the plane's body structure as much as possible. For the Kikka's front and main wheels, windscreen, and flap actuators, they chose components that were already in production for other aircraft types. To further reduce production costs, wherever possible the engineers substituted aluminum alloy with steel, tinplate sheets, and wood.[57] Even though the airframe design followed the general structure of the German Me 262 jet fighter, the Japanese engineers made several innovative modifications for the Kikka. Folding wings allowed the aircraft to be stowed in a tunnel-type air-raid shelter. Two auxiliary rockets, one under each wing, provided additional thrust during the takeoff run to allow the aircraft to be deployed from short airfields.

The navy's ambitious plan for the Kikka's mass production reveals the high hopes invested in the new weapon. Until the end of 1945, Nakajima and Kyūshū Hikōki, together with naval arsenals Sasebo and Kūgisho, were to produce 590 jet fighters.[58] However, prototype construction met with considerable difficulties. In February and March 1945, heavy air raids on Nakajima's Ōta plant forced the company to disperse its production. The Kikka design team evacuated to Sano, a small city twenty kilometers west of Ōta, and continued its work in the city's school and nearby factory buildings. The assembly of the prototype took place in Serada, a nearby silkworm-raising village. On May 3, 1945, the fuselage of the aircraft that was to herald Japan's jet age was completed in a silkworm nursery hut.[59] Lieutenant Commander Kofukuda Mitsugi, in charge of the final airframe inspection, remembered: "I carried out the [Kikka's] examination in a farmer's shabby silkworm shack. To me this scene seemed to symbolize Japan's decline."[60]

The Maiden Flight of the Shūsui

In summer 1945, two eventful flights went down in Japan's aviation history. By early July 1945, less than a year after the start of the rocket program, the first Shūsui was ready for its maiden flight. For many of us, who now take safe air travel for granted, it is important to realize the challenges and dangers of new rocket technology. The Shūsui was a pilot's nightmare. Picture the following scene: surrounded by tanks in the fuselage and wings that contained highly explosive rocket fuel, the pilot had to start his takeoff run with an aircraft that was notoriously difficult to steer. After becoming airborne, he would initiate a steep climb of sixty degrees to reach an altitude of 12,000 meters in less than three minutes, with a climb rate more than four times that of a conventional interceptor. Then five more minutes of powered flight were available, during which the attack on enemy aircraft had to be completed before the fuel was exhausted. After the rocket motor had stopped, the pilot would return in a glide back to the airfield. He had to be sure to land with completely empty tanks. Any remaining fuel was likely to explode during a hard touchdown.

Even though the unpowered version of the Shūsui had flown successfully and the rocket motor passed its ground test, the rocket engineers remained uneasy. Lieutenant Nakaguchi Hiroshi, who oversaw the airframe development, and his colleagues opposed a "forced test flight with machinery that cannot be fully trusted yet." In his memoirs Nakaguchi blamed the fanaticism of the navy's upper ranks for the engineers' enormous effort to achieve quick results.[61] Indeed, the impatience of the navy's officials, together with a stopgap measure to soothe the worries of the engineers about possible engine troubles, led to two fateful decisions. So as not to lose even a single day, the navy chose the test site that was closest to the rocket laboratory. However, the small, narrow Oppama Airfield significantly constricted the pilot's margin of error. Furthermore, Commander Shibata Takeo, who was in charge of the test flight, decided on a short flight time so that the engine would not be pushed to its limits. He ordered the tanks to be filled to only a third of their capacity, which would limit the full-power phase of the test flight to two minutes instead of seven.[62] Shibata appointed Captain Inuzuka Toyohiko, who had already carried out flights with the unpowered version of the Shūsui, to be the test pilot for the maiden flight.

July 7 was the decisive day. At 4:55 PM, Inuzuka started the takeoff run of his Shūsui rocket aircraft. Discharging a blue-green flame, the engine accelerated to maximum thrust (figure 8.4). After 220 meters, the

FIGURE 8.4 This dramatic photograph shows Japan's first rocket aircraft shortly before takeoff in July 1945. Two crewmen are still holding the Shūsui's wing. Water is being sprayed below the fuselage to prevent any chemical reaction caused by spilled rocket fuel. (Source © Japan Aeronautic Association)

aircraft took off and started a forty-five-degree climb. At an altitude of around 400 meters, the engine spewed out smoke, made a popping sound, and stopped. The skilled pilot immediately dropped the plane's nose to prevent a stall. Aware of his low altitude, Inuzuka abandoned the prearranged emergency plan to ditch the aircraft in Tokyo Bay. Instead, he decided to return to the airfield while starting to dump the remaining fuel. Presumably not considering that his aircraft had a higher sink rate than the much lighter glider used for training, Inuzuka initiated his final turn toward the runway too late and scraped an observation tower that badly damaged his plane. The aircraft touched down, and both fuel tanks burst. The liquid fuel sprayed in all directions but miraculously did not explode. Inuzuka survived the immediate crash but suffered a skull fracture. He died the next morning.[63]

In hindsight, we can see that two factors led to the unfortunate failure of the test flight. With tanks only partially filled, a wrongly positioned fuel outlet resulted in a fuel cutoff during the steep climb. The choice of a second-rate airfield with obstacles in close proximity contributed to the pilot's failure to perform a safe landing after the engine failed. Even considering the tragic death of the pilot, it is safe to concur with the assumption of the US Air Technical Intelligence Group, published in October 1945, that "the power plant failure terminating the first and only flight had been the result of a readily correctible defect."[64]

The Kikka's Maiden Flight

In June 1945, Nakajima and the Naval Air Arsenal installed the recently developed jet engine on the new airframe. In mid-July, the first ground tests were carried out at the Kisarazu naval airfield in Chiba prefecture. Lieutenant Commander Takaoka Susumu, a test pilot of the Yokosuka Naval Air Force flying corps, was to be the pilot for the first test flight.

Takaoka prepared for his task in a truly professional way. He was aware that the Kikka's engine was very different from a conventional motor. It was slow to respond to the pilot's input, needing more than ten seconds to accelerate from idle thrust to full power. He was mindful of moving the throttle lever with great care and not letting the engine run slower than 6,000 revolutions per minute to avoid a potentially disastrous flame-out.[65] Upon closer study, Takaoka began to worry about several potential flaws in the Kikka's design. The aircraft's wheels and brakes were designed for the Zero-sen fighter's relatively slow takeoff speed and might not be adequate for the Kikka, which had to accelerate to at least 180 kilometers per hour before lift-off. Takaoka also learned about the inherent problems of the auxiliary rockets, which were to increase acceleration during the takeoff run but also generated a nose-up momentum. Finally, he was well aware that the light fuel load during the first test flight allowed only one landing approach, with no chance for a go-around after an aborted landing approach.

Arguing that safety should be the top priority, Takaoka requested that the main wheels be equipped with more powerful brakes. He also suggested mounting the auxiliary rockets closer to the plane's center of gravity to balance their nose-up momentum. The engineers responded that carrying out these requests would take another six months. Even though this assertion was clearly exaggerated, it seemed to Takaoka that he had no choice but to accept the aircraft's obvious shortcomings.

On August 7, 1945, exactly one month after the Shūsui's first flight, Takaoka took off for the maiden flight (figures 8.5 and 8.6). He climbed with his Kikka to an altitude of 600 meters and made a safe landing after ten minutes. The airplane had shown stable flight characteristics and adequate maneuverability. An examination and overhaul of the jet engines did not reveal any anomalies.

The second test flight was scheduled for August 11. This time the successful use of the two wing-mounted acceleration rockets was to be demonstrated. Takaoka had calculated that to let the auxiliary rockets deliver their thrust only while the aircraft was still accelerating on the runway, he would have to ignite them exactly three seconds after starting his takeoff run. Here is his account of the fateful flight:

> After all preparations were done, I applied full engine power and checked the engine parameters. I released the brakes, and after counting "one thousand, two thousand, three thousand" I pressed the rockets' ignition switch. I heard a thundering noise and felt the strong acceleration that pushed me from behind. At the same time, the aircraft's nose went up about twenty-five degrees and its rear part was grazing the runway. While the

FIGURE 8.5 Ground crew members are getting Japan's first jet aircraft, the Kikka, ready for flight in August 1945. The black cylindrical container left of the starboard engine houses one of the two rocket boosters that were to increase the aircraft's acceleration during takeoff. (Source © Japan Aeronautic Association)

FIGURE 8.6 This photograph captures a historical moment in Japan's aviation history. It shows the Kikka during its takeoff run with the nose wheel already having left the ground. (Source © Japan Aeronautic Association)

> aircraft accelerated rapidly, I could not see the runway anymore. Instinctively I pushed the control stick forward. I have never felt that eleven seconds could be so long.
>
> Suddenly the rockets stopped, and the nose bounced down, clearly because I was still pushing the control stick fully to the front. At that time, I felt something odd. I checked the engine instruments, but they did not show anything abnormal. I then heard a strange banging noise—did the front wheel burst? I had only one second to judge and act. With only half of the runway length left I decided to abort the take-off and to cut the engine. However, the brakes failed, I overshot the runway, a ditch sheared off the aircraft's wheels, and I slid into the sand.[66]

The combined—and foreseeable—shortcomings of the booster rockets and inefficient brakes led to the loss of the aircraft. Fortunately, the pilot survived this test flight. The parallels to the first Shūsui flight are obvious: once more a test had failed not because of any flaws in advanced technology but because engineers and operators ignored some seemingly trivial but essential details.

Even during the final stage of the war, engine experiments continued. Japanese engineers conducted tests running the Ne-20 on different types of fuel, which included diesel oil, alcohol, and even an extract made from crushed pine roots. The navy ordered the experimental production of advanced versions of the Ne-20. Ishikawajima and Nakajima were to build the Ne-130 and Ne-230, which were to deliver a thrust of 900 kilograms, roughly twice as much as the Ne-20. Mitsubishi was to develop the even more powerful Ne-330, which was to have twice the Ne-20's diameter and three times its thrust. All these projects had to cope with air raids and production dispersal. None of the engines completed the experimental phase, let alone entered production.

The Shūsui's failed test flight did not stop the navy from working on an improved version of the interceptor. On the morning of August 15, a number of navy officials discussed a plan for the J8M2, a second-generation rocket fighter with more stable flight characteristics, less armament, and greater fuel capacity.[67] A few hours later, Japan officially surrendered, with the emperor's Imperial Rescript on the Termination of the War broadcast nationwide.

Suicidal Cherry Blossoms: The Ōka Attack Aircraft

The promises of rocket and jet technology enticed Japanese Navy officers into visions of flying torpedoes for the ultimate defense of the homeland. In summer 1944, Naval Ensign Oota Shōichi presented his concept of a suicide aircraft that was to be released from a bomber at a high altitude. Its pilot would ignite the engine and start a high-speed crash dive onto its target. We might expect that such a proposal would be flatly rejected as the absurd idea of a low-ranking officer. But Oota received full support from his superiors, who even helped him get cooperation from some of Japan's most distinguished aeronautical experts. Aeronautical engineer Yamana Masao led a design team that included Ogawa Taichirō (1899–1952), Tani Ichirō, and Kimura Hidemasa (1904–86), all high-profile academic researchers at the Aeronautical Research Institute.[68] The airplane was to be named Ōka ("cherry blossom"), a term that traditionally carried the connotation of short-lived transience.

The actual aircraft was the antithesis of such flowery melancholia. The Ōka's design specifications called for a cheap, disposable flying bomb. The explosive head made up 80 percent of the plane's weight. Its short wings were to be made from plywood that could be worked by unskilled laborers. The aircraft's first version, the Model 11, was powered by three solid fuel rockets. Soon the severe operational flaws became obvious. Due to its short cruise range, the bomber carrier had to fly deep into the target's air defense zone, exposing itself to enemy fire. Because the Ōka added to the carrier's weight and drag, it significantly reduced the mother plane's range and cruise speed.[69] On March 21, 1945, the first major deployment of the Ōka flying bombs off the Okinawa coast ended in disaster. Before coming into close enough range with the US fleet, the Ōka-carrying bombers were either shot down by American fighters or had to drop their load to escape from enemy fire.[70] The failed attack made it clear that Japan's loss of air superiority had rendered the rocket-driven Ōka useless.

To make up for the obvious shortcomings of the Model 11 Ōka, Yamana conceived of the idea to equip the aircraft with a Campini-type motorjet, a combination of a piston motor for air compression and a thrust-producing combustion chamber.[71] The Tsu-11 engine and the new Ōka Model 22 (figure 8.7) underwent one unsuccessful test flight in

FIGURE 8.7 The Imperial Japanese Navy's MXY7 suicide attacker Ōka 22. Note the cherry flower symbol at the aircraft's nose section. Photo by Eric Long, Smithsonian National Air and Space Museum (NASM 2000-9387).

July 1945. An even more powerful version, the Ōka 43B, was to be equipped with Japan's new Ne-20 turbojet. These aircraft had an extended flight range of up to 270 kilometers and were to be launched from catapults. Wada Misao, chief of the Naval Air Technical Arsenal, drummed the decisive importance of this new weapon into his engineers:

> When the enemy assumes that all Japanese airplanes, warships, and every single defensive weapon had been completely destroyed, he will mass his large fleet around Japan proper to begin an invasion. Then these planes, hidden along the coast, will be all launched together from catapults and send the enemy's invasion troops to the ocean floor.[72]

Catapult construction began at Japanese beaches and even on sacred Mount Hiei, the home of one of Japan's most important monasteries. However, in early August, after six failed test flights, the project was abandoned. The apocalyptic vision of a fleet of jet-powered suicide bombers catapulted out of their secret hideouts had evaporated.

Conclusion

In 1944, shortly before transport routes between Germany and Japan completely broke down, German aviation technology once more wielded substantial influence on Japan. However, with the Pacific War entering its final stage, the nature of technology transfer had changed fundamentally. Japanese expertise had developed to such a degree that even very limited and incomplete information from Germany could catalyze a major technological breakthrough.

Japan's ambitious jet and rocket projects led to remarkable technological advances. At the same time, they exposed with striking clarity all the fundamental shortcomings of the country's wartime aviation. Production managers had to comply with absurd production plans while coping with material shortages, bombing damage, and plant dispersals. Military leaders ignored the country's crumbling industrial base and put their faith in the latest technology, which made them believe they could win a final decisive battle. Engineers dutifully put their skills to work. At the same time, as epitomized in the case of the suicide bomber Ōka, they did not object to the operationally flawed concepts and unreasonable demands of their military clients.

Any critical evaluations of the Shūsui and Kikka project, then, will lead to vastly diverging results. From a manufacturer's point of view, any aspiration for significant mass production of the two aircraft was illusory. With counterproductive overlaps of the design and production phase, wide dispersion of factories, and shortages of material and manpower, there was no prospect of producing these planes in large numbers. Furthermore, especially in the case of the Shūsui, fuel production bottlenecks made it impossible to sustain the planned operation of a fleet of several hundred aircraft. According to US postwar estimates, Japan's monthly hydrogen peroxide output amounted to 100 tons by 1945.[73] The unrealistic aim to produce 3,000 tons of the rocket fuel each month could never be met.

Limits in production and fuel supply already put the strategic value of Japan's jet and rocket program in question. The Kikka and Ōka arrived too late to make up for the shortcomings of the nation's air defense. In addition, it was doubtful whether the Shūsui would have been effective as an interceptor against the B-29 bombers.[74] With its short flight time and limited radius of action, the aircraft would have had to locate and destroy its targets within minutes. Such a task could have been carried out only with precise radar guidance and effective communication, neither of which was available to the Japanese military at the time.

From an engineering point of view, the development of Japan's first rocket and jet aircraft was a remarkable achievement. Within less than a year, the Japanese specialists had designed and built the Shūsui's exotic airframe and mastered a revolutionary new propulsion technology—when three years between the early design phase and test flight was the norm even for a conventional aircraft. Even though a German predecessor was available, we have to credit the Japanese engineers for their own inventiveness in overcoming countless technological challenges for which the very limited German material provided no answers. The Shūsui's successful takeoff had proven the airworthiness of the engine and the airframe. Once again Japanese aeronautical engineers proved their ingenuity. In a similar way, the rapid development of the Ne-20 jet engine was a technological success story. Based on a single blueprint, the Japanese engineers skillfully modified their earlier designs. In less than a year, they developed new engine components and systematically solved all the technical problems that occurred during the test runs. Considering that the Allies had no rocket aircraft and only one jet aircraft (the British Gloster Meteor) in operational use, Japan's aviation technology once more had overtaken that of their wartime enemies.

Why, then, did Japan's remarkable advances in jet and rocket technology fail to impress the Allies? The most plausible explanation could be the destruction of most of the related classified documents immediately after the end of the war.[75] According to one member of the Shūsui project, the engineers and technicians destroyed their documents out of fear of being captured and brought to the United States.[76] US inspection of remaining material and interrogation of Japanese experts further reinforced the Americans' stock image of Japan's backward research and development.[77] After Japan's surrender, the US occupying forces lost no time in entering the aircraft manufacturing plants and military arsenals and confiscating remaining documents and matériel.[78] Already on August 31, 1945, US Military Intelligence presented its classified report on "German Technical Aid to Japan." The study reveals how initial astonishment over the "presence there of bat-like flying wings" gave way to the hasty conclusion that the Shūsui was nothing but a copy of the German Me 163. The document also discloses US ignorance of the advanced state of Japanese jet technology. It arrived at the flawed evaluation that "development of a turbo-jet unit was underway in Japan as far back as 1941; the evidence suggests that little success was achieved."[79]

The interrogation of two Messerschmitt engineers consolidated the US Intelligence Service's low regard for Japan's aviation industry. The German engineers, who at the time of their surrender did not know about the successful Kikka flight, told their interrogators they were convinced that without complete documentation and without the guidance of German experts, Japan's jet fighter project would inevitably fail. Even with Germany's help, it would take the Japanese a minimum of eighteen months to build the first jet fighter. These statements led the US Office of Naval Intelligence to conclude that "apparently the Japanese were having difficulties in this field [jet and rocket propulsion] since they were constantly pressing for additional information."[80] Nevertheless, the US Navy took the effort to ship three Shūsui and three incomplete Kikka airframes along with one prototype to the United States. However, it seems that these aircraft were of little interest to the US military, which did not carry out any flight evaluations of the Japanese rocket or jet interceptors. Instead, it was decided to display one of the three Shūsui in Hollywood. Another was on exhibit at the Naval Air Station in Glenview (Illinois) for a time and scrapped thereafter.[81]

Such was the lackluster end of Japan's miracle weapons. Twelve years passed before another made-in-Japan jet aircraft took off. On January 19, 1958, Japan's first postwar jet aircraft, the T1 trainer, successfully made its first test flight. It was flown by Takaoka Susumu, who had been the pilot of the first Kikka. The T1 was designed and built by Fuji Heavy Industries, which had emerged from the former Nakajima Aircraft Company. The trainer was initially driven by a British motor. Its successor, the T1B, featured the first made-in-Japan jet engine in the postwar period, the Ishikawajima-Harima J3. It was designed and produced by Ishikawajima, the same company that during World War II had played a central role in the Ne-20 development.

EPILOGUE

The history of Japanese aviation followed a transnational trajectory of aeronautical advancement driven by technological development, public air-mindedness, and aerial armament. Japanese airmen, aeronautical engineers, and strategists participated in the transnational flow and exchange of people, matériel, and ideas. Beginning in the 1910s, when Japanese aviation enthusiasts thronged the Yoyogi Parade Ground to see those magnificent flight demonstrations, they shared their fervent air-mindedness with the exhilarated crowd of the 1909 Reims Air Meet and cheering spectators at the Johannisthal airfield in Germany. While the rise of popular aviation enthusiasm was a transnational phenomenon, the close connection among the industry, military, and public became a unique feature that defined the development of aviation in Japan. New avenues of technology transfer, a peculiar mix of visions and anxieties, and an ever-changing geopolitical landscape generated pressures and challenges along with windows of opportunity that allowed Japanese military officials, diplomats, and industrialists to skillfully access the latest foreign aviation technology and build an autonomous domestic industry.

Technology Transfer: Causes, Conduits, and Consequences

Persistent worries about Japan's backwardness and dependence on foreign technology drove the fast buildup of the country's airpower and aviation industry. This fear materialized for the first time during the Russo-Japanese War of 1904–5 after the late and damaged delivery of French observation balloons. It gained strength during World War I, when Western exports to Japan all but dried up; it escalated after 1918, when Japanese observers and air strategists argued that World War I had shown that any future war would be a battle between each country's factories. These anxieties resulted in a massive buildup of Japan's aviation industry—a buildup that provided the base of airpower expansion and ultimately made the country fully independent of imports.

In their quest for technological autonomy, the Japanese military and domestic aircraft makers built on a tried-and-tested practice of exchanging foreign and Japanese experts that reached as far back as the Meiji era. Japanese engineers, military observers, and airmen were dispatched to Britain, France, Germany, and the United States. They returned to Japan with detailed information about the latest aeronautical developments and an almost missionary zeal for advancing aviation at home. In return, Japan invited a considerable number of foreign advisers, most notably the French Aeronautical Mission (1918–20) and the British Aviation Mission (1921–23). French, British, German, and US influence greatly accelerated the reshaping of Japan's aviation. Yet Japan's military and the country's emerging aviation industry deftly selected from a wide range of accessible technologies and thus avoided sole reliance on a single country's expertise.

The direct purchase of hardware and production licenses was another straightforward—if expensive—conduit for technology transfer. For a transition period, foreign imports continued on a limited and very selective base; they allowed Japan to absorb and diffuse the latest foreign aviation technology. Soon aircraft makers left behind the simple assembly of imported, prefabricated parts and managed to take control of the complete production process. Japanese engineers replicated the new technologies and, within a short period, devised their own original innovations.

By the late 1920s, they had left behind earlier design methods of trial and error and adopted a scientific approach that transformed aircraft design from a craft into an engineering discipline. Their expertise reached a level that allowed them to fully engage in the independent design of outstanding aircraft types that were among the best in the world.

This historic achievement depended not only on the engineers' expertise and skill but also on their creativity and originality. What were the sources of such technological creativity and originality as the driving forces behind technological change? There is serendipity or—even more mysteriously—the inventor's accidental flash of inspiration. Memoirs and hagiographies of heroic designers tend to downplay the chance character of discoveries while connecting the proverbial "eureka moment" with the protagonist's genius.[1] Such a view seems to suggest that creativity is driven by an autonomous (that is, context-free) dynamic that defies explanation.

In this book I tried to frame the phenomenon of creativity with an institutional and social context. Indeed, the emergence of an organizational structure for sustained technological inventiveness was a critical juncture in Japan's aviation history. Such a "technological support network"[2] had a significant practical implication: it institutionalized innovation. For many who entertain the idea of the brilliant solo inventor, the phrase "institutionalized innovation" might sound like an oxymoron. Yet it is important to realize the role of the state, the military, and the industry that provided the organizational structure along with the capital and equipment to motivate and stimulate innovators. As we have seen, the academic departments of aeronautics and laboratories and design sections of civil aircraft makers—together with the military's arsenals—developed into venues of original research and creative development that propelled Japanese aviation to the forefront of aeronautical advance.

Obviously, the institutional model covers only the functional aspects of creative innovation. Any exclusive reliance on such a framework might result in a "total systematization [that] excludes all adventure."[3] To steer clear of any institutional determinism, this account of Japan's aviation history paid careful attention to the creative historical actors: ingenious and resourceful aeronautical innovators. These engineers, with formal scientific and technical training, formed well-organized groups of "rational inventors."[4] They embodied the decisive transition from a purely empirical to a truly scientific approach or—to put it more figuratively—from the solitary tinkerer in oil-stained overalls to the creative team of engineers in white coats. This new breed of skilled specialists abandoned the aviation pioneers' trial-and-error approach. They put forward new theories and refined or rejected them after careful and accurate experimentation and only then implemented new findings into a prototype. In other words, the engineers redefined the creative effort as an endeavor to get an idea from the ivory tower down to the drawing board and then materialize it on the shop floor.

Japan's aeronautical engineers might have followed the motto "the sky is the limit." Yet even the most brilliant innovators could not shun the gravitational pull of institutional inertia and rivalry. Even though these rational inventors benefited from institutional support, they often faced fierce resistance to innovation at the same institution. Paradoxically, even though the military was the driving force behind the development of Japan's aviation, a significant number of high-ranking officials obstructed the advancement of the air arm. The notorious rivalry between the Imperial Japanese Army and the Imperial Japanese Navy resulted in costly parallel developments and effectively interdicted technological transfer between the two services. Furthermore, military innovators often faced a rigid environment that showed little inclination to consider unorthodox ideas or experiment with new concepts. In the 1930s, the army's traditionalists still valued the soldier's fighting spirit over modern technology and were loath to attribute a significant role to the new aerial weaponry. As a result, the army failed to implement advanced airpower doctrines that stressed the importance of fighter aircraft for gaining air superiority and a strike force of long-range bombers for strategic bombing missions into an enemy's hinterlands.

The modernizers at the Imperial Japanese Navy had to fight similar battles with the advocates of a big-ship, big-gun policy: these advocates pledged the navy's strike force to giant battleships rather than powerful naval aviation. Yet the remarkable advancement of Japan's naval aviation technology strengthened the position of the navy's airpower faction. As a result, and in marked contrast to the army, technological innovation had a profound effect on the navy's battle doctrine and air strategy. A new generation of flying boats enabled the navy to independently carry out long-range air missions in the Pacific. Even more important, carrier-launched fighters and dive bombers enticed the navy's strategists to devise an aggressive doctrine of preemptive air strikes at the opening phase of what they envisioned to be a short, decisive war.

The Media and the Public: Anxieties, Exhilaration, and Fervent Nationalism

Japanese media became the vital link between the military's aviation project and the public. The press repeatedly echoed the military's concern about foreign aeronautical superiority and Japan's technological backwardness. A pattern emerged where newspaper articles articulated, intensified, and at the same time soothed their readers' worries. Detailed accounts informed the readers about the latest advances in Japanese aviation. These reports emphasized the prowess of the country's engineers, workmen, and pilots and celebrated Japan's apparently unstoppable progress in the aviation world. A topic of similar national importance was the quest for attaining technological independence. The Japanese press continuously supported the development of a thriving national aircraft industry and praised the appearance of more and more aircraft that had been built and designed by domestic manufacturers.

Already in the 1870s, the military had become aware of the flying machines' potential to capture the public's attention and burnish the army's and the navy's images. Public balloon launches received intense press coverage and attracted huge crowds of spectators. In December 1910, Japan's first motorized flight became a mass spectacle of epic proportions. The event that the military carefully staged in the center of Tokyo attracted more than 100,000 spectators. When Japan's first pilots took off in their flimsy biplanes, an exhilarated crowd celebrated them as heroes, and the press praised the airmen for having secured Japan's membership in the exclusive club of airfaring nations.

During the early months of World War I, a curious mix of aerial sensationalism and nationalistic war fever made its first appearance in Japan. When the country fought its first aerial warfare over the German base at Qingdao, the press glorified the fearless Japanese pilots who proved equal or superior to a powerful enemy. In the following years, Japanese air-mindedness took on an increasingly nationalist overtone that became a leitmotif for covering future air battles. In the early 1930s, the press applauded the early successes of Japan's military aircraft in the Manchurian conflict and continuously urged support for "our young Japanese soldiers in the cold of Manchuria." The resulting donation fever enabled the Japanese Army to substantially reinforce its airpower.

Close cooperation between the press and the military resulted in a series of transcontinental flights that became a strikingly efficient instrument to forge national pride and identity in foreign air space. In July 1925, more than 150,000 Japanese watched two aircraft take off for the "visit Europe flight," a venture sponsored by the *Asahi* newspaper and carried out largely by the Japanese Army. In April 1937, a converted Mitsubishi bomber left Japan for a flight to London, setting a new flight-time world record. Just two years later, Japan celebrated its first around-the-world flight: the *Nippon* covered a distance of more than 52,000 kilometers before safely returning to Tokyo. These endeavors were advertised as goodwill flights and manifestations of Japan's efforts for international understanding. They further stirred Japanese air-mindedness and propagated the high standard of Japan's aviation at home and abroad.

International Relations: From Cooperation to Alienation and Conflict

The interaction between Japan's aviation and the country's international diplomacy and military strategy is a tale full of ironies, unintended consequences, and devastating outcomes. Post–World War I diplomacy had a profound impact on Japanese aviation, leading to a massive transnational flow of matériel, experts, and ideas. The 1919 Treaty of Versailles provided the Japanese unprecedented access to German aircraft, aircraft engines, and equipment and unintentionally paved the way for German aircraft makers to do business with Japan. This illicit move, which was sanctioned by both governments, led to a Japanese–German rapprochement and alienated Japan from its erstwhile allies. At the same time, the Japanese benefited from the postwar slump of the Allied aviation industries. They welcomed French and British willingness to export aeronautical hardware and know-how while deftly keeping foreign powers from establishing a monopoly in the development of Japan's aviation.

During the interwar period, two important diplomatic initiatives led to what can be called a paradoxical disarmament. The 1922 Washington Naval Treaty and the 1930 London Naval Treaty were calls for arms reduction. While the Japanese Navy agreed on limiting the number of its warships and submarines, it redirected appropriations toward a significant expansion of naval airpower. Even though the conclusion of the treaties still nurtured the hopes of multilateral cooperation, Japan's international relations soured. A revised Imperial Defense Policy promoted the United States to "Japan's number-one hypothetical enemy," and military planners initiated a hugely expensive project for a fleet of long-range bombers to attack US bases in the Philippines.

During the takeover of Manchuria, Japanese bombers engaged for the first time in air raids not only against combatants but also against civilians. These campaigns met with international criticism and accelerated Japan's isolation, which became manifest with the country's withdrawal from the League of Nations in 1933. In 1934 the Japanese government decided to abrogate the Washington Naval Treaty; in the following year, it withdrew from the London treaty as well. As a result, the pace of Japanese aerial armament accelerated dramatically.

After the start of the Second Sino-Japanese War in 1937, the Japanese resumed bombing civilians on an unprecedented scale. This time the airpower deployment caught the Japanese in a fundamental dilemma. The further expansion of airpower crucially depended on the import of US fuel, aluminum, and production tools. But US embargoes set up in response to Japan's bombing campaigns severely jeopardized the advancement of Japan's aviation industry. A ticking-clock scenario emerged, enticing Japan's military officials to bet the country's fate on the technological excellence of their airpower and on—what they believed to be—a uniquely Japanese fighting spirit of their pilots to win against a numerically superior enemy. Cut off from essential resources and technology, Japan's reliance on international diplomacy gave way to a dogged confidence in military power to fight what they imagined would be a short, decisive war.

Transwar Continuities and Postwar Disruptions: Japanese Aviation after 1945

Even a rough sketch of Japan's postwar aviation will allow us to identify old acquaintances, long-lived doctrines, and familiar rivalries—but will also reveal radical ruptures and new departures. The formation of Japan's postwar aviation was in many ways a replay of its pre-1945 history, when the country evolved from a late starter to a major player. Once more, fears of technological backwardness and dependency gave rise to aspirations for an autonomous high-tech industry that would allow Japan to join the club of advanced aviation nations. Once more the military strove to include the public its aviation project; once more the country's leadership deftly gasped the opportunities arising from major geopolitical realignments to rebuild Japan's airpower. Yet some transwar legacies greatly accelerated Japan's aerial rearmament and the buildup of an advanced aviation industry, whereas others became formidable obstacles that were difficult to overcome.

Soon after their country's defeat in World War II, Japanese aeronautical engineers reexamined the fundamental deficiencies of the country's wartime aviation industry. Many of these specialists arrived at the view that—apart from the obvious shortage of resources and labor—Japan had lost the war in its research departments and design offices.[5] Some also declared that Japan's science and technology had been driven by a mistaken war of aggression and compared the results of their wartime efforts with "jewels sunk to the bottom of disgrace." Now it would be necessary to restore science and technology to its "original position as a foundation of peace."[6] Virtually from day to day, scientists and engineers repurposed their trade from a means of imperial expansion to an instrument for rebuilding a democratic Japan.[7] In January 1946 the Federation of Democratic Scientists (Minshushugi Kagakusha Kyōkai) was founded.[8] In the same year, the Foreign Ministry issued a report that contended a mutual dependence between democracy and technological development: "Without democratization it will be difficult to achieve technological advance, and at the same time without technological advance we will be unable to achieve true economic democratization."[9]

THE DEMISE OF JAPAN'S AIRCRAFT INDUSTRY

After World War II, the remnants of Japan's air fleet and aviation industry rapidly disappeared. On August 24, 1945, nine days after Japan's surrender, the Supreme Commander for the Allied Powers (SCAP), General Douglas MacArthur (1880–1964), issued a ban on all Japanese aircraft flights. In September US forces took control of aircraft manufacturing sites, both civil and state-run, and began the large-scale destruction of all remaining civilian and military airplanes (figure E.1). A November 18 directive prohibited "anyone from owning, purchasing, possessing, or operating any aircraft, conducting research or teaching, or experimenting in aeronautical science, aerodynamics or other subjects related to aircraft or balloons."[10]

Determined to completely break up Japan's aircraft industry, SCAP's General Headquarters went even further. In January 1946 they took into custody around 390 research laboratories, arsenals, and aircraft manufacturing sites. In November 1947, the dismantling of Japanese factories began with a priority on the former armament industry. There was only one way for erstwhile arms manufacturers to avoid the demolition of their factories: they could apply for armament conversion. By November 1948, 182 out of 300 aircraft manufacturing plants received a permit for the "conversion" of their production from military supply to civilian goods.[11]

The workforce of the former aviation industry also faced tough times. By the end of the war, the number of employees involved in design or production of aircraft had exceeded 1 million, including about 100,000 engineers and highly qualified technicians. Confronted with the General Headquarters' draconian measures, many former aviation engineers moved into the industries of precision mechanics and optics or went into automobile manufacturing. Others joined the Transportation Ministry's Railway Technology Research Institute (Tetsudō gijutsu kenkyūjo), where they played a key role in the development of the Shinkansen high-speed passenger train.[12]

A CHANGING GEOPOLITICAL CLIMATE AND THE REVIVAL OF JAPAN'S AVIATION

With aircraft manufacturing coming to a complete stop and engineers finding employment elsewhere, it was difficult to imagine a return of Japan's aviation industry. However, in the early 1950s the escalation of

FIGURE E.1 "The mighty falls at last, to be no more than dust before the wind." This bleak photograph shows some remainders of a once-proud air fleet crammed in a scrapyard. (Source © Japan Aeronautic Association; quote from *Tale of the Heike*, translation by Helen Craig McCullough)

the Cold War increased Japan's geopolitical importance, accelerated the country's economic recovery, and led to its eventual rearmament. During the Korean War (1950–53), Japanese companies received orders for the maintenance and repair of US military aircraft. This unexpected opportunity exposed Japanese specialists to the latest advances in US aviation and prepared them for taking up aircraft production again. The ongoing war in Korea also made the US Department of State change its occupation policy. It advocated a formal peace treaty that would integrate Japan in America's Cold War strategy. When the Peace Treaty of San Francisco came into effect on April 28, 1952, Japan regained air sovereignty, and the prohibition on aircraft manufacturing came to an end. Five companies emerged as the major players in Japan's postwar aerospace industry: Kawasaki, Mitsubishi, Fuji, Shin Meiwa, and Ishikawajima. They all relied on decades of experience in aircraft manufacturing that dated back to well before World War II.

While the Korean War hastened US plans for Japan's rearmament, large segments of the Japanese public opposed the country's remilitarization and any reinstatement of a national armament industry. Even though Prime Minister Yoshida Shigeru (1878–1967) was well aware of the unpopularity of a fast and comprehensive military buildup, he reluctantly gave in to US demands. In March 1954 Yoshida signed the US-Japan Mutual Defense Assistance Agreement; the same year, Japan Self-Defense Forces were established under the direction of the Japan Defense Agency.

Soon a sweeping interministerial rivalry emerged that bore a remarkable resemblance to the notorious pre-1945 army-navy antagonism. Unlike Japan's powerful World War II Imperial General Headquarters, the Defense Agency had no direct authority over the domestic aviation industry. Neither was the agency in a position to counter the Ministry of Finance's strong hold on the defense budget. Moreover, the powerful Ministry of International Trade and Industry (MITI) also showed a strong ambition to be in control of the newly reemerging aviation industry. Finally, a compromise was reached that put the Defense Agency in charge of setting up plans and guidelines for the Air Self-Defense Force's equipment, while MITI controlled the aircraft manufacturers' production facilities and channeled government subsidies and investments.[13]

THE ROCKY ROAD TO JAPAN'S FIRST POSTWAR JET ENGINE

Jet engine development in early postwar Japan highlights the struggle to revive the prewar doctrine of independent engine production. Ironically, the rather earthbound Railway Technology Research Institute played a major role in this process. In January 1950, the institute's engineers completed their first gas turbine. The engine's successful test run evoked memories of past achievements. According to one observer, the Tekken 1-gō gas turbine became "a fine thread that connected the golden age of [Japan's] prewar aviation industry with the era of independent postwar jet engine technology."[14]

Ishikawajima Heavy Industries, the company that developed the wartime Ne-20 jet engine, also endeavored to build on past successes. In July 1953, the firm established—along with Shin Mitsubishi, Fuji Precision Industry, and Ōmiya Fuji Industry—the Nihon Jet Engine Company (NJE).[15] Yet it turned out that NJE's first engine, the JO-1, was no match for its US counterparts, which were more than twice as powerful. The JO-1 project was abandoned in December 1955. In the same year, the Defense Agency ordered NJE to develop a new engine for Japan's first jet trainer. The company could not meet the delivery deadline, so—contrary to the Defense Agency's ambitious plans—the new Fuji T-1 jet trainer made its first flight in January 1958 with a British Orpheus engine. To make matters worse, when NJE finally delivered four jet engines, none met the required performance levels. With little prospect for setting up a successful series production, NJE ceased operations in 1959.

NJE's failure revealed the difficulties in closing the gap between wartime technology and the advances of the 1950s. It also confirmed the view of those who argued that Japan should give up its expensive aspirations for indigenous engine development and engage in the licensed production of foreign designs. Such an approach would be more helpful for advancing the jet engine industry, especially after the United States announced easier access to production licenses.

JAPAN RETURNS TO THE JET AGE

Indeed, in an astoundingly close parallel to the years preceding the Pearl Harbor attack, US technology transfer once more became crucial for advancing Japan's aircraft manufacturing. After the Korean War ended, US manufacturers had to cope with shrinking defense budgets, causing them to welcome the opportunity to sell production licenses to Japan. Their aspirations materialized with the 1954 US and Japan Mutual Defense Assistance Agreement. The treaty was a response to an increasingly confrontational policy adopted by the Soviet Union and its close ally, the People's Republic of China. The arrangement, known as the Mutual Security Agreement, endorsed further rearmament of Japan and the reconstruction of Japan's air defense industry.

Kawasaki and Mitsubishi were the main beneficiaries of the Mutual Security Agreement. Kawasaki set up a production line for Lockheed's T-33 jet trainer, and Mitsubishi engaged in building the North American Aviation's F-86 Sabre fighter aircraft. The US government agreed to cover about 45 percent of the US$155 million production cost, clear proof of the importance the United States attached to Japan's rapid aerial rearmament.[16] The arrangement gave Kawasaki and Mitsubishi full access to US jet technology and US-style production. A distinctive pattern evolved that became characteristic of the transfer of US military aviation technology to Japan. Initially, a few aircraft were built and tested in the United States and then brought to Japan. As a second step, beginning in June 1955, Kawasaki and Mitsubishi started with assembling imported, prefabricated parts. Then the Japanese companies moved forward to licensed production, initially under the close guidance of US specialists. During this third phase, the Japanese manufacturers gradually increased the portion of domestically produced components. In sharp contrast to the country's go-it-alone jet engine program, the licensed production of jet aircraft was remarkably successful. By 1961 Kawasaki and Mitsubishi had built more than 500 trainers and fighters. During these production periods, both companies significantly expanded the portion of made-in-Japan parts to over 60 percent.[17]

BLUE IMPULSE: JAPAN'S ACROBATIC FLIGHT TEAM

The new F-86 fighter aircraft not only promoted Japan's airpower and technological expertise, it also was instrumental in reviving a long-standing tradition of fostering air-mindedness among the Japanese public. The formation of a new acrobatic team received considerable support from Genda Minoru, who established an acrobatic team for the Imperial Japanese Navy already in 1933 before gaining fame as one of the key figures behind the Pearl Harbor attack. When Genda became the chief of staff of the Japan Air Self-Defense Force (JASDF) in July 1959, he strongly advocated for establishing an official JASDF acrobatic flying team. As in the 1930s, he was convinced that Japanese pilots performing acrobatic flights—even with US-designed fighter aircraft—would boost the JASDF's public image and self-esteem.[18] In 1960 the Aerial Maneuvers Research Squad was officially established. The acrobatic team became known under the name Blue Impulse to a global audience during the opening ceremony of the 1964 Summer Olympics in Tokyo when, after the release of 8,000 doves of peace, five F-86 fighters painted the five-ring Olympics symbol with colored smoke in the sky (figure E.2). Over a period of twenty-three years, the Blue Impulse team flew more than 540 shows with their F-86s. Their "aviation festivals" (*kōkū matsuri*) still attract huge crowds, as on November 3, 2013, when in a single day the flight show at Iruma Air Base in Saitama prefecture reached a record turnout of more than 320,000.[19]

AN ENGINEERING MASTERPIECE TURNS INTO A SALES DEBACLE

In 1957 some of Japan's most experienced engineers went to their drawing boards to design Japan's first turboprop passenger aircraft, the YS-11.[20] All of them had established their reputation with the development of military aircraft before and during World War II. Kimura Hidemasa, designer of the rocket-driven Ōka kamikaze attack plane, chaired the technology committee that consisted of former Nakajima designer Ōta Minoru; Kikuhara Shizuo, who had developed Kawanishi's H8K flying boat; Horikoshi Jirō, who had risen to fame with his Zero-sen fighter;

FIGURE E.2 The opening ceremony of the 1964 Tokyo Summer Olympics was a defining moment of Japan's postwar transformation. The picture shows Japan's acrobatic flight team Blue Impulse painting the five-ringed Olympics symbol in a brilliant blue sky. (Source © International Olympic Committee)

and Doi Takeo, well known for his high-performance Hien fighter. The group had strong support from industrialists and government officials who aspired to "dream once more of Japan as an aircraft empire."[21] A new Law for the Promotion of the Aircraft Manufacturing Industry allotted ¥300 million for the YS-11 project to the newly founded Nihon Airplane Manufacturing Company (NAMC).[22] The generous funding boosted the YS-11's development but ultimately did not provide much incentive for cost-efficiency. Echoing their pre-1945 experience, engineers focused on sophisticated design features while giving little consideration to cost control and efficient manufacturing.

The YS-11 made its first flight at Nagoya Airport on August 30, 1962. However, intractable flight control problems caused a major delay for series production. As a result, an impatient All Nippon Airways did not respond to patriotic appeals to "buy Japanese." In 1963 the airline canceled its YS-11 orders and purchased twenty-five airplanes from Dutch aircraft manufacturer Fokker instead. The aircraft's complicated design and poor coordination between NAMC and the six different aircraft manufacturers greatly hampered efficient production. In addition, due to NAMC's lack of experience in international sales and marketing, the plane failed to attract a large enough number of foreign customers.[23] By 1970 NAMC's soaring deficits increasingly came under attack by opposition politicians and the media. In 1971 the Japanese government decided to cease production of the YS-11.

Japanese ambition to design, build, and sell the country's first postwar passenger aircraft suffered from a prewar legacy of overambitious engineering that neglected the requirements of efficient production. Furthermore, a fiercely competitive international environment—and a public response that turned from lukewarm support into open criticism—turned a patriotic dream into an economic crash landing.

BEYOND THE SOUND BARRIER

While Japan advanced into the supersonic age, tensions with its US ally heightened. In December 1959, Japan's National Security Council opted for the Lockheed F-104 Starfighter as Japan's first supersonic fighter aircraft.[24] As it turned out, Lockheed had paid about US$1.5 million to Japanese government officials to change their previous decision which favored Lockheed's competitor Grumman.[25] Presumably unaware of these dealings, Mitsubishi, Kawasaki, and engine maker Ishikawajima-Harima welcomed the opportunity to engage in producing a supersonic aircraft.

The Starfighter's successor also caused intense controversy. In November 1966, the Japanese cabinet approved the procurement of the McDonnell Douglas F-4 Phantom. In the Diet, this decision for an aircraft that already had been deployed in large numbers in the Vietnam War led to fierce discussions. Opposition politicians argued that "the aircraft with its long flight range and its bombing equipment poses an aggressive threat to neighboring countries."[26] In response to these apprehensions, the Defense Agency ordered the aerial refueling devices and bombing computers to be removed from all F-4 aircraft. In addition to these domestic disputes, the licensed production of the F-4 made it obvious that the United States had become less willing to share advanced military technology, rekindling Japanese fears of technological dependency. In response, Japanese companies, such as Nihon Keiki and Hitachi, developed and produced essential components like the radar-warning device and data-link equipment.[27]

Before the start of the F-4 licensed production, the military ordered Japanese manufacturers to engage in independent design and construction for a supersonic training aircraft. In 1971 a prototype of the trainer made its first test flight. As a next step, Japanese engineers modified the basic design to turn the trainer into a fighter, the Mitsubishi F-1. The two aircraft types bolstered Japanese self-confidence as the country became the sixth member in the exclusive club of nations that had developed their own supersonic aircraft.[28]

The quest for a successor to the F-1 fighter became notorious for causing considerable contention at home and abroad.[29] Both the Defense Agency and Japan's aircraft manufacturers favored a made-in-Japan fighter rather than relying on foreign technology. In 1983 the National Security Council approved the development of a new fighter, the FS-X, and in 1985 Mitsubishi and Kawasaki presented their detailed design proposals. However, US politicians opposed Japan's initiative. Expressing their worries about the United States' soaring trade deficit with Japan, they argued that Japan should buy US aircraft to support US manufacturers and reduce the massive trade imbalance. In October 1987, both governments arrived at a compromise. Mitsubishi would design the FS-X based on General Dynamics's F-16 fighter. Furthermore, the Japanese had to agree to a bidirectional but lopsided technology transfer. While the United States would not disclose all technological details of the F-16, any new Japanese developments would be made fully available to the US aviation industry. The denial of advanced technology proved to be an opportunity rather than a handicap. Crucial components of the aircraft's electronic equipment—such as the fighter's radar, head-up display, and navigation system—were made in Japan. Yet for many Japanese, the unequal partnership was difficult to swallow. Politician Ishihara Shintarō famously protested that "[Prime Minister] Nakasone gave away our technology; all he got in return was Reagan's friendship."[30]

FROM "MADE IN JAPAN" TO "MADE WITH JAPAN"

In contrast to the US-imposed technological cooperation in the military sector, Japan's commercial aviation industry engaged on its own accord in international collaboration. After the financial debacle of the "purely Japanese" YS-11 turboprop airliner, participation in international joint ventures became the new strategy for survival in a fiercely competitive market. This was a fundamental move away from a doctrine of "techno-nationalism" to what might be called "technointernationalism."[31]

Japan's participation in developing and producing the Boeing 767 wide-body twinjet was the country's first commercial aviation project in cooperation with a foreign manufacturer. Mitsubishi, Kawasaki, Fuji, Nippi, and Shin Meiwa collectively received a 15 percent share of the Boeing 767's overall production. Production began in 1981, and a satisfied Boeing official confirmed that "Japanese companies have developed into a competent source for the manufacture of commercial aircraft parts."[32] The success of the joint venture led to an important change in Japan's official aviation policy. According to the 1986 revision of the Law for the Promotion of the Aircraft Manufacturing, subsidies would now be extended exclusively to projects based on international cooperation.[33] In early 1989, Boeing showed interest in Japan's cooperation for the Boeing 777, a 350-seat wide-body airliner. This time Japanese manufacturers secured a record 21 percent share of the new airplane's development and production.[34] The "Triple Seven" made its first flight in 1994 and became a remarkable commercial success, with more than 1,100 sold by 2017.

Japanese participation in designing and manufacturing US airliners reached an unprecedented scale with Boeing's 787 "Dreamliner." Mitsubishi, Kawasaki, and Fuji built 35 percent of the Boeing 787. Japan played a key role in developing a light airframe and a wing with advanced aerodynamic properties, significantly reducing fuel consumption and operating costs. Boeing branded the 787 as part of its "made with Japan" program, whereas the Japanese companies proudly marketed the aircraft as a "quasi-national production."[35]

A RETURN TO A "MADE-IN-JAPAN" POLICY?

After having cooperated for more than two decades with Boeing, Mitsubishi's engineers became confident they could develop a commercial jet aircraft by themselves. In 2008 the company started developing a twin-engine aircraft that could carry between eighty and ninety passengers. On November 11, 2015, more than half a century after the YS-11's first takeoff, the Mitsubishi Regional Jet (MRJ) made its successful maiden flight. Even with strong international competition in the market for regional jets, Mitsubishi received more than 400 orders for its new aircraft by late 2017.[36]

Soon after the start of the MRJ development, Japan's government launched a parallel project that was to establish Japan's technological independence in cutting-edge military aviation. Japanese officials were frustrated by the continuing refusal of the United States to sell its fifth-generation stealth fighter, Lockheed Martin's F-22 Raptor, to Japan. In 2009, the Japanese government earmarked ¥39 billion for developing an all-Japanese stealth fighter, known under the name X-2 Shinshin ("spirit"). The aircraft's shape and its carbon-fiber composite material minimize detection by radar. With the X-2's first flight in April 2016, Japan became the fourth nation after the United States, Russia, and China to have mastered advanced stealth technology.[37]

Conclusion

A fast-changing geopolitical environment became the key factor in the fate of Japan's postwar aviation sector. Whereas the end of World War II had brought about the eventual demise of Japan's aviation, the Cold War sparked and then accelerated its unlikely revival. US–Japan relations became the single most important feature driving this development. Against a background of increasingly strained US relations with the Soviet Union and communist China, Japan transformed from a former enemy into a close ally. This process led to an equally radical reversal of US policy from a complete aviation ban to the full support of Japanese aerial rearmament via generous provision of funds and know-how.

The country's aircraft manufacturers deftly adapted to the new situation by referring to a pattern of technology transfer that had been already tried and tested in the prewar period. They soon replaced imports with licensed production, which they followed with the ultimate aim of establishing an independent domestic industry. The rapid and efficient transfer of technology made it possible to catch up in the crucial fields of jet aircraft and make the fast transition to supersonic and—most recently—stealth technology.

While this process can be seen as a prime example of Japan's transwar continuities, not all transwar legacies were successful. World War II engineers who aimed to revive the wartime pioneer jet research failed to establish large-scale production of a domestically made jet engine. In a similar way, a selected group of Japan's top World War II aeronautical engineers managed to design and build the YS-11, Japan's first commercial turboprop airliner. However, largely due to a wartime legacy of neglecting cost-efficient production, the YS-11 project ended up as a commercial disaster. As a response, Japan's aircraft industry and lawmakers abandoned an "all-made-in-Japan" doctrine. They turned toward international cooperation, which culminated in the Boeing 787 being "made with Japan."

Japanese aviation enthusiasm outlived the country's defeat in World War II, though with a decidedly nonmilitaristic note. The Blue Impulse acrobatic team owes its popularity to upbeat airmanship rather than a display of military airpower. Political and economic leaders continue to emphasize a strong and independent aviation industry and connect it with the country's prestige and international standing. However, such nationalist rhetoric was faced—and is still facing—an increasingly critical public that, along with the press and opposition parties, has spoken out against rapid rearmament and offensive airpower as much as against bribes and wasteful government spending.

A strong US–Japanese military alliance, well-established technological internationalism, and a government under close public scrutiny suggest a rather down-to-earth development of Japan's aviation with no high-flying dreams of military supremacy and little risk of getting lost in technological extravagancies. Yet recent developments, such as the 2014 end of the arms export ban, new security legislations in 2015, and the 2018 plan to upgrade two *Izumo*-class destroyers to aircraft carriers indicate that the future of aviation in Japan remains highly unpredictable. Looking back at the turbulent history of Japan's aviation allows us to imagine that major shifts in the geopolitical environment, the emergence of disruptive technologies, or wavering public opinion could easily result in yet another radical change that will open up a new chapter in Japan's aviation history.

Notes

Introduction

1. Isabel Hofmeyr in Bayly, "AHR Conversation," 1444.

2. Garon, "Transnational History and Japan's 'Comparative Advantage,'" 68.

3. Ruttan and Hayami, "Technology Transfer and Agricultural Development," 119–51.

4. For details about such a diffusion process, see Goldman and Eliason, *Diffusion of Military Technology and Ideas*, 11–22.

5. Samuels, *"Rich Nation, Strong Army."*

6. Mizuno designates this ideology as "scientific nationalism" in *Science for the Empire.*

7. Jones, *Live Machines*, 145.

8. Muramatsu, *Westerners in the Modernization of Japan*, 24.

9. For a detailed discussion of foreign influence on the formation of the Japanese Army and Navy, see Presseisen, *Before Aggression*, and Shinohara, *Nihon Kaigun oyatoi gaijin.*

10. For details on the institutional struggles over the establishment of the Aeronautical Research Institute, see Bartholomew, *Formation of Science in Japan*, 217–23.

11. See especially Fritzsche, *Nation of Fliers*; Palmer, *Dictatorship of the Air*; Van Vleck, *Empire of the Air*; and Young, *Aerial Nationalism.*

12. Kern, *Culture of Time and Space*, and Schivelbusch, *Railway Journey.*

13. Corn, *Winged Gospel*; Wohl, *A Passion for Wings*; Wohl, *Spectacle of Flight.*

14. Included in Apollonio (ed.), *Futurist Manifestos*, 22.

15. Le Corbusier, *Aircraft*, 6, 10.

16. Bijker, Hughes, and Pinch, *Social Construction of Technological Systems*, 40. See also Law, *Aircraft Stories.*

17. Kasza, *State and the Mass Media.*

18. Barnhart, *Japan Prepares for Total War*, and Mimura, *Planning for Empire.*

19. Etherton and Hessell, *Manchuria, the Cockpit of Asia*, 163, 307, 312.

20. Young, *Japan's Total Empire*, 88.

21. Quoted in Naitō, *Ōka: Hijō no tokkō*, 163.

1. Powerful Images and Grand Visions

1. For an inventory of publications that cover various aspects of technology transfer, see Staudenmaier, *Technology's Storytellers*, 123–33.

2. I am aware of the anachronistic use of the expression "air-minded." According to the *Oxford English Dictionary*, the word entered the English language only in 1927, nearly a quarter of a century after the Wright brothers' first powered flight in 1903.

3. Saaler has argued that this "independence from civilian control," rather than the defeat of France in the 1890–71 Franco-Prussian War, was the main motive for a move away from the French toward the German model. See Saaler, "The Imperial Japanese Army and Germany."

4. Yet during wartime it would still be an army general who issued imperial orders to the army and the navy. See Drea, *Japan's Imperial Army*, 76–77.

5. Evans and Peattie, *Kaigun*, 1–31.

6. Akimoto, *Nihon hikōsen monogatari*, 99.

7. Reconnaissance balloons had already been successfully deployed by the French Aerostatic Corps (Compagnie d'aérostiers) in 1794 and during the US Civil War (1861–65). The British army started experimenting with them in 1863.

8. Ōura, *Saikin sekai no hikōsen*; and Ono Shōzō, *Kūchū hikōki no genzai oyobi shōrai*, 8–9.

9. *Yomiuri Shinbun*, May 22, 1877.

10. A quite appropriate expression, considering that the word *appare* can be written with the ideographs for "bright sky."

11. Ōura, *Saikin sekai no jikōsen*, 87–88, and Ono, *Kūchū hikōki*, 10–11.

12. Markus, "The Carnival of Edo."

13. Ōtani, "Kyōto to Shimazu Genzō fushi," 18–19.

14. Akimoto, *Nihon hikōsen monogatari*, 143.

15. Akimoto, *Nihon hikōsen monogatari*, 141.

16. Minister of the Army Viscount Terauchi Masatake, "Yamada Isaburō," 1909, JACAR ref. A10112670000.

17. The original report is reproduced in Nihon Kaigun Kōkūshi Hensan Iinkai, *Nihon Kaigun kōkūshi 1 Yōhei hen*, 50–53.

18. Bōeichō bōeikenshūjō senshi shitsu, *Rikugun kōkū no gunbi*, 1:12–13.

19. Drea, *Japan's Imperial Army*, 125–31; Schencking, *Making Waves*, 137–65.

20. Evans and Peattie, *Kaigun*, 159–60.

21. Bōeichō bōeikenshūjō senshi shitsu, *Rikugun kōkū no gunbi*, 1:15–16.

22. French pilot Henri Fabre (1882–1984) is generally credited with having carried out the first successful seaplane flight in March 1910.

23. Nihon Kaigun Kōkūshi Hensan Iinkai, *Nihon Kaigun kōkūshi 3*, 9.

24. Bōeichō bōeikenshūjō senshi shitsu, *Rikugun kōkū no gunbi*, 1:11–14.

25. Roach, *The Wright Company*, 55–71.

26. Headline of the *Daily Mail*, July 27, 1909. The same phrase appeared in the July 29 edition of the *Asahi* newspaper.

27. *Yomiuri Shinbun*, July 28, 1909, morning edition, 1; and *Asahi Shinbun* (Tokyo), July 29, 1909, morning edition, 2.

28. A French newspaper article quoted in Demetz, *Die Flugschau von Brescia*, 50.

29. Hartmann, "Les premiers Farman."

30. For more biographical details, see Shibuya, *Hino Kumazō den*.

31. Bōeichō bōeikenshūjō senshi shitsu, *Rikugun kōkū heiki*, 12.

32. Tokugawa, *Nihon kōkū koto hajime*, 241.

33. Nihon Kōkū Kyōkai, *Nihon kōkūshi*, 2:186–87, 241–44.

34. Yokota, *Kumo no ue kara*, 15–16.

35. Gunmukyoku gunmu ka, "Hino Tokugawa ryōtaii ōshū haken no ken" [The dispatch of the two officers Hino and Tokugawa to Europe], 1910, JACAR ref. C06084949000.

36. Tokugawa, *Nihon kōkū koto hajime*, 52–53.

37. Nagaoka Gaishi, "Request for the Release of the Contract Guarantee Fund," 1910, JACAR ref. C06084948200. Nohara, *Nihon gun'yōki jiten, rikugun hen*, 10, gives a sum of ¥18,835, which includes two spare propellers and the transport cost. Nohara calculates that this sum would have amounted to ¥94 million (around US$920,000) in 2005.

38. Tokugawa, *Nihon kōkū koto hajime*, 54–58.

39. Ader, *L'aviation militaire*, 79.

40. Nagaoka Gaishi, "Request for the Release."

41. Provisional Committee for Military Balloon Research, "On the Use of the Yoyogi Parade Ground," 1910, JACAR ref. C06085048200.

42. For more on the new rituals of the Meiji era and how the leaders exploited them for nation-building and social disciplining, see Fujitani, *Splendid Monarchy*, 105ff.

43. Tokugawa, *Nihon kōkū koto hajime*, 63.

44. The French seven-cylinder Gnome motor was a rotary engine. Its entire engine block rotated together with the propeller around a fixed crankshaft.

45. Nihon Kōkū Kyōkai, *Nihon kōkū shi Meiji Taishō*, 35.

46. The question about who Japan's first pilot actually was has received considerable attention. The discussion is further complicated by contradictory newspaper reports. The *Nichi Roku Shinpō* of December 15, 1910, reported Hino's flight, and the *Kokumin Shinbun* of December 20, 1910, announced Tokugawa's flight as being the first in Japan. A more recent publication by Hiraki Kunio, *Baron Shigeno no shōgai*, and the exhibits of the aviation museums in Tokorozawa and Kakamigahara give the credit to Hino.

47. *Asahi Shinbun*, morning edition, December 18, 1910, 5.

48. *Yomiuri Shinbun*, December 20, 1910, 3.

49. Nihon Kōkū Kyōkai, *Nihon kōkū shi Meiji Taishō*, 91.

50. Quoted in Nihon Kōkū Kyōkai, *Nihon kōkū shi Meiji Taishō*, 34.

51. Article published in the *Yomiuri Shinbun* on October 28, 1912, the day after Tokugawa's flight.

52. Tokugawa, *Nihon kōkū koto hajime*, 115.

53. Takano, *Hino Tokugawa ryō taii*.

54. Iwaya, *Hikō shōnen otogi ebanashi.*

55. Such conflation of youthful enthusiasm and militaristic flag-waving persisted. Hiromi Mizuno has pointed out in *Science for the Empire* how, starting in the 1920s, popular magazines promoted interest in science and technology among Japanese children. When these journals increasingly turned toward military science and technology, they became essential tools for wartime mobilization.

56. Arguably Jules Verne, in his 1904 novel *Master of the World*, was one of the first writers to develop a narrative of aircraft inspiring visions of world domination.

57. For more on these novelists see, for example, Wohl, *A Passion for Wings.*

58. *Yomiuri Shinbun*, April 6, April 9, April 15, 1911, October 25, October 28, 1912.

59. *Kai* refers to the last syllable of the Balloon Committee's name, Rinjigunyō Kikyū Kenkyūkai.

60. For a detailed account of Hino's flagging career, see Shibuya, *Hino Kumazō den*, 206–47.

2. The French Decade

1. *Notre avenir est dans l'air* (*Our future is in the air*) was the title of a 1912 pamphlet published by industrialists André and Édouard Michelin with a print run of over a million copies.

2. Quoted in Hiraki, *Baron Shigeno no shōgai*, 45.

3. For more on Japanese World War I volunteers, see Melzer, "Warfare 1914–1918 (Japan)."

4. Decree of the French War Ministry published in *L'Aérophile*, no. 21 (1915): 253.

5. For details, see Nish, "Japan and the Outbreak of War in 1914," 173–88.

6. Unattributed quotation in Nish, "Japan and the Outbreak of War in 1914," 182.

7. The Siemens Scandal exposed that high-ranking naval officers had accepted bribes for granting German company Siemens a monopoly for naval procurements.

8. As yet another example of the labyrinthine Japanese designation system for military aircraft, "Mo" stood for the first syllable in Maurice Farman, the aircraft designer.

9. Nihon Kōkū Kyōkai, *Nihon kōkū shi Meiji Taishō*, 142.

10. According to Bōeichō bōeikenshūjō senshi shitsu, *Rikugun kōkū heiki*, 15–16, out of the army's sixteen airplanes, these five aircraft were the only operational ones, so the army had mobilized its entire airpower available.

11. *Yomiuri Shinbun*, morning edition, September 7, 1914, 7.

12. *Osaka Mainichi Shinbun*, extra edition, September 27, 1914.

13. Nihon Kōkū Kyōkai, *Nihon kōkū shi Meiji Taishō*, 145–47.

14. *Asahi Shinbun*, extra edition, September 28, 1914.

15. Nihon Kōkū Kyōkai, *Nihon kōkū shi Meiji Taishō*, 143.

16. Kōri, *Aireview's the Fifty Years*, 14; Bōeichō bōeikenshūjō senshi shitsu, *Rikugun kōkū heiki*, 15; Inoue, *Inoue Ikutarō den*, 182; Kimura and Tanaka, *Nihon no meiki hyakusen*, 27.

17. Kōri, *Aireview's the Fifty Years*, 14.

18. Bundesarchiv RM/16/19, *Kriegstagebuch zur Marine-Fliegerstation Tsingtau und Bericht des Oberleutnants zur See Plüschow zur Belagerung Tsingtaus.*

19. *Asahi Shinbun*, morning edition, October 31, 1914, 5.

20. Plüschow, *Die Abenteuer des Fliegers von Tsingtau*, 53. The book ranked among the ten bestselling German books between 1915 and 1940. It was translated into English and—despite its racist undertones—even into Japanese.

21. Plüschow, *Die Abenteuer des Fliegers von Tsingtau*, 53, 56.

22. These numbers include 49 flights and 199 bombardments carried out by the navy.

23. *Yomiuri Shinbun*, morning edition, November 29, 1914, 7.

24. *Asahi Tokyo*, morning edition, December 24, 1914, 3.

25. Dickinson, *War and National Reinvention*, 85.

26. Minister of the Army Oka Ichinosuke, "Officers to Be Dispatched for Aviation Research Purposes," 1914, JACAR ref. B07090456200.

27. The diary refers to visits of aircraft makers LVG, Rumpler, Albatros, AEG, and Otto; engine manufacturers Mercedes and Maybach; and airship builders Parseval and Zeppelin. Kusakari, *Taiō nikki*, 8–28.

28. Diary entry of June 4, 1914. Kusakari, *Taiō nikki*, 8.

29. Kusakari's diary mentions the examination of French Rhône, Clerget, Renault, and Nieuport engines and German Benz and Daimler motors. Kusakari, *Taiō nikki*, 188.

30. Kusakari, *Taiō nikki*, 194.

31. Kusakari Shirō, "An Outline of the Present Condition of the European Countries' Aviation," 1917, JACAR ref. C08020891800.

32. On August 5, 1908, the Zeppelin airship *LZ 4* exploded during a stopover at Echterdingen near Stuttgart. This national disaster triggered an unprecedented donation campaign that became known as the "miracle of Echterdingen." The movement raised capital for the construction of a new Zeppelin ten times the value of the lost airship.

33. "Sekai kōkū genkyō," *Fukuoka Nichinichi Shinbun*, February 1, 1918.

34. Ikutarō, "Kōkūnihon no sōsho," reprinted in Nihon Kōkū Kyōkai, *Nihon kōkūshi*, 1:248–58.

35. Inoue, *Inoue Ikutarō den*, 49–51.

36. The Japanese Navy established its own aviation department in its Bureau of Naval Affairs (Gunmukyokukōkūbu) in 1919.

37. After the Russo-Japanese War, in an effort to establish closer ties between the army and the populace, the slogan "Good Soldiers and Good Citizens" replaced the previous catchphrase, "Rich Nation Strong Army." For more on this phenomenon, see Fujii, *Zaigō Gunjinkai*, 10–11.

38. Nihon Kōkū Kyōkai, *Nihon kōkūshi*, 1:258.

39. Tanaka Giichi, "Rikugunkōkūbu chokurei" [Army flight department ordinance], 1919, JACAR ref. A03021185100.

40. "Tanaka rikugundaijin no kunji chihōchōkan kaigi ni oite" [The Minister of the Army Tanaka's instruction at meeting of local governors], *Yomiuri Shinbun*, September 26, 1920, 2.

41. Bōeichō bōeikenshūjō senshi shitsu, *Rikugun kōkū no gunbi*, 1:168.

42. Inoue, "Kōkūnihon no sōsho," 1:248–58.

43. Hiraki, *Baron Shigeno no shōgai*, 204–6; and Bōeichō bōeikenshūjō senshi shitsu, *Rikugun kōkū heiki*, 35.

44. For more details on these missions, see Presseisen, *Before Aggression*, 1–68; Polak, *Kinu to hikari*, 53–77; and Polak, *Fude to katana*, 10–61.

45. National Archive of Japan, JACAR ref. C03024710400, manuscript numbers 2061 and 2062.

46. Quoted in Polak, *Fude to katana*, 140.

47. Quoted in Polak, *Fude to katana*, 140.

48. "Kōkūkai no gensei jō," Tokyo *Asahi Shinbun*, January 16, 1919; "Kōkūkai no gensei ge," Tokyo *Asahi Shinbun*, January 17, 1919.

49. Quoted in Porte, "L'échec de la Mission militaire française," n7.

50. A tragic event happened in the midst of the aviation craze. On April 16, 1919, several thousand spectators gathered at the ferry terminal on the Kiso River. In the subsequent scramble to get into the ferry, the boat sank, and seven people drowned. Kakamigaharashi kyōiku iinkai, *Kakamigaharashi shi*, 498.

51. A detailed account of the training program can be found in Polak, *Fude to katana*, 106–23.

52. National Archives of Japan, ref. 03011244800, manuscript numbers 1060–65.

53. Jones, *Live Machines*, 11.

54. Porte, "L'échec de la Mission militaire française," n26, Rapport du COL Faure: "The *Spad Lorraine* aircraft have not arrived and cannot be used. . . . The *Spad* monoplanes which we requested for demonstration flights have not arrived . . . not even the silicon which we need for hydrogen production arrived."

55. Bōeichō bōeikenshūjō senshi shitsu, *Rikugun kōkū no gunbi*, 109–10.

56. Report dated February 4, 1920, quoted in Polak, *Fude to katana*, 126.

57. Bōeichō bōeikenshūjō senshi shitsu, *Rikugun kōkū no gunbi*, 91–96.

58. Faure pointed out that even in peacetime the average life span of an aircraft is six months. His figures for required aircraft and pilots concur with those given by Kusakari in his lecture. Kusakari attributes an average life span of only three months to a military aircraft during wartime.

59. Providing target information for the artillery was one of the first relevant uses of military aircraft.

60. Bōeichō bōeikenshūjō senshi shitsu, *Rikugun kōkū no gunbi*, 94.

61. Nihon Kōkū Kyōkai, *Nihon kōkūshi*, 1:255.

62. Quoted in Polak, *Fude to katana*, 126.

63. Quoted in Polak, *Fude to katana*, 124.

64. Bōeichō bōeikenshūjō senshi shitsu, *Rikugun kōkū heiki*, 57.

65. Nohara, *Nihon gun'yōki jiten: Rikugun hen*, 32.

66. Nohara, *Nihon gun'yōki jiten: Rikugun hen*, 33.

67. Nohara, *Nihon gun'yōki jiten: Rikugun hen*, 34.

3. *Japan's Army Aviation in the Wake of World War I*

1. For a closer examination of these competing concepts, see Drea, *Japan's Imperial Army*, 146–62.

2. Hiromi Mizuno discusses the impact of World War I on Japanese science and technology in *Science for the Empire*, 13–14.

3. Barth, *Dolchstoßlegenden und politische Desintegration*.

4. Besides the Hans Grade plane, discussed in the first chapter, imports of German aviation technology had been limited to a Parseval airship in 1912 and a Rumpler Taube monoplane in 1913.

5. Katō Hiroharu, "Ōbei kakkoku gunji shisatsu jōkyō (tōmiyadenka e gokōen 6 gatsu 29 nichi)" [An inspection of the military's condition in each Western Country—lecture to Crown Prince on June 29], 1920, JACAR ref. C11081069100.

6. A diagram of the Supreme War Council Military Representative gives the following numbers of officers in the Inter-Allied Aeronautical Commission of Control: Great Britain (sixty-six), France (sixty-two), Italy (twenty-six), United States (eighteen), Belgium (ten), and Japan (five). Supreme War Council Military Representative, "Regarding the Principles Which Should Govern the Distribution of the Aeronautical Material, Given Up or to Be Given Up by the Central Powers," 1919, JACAR ref. B06150307700.

7. Hara Takeshi, "Tokumei zenkentaishi Matsui Keishirō ika 74mei o heiwajōyaku jishi iin toshite jōsō" [The ambassador extraordinary and plenipotentiary Matsui Keishirō and the following 73 members of the peace treaty enforcement committee report to the emperor], 1919, National Archives of Japan: Honkan-2A-019-00 • B00910100.

8. Members of the Supervisory Committee on German Aviation, "Ō dai 2 gō shi 30 Taishō 14 nen 4 gatsu 25 nichi Doitsu kōkūkai genkyō no ken hōkoku" [Europe No. 2-30: Report on the current situation of the German aviation industry, April 25, 1925], 1925, JACAR ref. C08040446100.

9. Koiso, *Katsuzan kōsō*, 402–10.

10. Junkers-Japan Archivmaterial Deutsches Museum, "Verwertung Flugzeuge Asien Japan," folder 0705 T1 1919–1921.

11. Koiso, *Katsuzan kōsō*, 410.

12. "Doitsu kōkūkai no genkyō" [The current condition of the German aviation industry], 1922, JACAR ref. C08040371600.

13. "Arumi yori karuku hagane yori kataku" [Lighter than aluminum, harder than steel], *Osaka Asahi Shinbun*, October 1, 1926.

14. It should be noted that the alleged long-term durability of duralumin was put into question in 1924 when scientists at the US Bureau of Standards discovered a type of erosion inside the metal that reduced its strength. For details, see Schatzberg, *Wings of Wood, Wings of Metal*, 54–56.

15. Watanabe Kōtarō, "Kinzokusei hikōki ni kan suru kenkyū" [Research on all-metal aircraft], 1925, JACAR ref. C08040449800.

16. Navy Ministry, "Fuzoku kokusai kōkūkaigi kankei 43 satsu" [Attachment to the 43 volumes on the international aviation conference], 1926, JACAR ref. C04015196500.

17. "Doitsu kōkūkai no genkyō."

18. Takada Yoshimitsu, "Doitsu kōkūkanshi geppō" [Monthly German aviation inspection reports], 1920, JACAR ref. C04015195400.

19. "Doitsu kōkūkai no genkyō."

20. Watanabe Kōtarō, "Ō dai 2 gō shi 92 Furiidorihhisuhaahen kōkūseizōkaisha kenetsu hōkoku" [Europe No. 2-92: Inspection report on the Friedrichshafen Aircraft Manufacturing Company], 1924, JACAR ref. C08040441100.

21. Supreme War Council Military Representative, "Regarding the Principles Which Should Govern the Distribution of the Aeronautical Material, Given Up or to Be Given Up by the Central Powers," 1919, JACAR ref. B06150307700, 198.

22. Article 224 provided the following numbers: France (30 percent), Great Britain (30 percent), United States (15 percent), Italy (15 percent), Japan (5 percent), and Belgium (5 percent).

23. Tanaka Giichi, "Senri kōkūki bunpai ni kan suru" [Distribution of war victory aircraft], 1919, National Archives of Japan honkan-2A-011-00, rui 01325100. The exchange rate is according to the Bank of Japan Institute for Monetary and Economic Studies.

24. Army Ministry, "Dai 49 gō Doitsu yori rengōkoku ni kōfu subeki kōkūki no ken" [No. 49 Aircraft to be handed over from Germany to the Allied Powers], 1919, JACAR ref. C08040384500.

25. Supreme War Council, "Regarding the Principles Which Should Govern the Distribution of the Aeronautical Material," 194.

26. Army Ministry, "Dai 50 gō Doitsu yori rengōkoku ni kōfu subeki kōkūki no ken" [No. 50 Aircraft to be handed over from Germany to the Allied Powers], 1919, JACAR ref. C08040384600.

27. Foreign Ministry, "Rikugun ni oite ōshū hikōkisaibu no buhinfuzoku kikaitō kōbaigata no ken" [Purchasing aircraft parts and accessories confiscated by the army], 1920, JACAR ref. B07090277400, 286.

28. Foreign Ministry, "Rikugun ni oite ōshū hikōkisaibu no buhinfuzoku kikaitō kōbaigata no ken" [Purchasing aircraft parts and accessories confiscated by the army], 1920, JACAR ref. B07090277400, 288.

29. Foreign Ministry, "Rikugun ni oite ōshū hikōkisaibu no buhinfuzoku kikaitō kōbaigata no ken" [Purchasing aircraft parts and accessories confiscated by the army], 1920, JACAR ref. B07090277400, 289.

30. "Doitsu kōkūshi yatoiire ni kan suru ken" [The employment of German aviation engineers], 1920 JACAR ref. C03025180000.

31. For more details on the role of trading companies, see Miyamoto and Yoshio, *Sōgō shōsha no keieishi.*

32. The above numbers can be found in Rikugun Kōkū bu, "Senri hikōkikanran no ken" [The exhibition of aircraft obtained as war reparations], 1921, JACAR ref. C03025216900. Nihon Kōkū Kyōkai, *Nihon kōkūshi*, 1:369, maintains that the delivery consisted of forty-five different types of aircraft and 320 engines.

33. Haddow and Grosz, *The German Giants*, 273–74.

34. Nihon Kōkū Kyōkai, *Nihon kōkūshi*, 1:369.

35. Parts of the catalog are reproduced in Tokorozawa kōkūshiryō, *Yūhi sora no makuake Tokorozawa*, 209.

36. Nihon Kōkū Kyōkai, *Nihon kōkūshi*, 1:314.

37. "Gozen hikō no yu-shiki teisatsuki" [The flight of a Junkers reconnaissance aircraft in the Imperial presence], *Osaka Asahi*, morning edition, October 25, 1922.

38. Kuwabara, *Kaigun kōkū kaisōroku*, 195.

39. "Doitsu kōkūshi yatoiire ni kan suru ken" [The employment of German aviation engineers], 1920, JACAR ref. C03025180000.

40. Foreign Ministry, "Rikugun ni oite ōshū hikōkisaibu no buhinfuzoku kikaitō kōbaigata no ken" [Purchasing aircraft parts and accessories confiscated by the army], 1920, JACAR ref. B07090277400.

41. "Doitsu kōkūshi yatoiire ni kan suru ken."

42. Army Air Section, "Gaikoku kōkūgijutsusha shōhei ni kansuru ken" [The invitation of foreign aeronautical engineers], 1923, JACAR ref. C03011780200.

43. Navy Ministry, "Kōkū ippan (8)" [Aviation general (8)], 1925, JACAR ref. C08051420300, 352.

44. "Mitsubishi zōsenjo ni oite Doitsu hikōki kōnyū no ken" [The purchase of a German aircraft by Mitsubishi Shipyard], 1924–25, JACAR ref. B07090278600, 328.

45. Navy Ministry, "Kōkū ippan (8)," 352.

46. "Mitsubishi zōsenjo ni oite Doitsu hikōki kōnyū no ken," 329.

47. For an analysis of Douhet's works, see Meilinger, "Giulio Douhet and the Origins of Airpower Theory," 1–41.

48. Bōeichō bōeikenshūjō senshi shitsu, *Rikugun kōkū no gunbi*, 233.

49. It is important to note that Douhet's new ideas were not universally appreciated. In 1916 Douhet was court-martialed and imprisoned for his open criticism of Italy's military leadership that in his view ignored the strategic advantage of large-scale aerial bombing.

50. Bōeichō bōeikenshūjō senshi shitsu, *Rikugun kōkū no gunbi*, 208–9.

51. Polak, *Fude to katana*, 140.

52. Jauneaud, *L'aviation militaire*, 208–14.

53. Jauneaud's enthusiasm for aerial armament was not shared by the French War Ministry, which in 1925 began its plans for the Maginot Line, consuming a huge part of appropriations at the cost of airpower expansion until 1940.

54. Ogasawara, *Kōkū senjutsu kōjuroku*, 1–31.

55. Bōeichō bōeikenshūjō senshi shitsu, *Rikugun kōkū no gunbi*, 224.

56. Watanabe Kōtarō, "Ō dai 2 gō shi 11 Taishō 13 nen 12 gatsu 15 nichi Doitsu hikōki seinō bessatsu no tōri gohōkoku" [Europe No. 2-11: Report on the performance of German and French airplanes as shown in the attached document, December 15, 1924], 1924, JACAR ref. C08040448700.

57. Watanabe Kōtaro, "Ō dai 2 gō shi 41 Taishō 14 nen 7 gatsu 14 nichi shoruisōfuzuke no ken hōkoku" [Europe No. 2-41: Report of sending documents, July 14, 1925], 1925, JACAR ref. C08040445700.

58. Sugita, *Nihon no seisenryaku*, 36–37.

59. Bōeichō bōeikenshūjō senshi shitsu, *Rikugun kōkū no gunbi*, 193.

60. Yokoyama, "Military Technological Strategy," 117.
61. Koiso, *Katsuzan kōsō*, 407.
62. Koiso and Musha, *Kōkū no genjō*, 74.
63. Ugaki, "Kokka sōdōin," 263–84.
64. Bōeichō bōeikenshūjō senshi shitsu, *Rikugun kōkū no gunbi*, 255–56.
65. Bōeichō bōeikenshūjō senshi shitsu, *Rikugun kōkū no gunbi*, 238.
66. Bōeichō bōeikenshūjō senshi shitsu, *Rikugun kōkū no gunbi*, 241.
67. Inoue, "Kōkūnihon no sōsho," 1:248–58.
68. See Kasza, *State and the Mass Media in Japan*, 28–53.
69. National Archive of Japan JACAR ref. C03012294800, 393.
70. Details of the "visit Europe flight" can be found in *Ōshū hōmon dai hikō* and in Maema, *Asahi Shinbun hōō daihikō*, vols. 1 and 2.
71. The term "black ships" refers to Commodore Perry's steamboats, which arrived in Japan in 1853. It became a synonym for foreign threat and national humiliation.
72. For the full text of "The Convention Embodying Basic Rules of the Relations between Japan and the Union of Soviet Socialist Republics," see *American Journal of International Law* 19, no. 2, Supplement: Official Documents (April 1925): 78–88.
73. Ōsaka Honsha Hanbai Hyakunenshi Henshū Iinkai, *Asahi Shinbun hanbai hyakunen shi*, 321, gives the numbers: the planned budget was ¥400,000; the actual costs amounted to ¥636,194; the proceeds from donations were ¥356,820.
74. Nihon Kōkū Kyōkai, *Nihon kōkū shi Meiji Taishō*, 479.
75. Nihon Kōkū Kyōkai, *Nihon kōkū shi Meiji Taishō*, 580.
76. For Abe's personal memoirs, see Abe, "*Hatsukaze, Kochikaze* hōōhikō no omoide," 69–72.
77. *Ōshū hōmon dai hikō kinen gahō*, 1:5.
78. Bōeichō bōeikenshūjō senshi shitsu, *Rikugun kōkū heiki*, 51.

4. *On the Way to Independent Aircraft Design*

1. Mikesh and Shorzoe, *Japanese Aircraft*, 144.
2. Taken from the memoirs of Tateyama Toshikuni, the former director of Kawasaki Heavy Industries. See, on this point, Kawasaki Heavy Industries, *Kawasaki Jūkō Gifu kōjō no omoide*, 5.
3. *Kawasaki jūkō Gifu kōjō gojūnen no ayumi*, 7.
4. Senba, *Hikōki ni miserarete*, 36.
5. van der Mey, *Dornier Wal*, 15.
6. EADS Corporate Heritage, Archiv Dornier, folder Lizenz Japan.
7. Abe, *Kawasaki Zōsenjo yonjūnenshi*, 313, gives a 1925 net profit of ¥5,790,151. For comparison, during the same year, the company spent ¥8 million on research and development. See, on this point, Kawasaki's semi-annual reports in the July 7, 1925, and January 4, 1926, *Asahi Shinbun* Tokyo, morning edition, 1.
8. EADS Corporate Heritage, Archiv Dornier.
9. Kawasaki Jūkōgyō kabushikigaisha, *Kawasaki Jūkōgyō kabushiki gaisha hyakunenshi*, 32.
10. Kawasaki Dockyard, "Manufacturing and selling of all-metal aeroplanes," 1924, JACAR ref. C03012003600, 933.

11. Bundesarchiv BA-MA RH8-3606.

12. "Kūchū no kaibutsu" [A monster in the sky], *Asahi Shinbun*, December 4, 1924; "Daiseikō ura ni migoto na hikōburi" [A very successful great flight performance], *Asahi Shinbun*, December 9, 1924.

13. "N" apparently standing for "Nippon" (that is, Japan).

14. EADS Corporate Heritage, Archiv Dornier, folder Lizenzvertrag.

15. Quoted in "Dornier Flugzeuge in Japan," 54–55.

16. Military historians have pointed out the "irrationality" of oversized weaponry while acknowledging its overawing psychological impact. See van Creveld, *Technology and War*, 67–78.

17. Kawasaki Kōkūkikōgyō kabushikigaisha, *Kōkūkiseisaku enkaku kitai no bu*, 7.

18. *Kawasaki Jūkō Gifu kōjō gojūnen no ayumi*, 8.

19. These bombers were the US Boeing YB-9 and the British Avro Anson.

20. "The Kawasaki Aircraft Factory," *Kōbe Yūshin Nippō*, April 9, 1932.

21. Board meeting, July 1924, Mitsubishi archive Tokyo, MHI-00339.

22. Mitsubishi archive Tokyo, MHI-00781. The contract was executed in English.

23. Bōeichō bōeikenshūjō senshi shitsu, *Rikugun kōkū heiki*, 68–70. See also Nihon Kōkū Kyōkai, *Nihon kōkūshi Shōwa zenkihen*, 3–38.

24. An elongated wing generates less induced drag, which is a result of air flowing from the lower wing surface to the upper wing surface. Less induced drag results in fewer vortexes, thereby increasing an aircraft's speed and flight range.

25. Nozawa, *Nihon kōkūki sōshū: Mitsubishi hen*, 44.

26. Nohara, *Zukai sekai no gun'yōkishi: 8*, 8.

27. Weckbach, *Heilbronner Köpfe*, 30.

28. Bōeichō bōeikenshūjō senshi shitsu, *Rikugun kōkū no gunbi*, 279.

29. Sanuki, *Sanuki Matao no hitorigoto*, 264.

30. *Aichi no kōkūshi*, 144–45.

31. Mitsubishi Jūkōgyō KK, *Ōjibōbō*, 1:145.

32. DFVLR Historisches Archiv Berlin, letter by Baumann's son Alex to Heinz Nowarra, September 30, 1982.

33. Kawasaki kōkūkikōgyō kabushikigaisha, *Kōkūkiseisaku Gifu kitai no bu*, 1946, Kakamigahara Aerospace and Science Museum, not cataloged, 7; and Doi, *Hikōki sekkei*, 44.

34. Vogt, *Weltumspannende Memoiren*, 65.

35. Vogt, *Weltumspannende Memoiren*, 60.

36. Anderson, *The Airplane*, 261.

37. Bōeichō bōeikenshūjō senshi shitsu, *Rikugun kōkū no gunbi*, 282.

38. Abe, *Kawasaki Zōsenjo*, 87.

39. Bōeichō bōeikenshūjō senshi shitsu, *Rikugun kōkū no gunbi*, 280.

40. Doi, *Hikōki sekkei*, 53.

41. Vogt, *Weltumspannende Memoiren*, 70.

42. Interview with Doi Takeo in "Sekkei wa aato obu conpuromaizu."

43. Tsuji and Kurita, *Hondo bōkū sakusen*, 9.

44. Inoue and Ugaki, "The competition of the trial single-seat fighter aircraft," 1926–27, JACAR ref. C01003741300, 9–10.

45. The specifications required the aircraft to be able to reach a speed of at least 250 kilometers per hour and climb to an altitude of 5,000 meters in less than twelve minutes.

46. Doi, *Hikōki sekkei*, 58.

47. Nakajima Chikuhei, "Request for having foreigners enter and leave the country," 1927, JACAR ref. C04015990700.

48. This aircraft was the Dornier Do H Falke (falcon), a 1922 design. *50 Jahre Dornier*, 21–25.

49. Doi, *Hikōki sekkei*, 62–63.

50. Mitsubishi Jūkōgyō KK, *Ōjibōbō*, 1:45.

51. Nowarra, *Die Flugzeuge des Alexander Baumann*, 131.

52. Vogt, *Weltumspannende Memoiren*, 71.

53. Doi, *Hikōki sekkei*, 66–73.

54. NACA was established in 1915 to coordinate and supervise aeronautical research. In 1920 NACA began its own research activities, making use of increasingly sophisticated wind tunnels: A 1.5-meter-diameter wind tunnel started operation in 1920, a variable-density wind tunnel in 1923, and a full-scale wind tunnel in 1930. For the early years of NACA, see Roland, *Model Research*, 1:1–98. More on the airfoil can be found in Higgins, "The Characteristics of the N.A.C.A. M-12 Airfoil Section."

55. With a load factor of fifteen, an aircraft can hold out against an acceleration force fifteen times greater than its weight.

56. Doi, *Hikōki sekkei*, 70.

57. "A World Record Set by a Military Aircraft," *Asahi Shinbun*, November 5, 1930.

58. "The Army Adopted Kawasaki's New Fighter Aircraft / Able to Fly within One Hour between Tokyo and Osaka / Special Speed Performance," *Yomiuri Shinbun*, October 22, 1931, 7.

59. "The Power of Our Air Force Will Protect the Skies," *Yomiuri Shinbun*, April 9, 1933, 9.

60. Nozawa, *Nihon kōkūki sōshū: Kawasaki hen*, 62; and Nozawa, *Nihon kōkūki sōshū: Nakajima hen*, 44.

61. The expression "Manchurian lifeline" was first used by Japanese politician Matsuoka Yosuke (1880–1946) in January 1931 and served as a metaphor for the vital importance of northeast China for Japan. For more on this topic, see Young, *Japan's Total Empire*, 88.

62. Doi, *Hikōki sekkei*, 58–62.

63. Francillon, *Japanese Aircraft of the Pacific War*, 114.

64. *Nihon no kōkū runesansu*, 81.

65. Kōno Fumihiko, *Giken Andō gishi yori chōshu shita Mitsubishi kankei jiki shisakuki no gaiyō* [An outline by the army engineer Andō about a new generation of experimental aircraft], 1939, Mitsubishi Nagoya Aerospace Museum, not cataloged.

66. "Rikukūgun ni shinei sekai saidai no chōjū bakugekiki" [The army air force's new and powerful machine, the world's largest super heavy bomber], *Yomiuri Shinbun*, June 9, 1932, 7.

67. Bōeichō bōeikenshūjō senshi shitsu, *Rikugun kōkū no gunbi*, 263; and Asada, *From Mahan to Pearl Harbor*, 55.

68. Bōeichō bōeikenshūjō senshi shitsu, *Rikugun kōkū heiki*, 81–83; and Jōhō, *Rikugunshō gunmukyoku shi, Jōkan*, 292–93.

69. Jōhō, *Rikugunshō gunmukyoku shi, Jōkan*, 287–88; and Kawata, "Report on the martial spirit among the US service personnel," 1923, JACAR ref. C03022632000.

70. Jōhō, *Rikugunshō gunmukyoku shi, Jōkan*, 292.

71. Bōeichō bōeikenshūjō senshi shitsu, *Rikugun kōkū no gunbi*, 263.

72. Koiso and Musha, *Kōkū no genjō to shōrai*, 43.

73. Koiso and Musha, *Kōkū no genjō to shōrai*, 8.

74. Bōeichō bōeikenshūjō senshi shitsu, *Rikugun kōkū heiki*, 81–83.

75. Army Special Bureau of Aviation, "Kyūnishiki kyūbakugekiki seishi ni kan suru ken" [Establishment of the standard of the Type-92 Heavy Bomber], 1933, JACAR ref. C01003983200, 649.

76. Army Special Bureau of Aviation, "Kyūnishiki kyūbakugekiki seishi ni kan suru ken" [Establishment of the standard of the Type-92 Heavy Bomber], 1933, JACAR ref. C01003983200, 658–62. For the army's 1928–31 budget, see JACAR ref. C12121666100.

77. Budraß, "Rohrbach und Dornier," 218; and Budraß, *Flugzeugindustrie und Luftrüstung*, 241.

78. Wagner, *Hugo Junkers*, 295–303.

79. Bōeichō bōeikenshūjō senshi shitsu, *Rikugun kōkū no gunbi*, 365.

80. Katsuragi, *Rekishi no naka no Nakajima hikōki*, 146; Army Special Bureau of Aviation, "Kyūnishiki kyūbakugekiki seishi ni kan suru ken."

81. As it turned out, the new German coalition government under Hermann Müller made substantial budget cuts in the Transocean Program in 1929 that also affected Junkers's G 38 project.

82. Junkers-Japan Archivmaterial Deutsches Museum, "Verwertung Flugzeuge Asien Japan," folder 0705 T4 1928 I; and Mitsubishi Tokyo Archive MHI-339.

83. *Mitsubishi shashi*, 35:246.

84. Junkers-Japan Archivmaterial Deutsches Museum, "Verwertung Flugzeuge Asien Japan," folder 0705 T5 1928 II and MHI 1642.

85. Junkers-Japan Archivmaterial Deutsches Museum, "Verwertung Flugzeuge Asien Japan," folder 0705 T6 1929 I.

86. Nohara, *Zukai sekai no gun'yōkishi, 6*, 43.

87. Army Special Bureau of Aviation, "Kyūnishiki kyūbakugekiki seishi ni kan suru ken," 683.

88. Army Special Bureau of Aviation, "Kyūnishiki kyūbakugekiki seishi ni kan suru ken," 683.

89. *Aichi no kōkūshi*, 152.

90. Junkers-Japan Archivmaterial Deutsches Museum, "Verwertung Flugzeuge Asien Japan," folder 0705 T9 1931–1933.

91. Kariya, *Nihon rikugun shisakuki monogatari*, 79–91.

92. Army Special Bureau of Aviation, "Kyūnishiki kyūbakugekiki seishi ni kan suru ken," 686–89.

93. "Sekai ni hokoriuru chōbakugekiki seisaku" [The production of a Superbomber of which we can be proud before the whole world], *Asahi Shinbun*, February 14, 1931, 3.

94. "Rikukūgun ni shinei sekai saidai no chōjū bakugekiki" [The army air force's new and powerful machine, the world's largest super-heavy bomber], *Yomiuri Shinbun*, June 9, 1932, 7.

95. Ministry of Army Newspaper Group, "Sending Issue on Crackdown of Article on Aircraft Examination," 1935, JACAR ref. C01004075900.

96. Army Special Bureau of Aviation, "Releasing Military Secret of Aviation Equipment Model 92 Heavy Bomber Construction," 1938, JACAR ref. C01001635900.

97. Nozawa, *Nihon kōkūki sōshū: Mitsubishi hen*, 39.

98. "Mitsubishi ni okeru kitaisekkei shisaku no hensen," 156.

99. Matsuoka, *Mitsubishi hikōki monogatari*, 176.

100. Kawasaki Heavy Industries, *Kawasaki Jūkō Gifu kōjō no omoide*, 54–55.

101. Mitsubishi Tokyo Archive MHI-339.

102. According to Nozawa, 118 Ki-1 and 174 Ki-2 were produced. Nozawa, *Nihon kōkūki sōshū: Mitsubishi hen*, 60, 93.

103. Bōeichō bōeikenshūjō senshi shitsu, *Rikugun kōkū no gunbi*, 329.

104. Bōeichō bōeikenshūjō senshi shitsu, *Rikugun kōkū no gunbi*, 337.

105. Bōeichō bōeikenshūjō senshi shitsu, *Rikugun kōkū no gunbi*, 341.

106. According to Japanese newspaper reports the army air force's activities had such an impact that Chinese general Ma Chan-shan (1885–1950) implored the Japanese not to let their aircraft fly anymore. *Yomiuri Shinbun*, November 15, 1931.

107. For the "news war" over covering the events in Manchuria, see Young, *Japan's Total Empire*, 57–88.

108. Fujii, "Shōwa shoki sensōkaishiji," 39–55.

109. An approximate translation; *gō* is a name suffix for vehicles.

110. The phonograph is available at the library of Showa-kan (National Showa Memorial Museum), Tokyo, item no. 009132.

111. *Yomiuri*, December 9, 1931.

112. *Yomiuri*, December 9, 1931.

113. *Chūgai Shōgyō Shinpō*, January 11, 1932.

114. An entire chapter in the popular book *Warera no kūgun* [Our airforce] (published in 1932) was devoted to the first thirty-six donors. See Shōnen kokubōkai, *Warera no kūgun*, 114–20.

115. For an account of how the army's promotion policy obstructed the buildup of professional expertise, see Ishida and Hamada, "Kyūnihongun ni okeru jinji hyōkaseido," 55.

116. Koiso, *Katsuzan kōsō*, 425.

117. Humphreys, *The Way of the Heavenly Sword*, 79–107. For details on Araki's clash with the army's "total war officers," see Barnhart, *Japan Prepares for Total War*, 34–37.

118. Akimoto, *Kyojinki monogatari*, 160.

119. Izawa, *Nihon Rikugun jūbakutai*, 51.

120. Akimoto, *Kyojinki monogatari*, 160.

121. The only other country that made use of its bombers before 1931 was Great Britain. Between 1919 and 1932, the Royal Air Force deployed their bombers for antiguer-

rilla missions in the Middle East protectorates. Cox, "A Splendid Training Ground," 157–84.

122. For more on the development of the army's air doctrine and its impact on aircraft procurement policy, see Mizusawa, *Gunyōki no tanjō*, 32–43.

123. Sawai, *Kindai Nihon no kenkyū kaihatsu taisei*, 233–54.

124. For critical comments on the incompetent leaders of the Army Aviation Headquarters, see Hata, Itō, and Hara, *Gensui Hata Shunroku kaikōroku*, 469–79.

125. In 1943 Nakajima also began two long-range bomber projects. The Nakajima G10N Fugaku, a six-engine aircraft designed to bomb the United States, never made it beyond the drawing board. The G8N Renzan, a four-engine bomber for the navy, made a successful test flight in 1944, but due to material shortage, it did not enter series production. See Torikai, ed., *Shirarezaru gunyōki kaihatsu*, 1:9–16, 1:73–124.

5. Navigating a Sea of Change

1. For more details on the debate about an independent Japanese Air Force, see Bōeichō bōeikenshūjō senshi shitsu, *Kaigun kōkū gaishi*, 74–79.

2. "Nachlass Kissel Flugmotoren," 30.12.38, Mercedes Benz Archive, Stuttgart.

3. According to Eiichiro Sekigawa, the naval arsenals produced a third of the navy's aircraft between 1920 and 1930. Sekigawa, *Pictorial History of Japanese Military Aviation*, 33.

4. Wieselsberger had already supervised the installation of a wind tunnel at the navy's Tsukiji Aircraft Test Laboratory in 1922. He led the construction of wind tunnels at Tokyo Imperial University's Aeronautical Research Institute, the army's Air Technical Research Institute, Mitsubishi, and Aichi Tokei. See Suikōkai, *Kaisō no Nihon Kaigun*, 476. For a description of six major Japanese wind tunnels in operation by 1931, see "Wind Tunnels in Japan."

5. For a thorough discussion of the issues shaping interwar military innovation, see Williamson, "Innovation: Past and Future," 300–328.

6. Millet, "Patterns of Military Innovation," 335–36.

7. I am using the term *doctrine* in the sense of "fundamental principles by which the military forces or elements thereof guide their actions in support of national objectives," as defined in Joint Chiefs of Staff, *Department of Defense Dictionary*, 78.

8. The Japanese Diet approved the eight-eight fleet program in 1920. However, budgetary restrictions during the postwar recession and the 1922 Washington Naval Treaty resulted in the program's cancellation. For more on the navy's eight-eight project see Schencking, *Making Waves*, 214–17.

9. Watabe, *Nihon no hikōkiō Nakajima Chikuhei*, 119–22.

10. Nihon Kaigun Kōkūshi Hensan Iinkai, *Nihon Kaigun kōkūshi 1*, 107–10.

11. Nihon Kaigun Kōkūshi Hensan Iinkai, *Nihon Kaigun kōkūshi 1*, 110.

12. JACAR ref. C11080598300, C11080598500.

13. JACAR ref. C11080598500.

14. The credit for the first landing on an anchored ship's deck goes to US pilot Eugene Ely (1886–1911), who landed on the Navy battleship *Pennsylvania* in 1911.

15. JACAR ref. C10100792400, C10100792500, C10100792600.

16. For details, see Nozawa, *Nihon kōkūki sōshū: Yunyūki hen.*

17. Nozawa, *Nihon kōkūki sōshū: Yunyūki hen*, 173.

18. Nozawa, *Nihon kōkūki sōshū: Nakajima hen*, 120.

19. Edagawa, *Aichi Tokei Denki 85-nenshi*, 90.

20. "Imperial Maritime Defense Volunteer Association purchased an all-metal aircraft from Germany," *Asahi Shinbun*, February 10, 1923, 5. For further details on the association, see *Giyū zaidan kaibō gikai no mokuteki shimei.*

21. Nozawa, *Nihon kōkūki sōshū: Yunyūki hen*, 86.

22. Asada, *From Mahan to Pearl Harbor*, 107.

23. Heinkel, *Stürmisches Leben*, 75.

24. Nozawa, *Nihon kōkūki sōshū: Yunyūki hen*, 89.

25. Duppler and Forsmann, *Marineflieger*, 18.

26. *Nihon no kōkū runesansu*, 29–30.

27. Miki Tetsuo, "Aichi Tokei Denki no hikōki," 101.

28. Edagawa, *Aichi Tokei Denki 85-nenshi*, 138.

29. Heinkel, *Stürmisches Leben*, 130.

30. Heinkel, *Stürmisches Leben*, 164.

31. Details on the development of Japanese catapults can be found in Nihon Kaigun Kōkūshi Hensan Iinkai, *Nihon Kaigun kōkūshi 3*, 772–74, and in Okamura, *Kōkū gijutsu no zenbō*, 2:309.

32. Acceleration by early catapults could result in g-loads in excess of four times an aircraft's weight.

33. An article about the Tondern raid appeared in *Asahi Shinbun* (Tokyo), July 23, 1918, morning edition.

34. There seems to be some irony in the fact that the first US aircraft carrier bore the name of Samuel Langley (1834–1906), an aviation pioneer whose attempts at manned flight ended up with him in the Potomac River.

35. Nihon Kaigun Kōkūshi Hensan Iinkai, *Nihon Kaigun kōkūshi 1*, 194.

36. Bach, *Luftfahrtindustrie im Ersten Weltkrieg*, 176.

37. Fearon, "The Formative Years of the British Aircraft Industry," 493.

38. Hoare, "Ernest Cyril Comfort," 190.

39. Mitsubishi Archives, Tokyo, MHI-00781.

40. Mitsubishi Jūkōgyō KK, *Ōjibōbō*, 3:67.

41. Ferris, "Armaments and Allies."

42. Quoted in Till, *Air Power and the Royal Navy*, 91.

43. JACAR ref. C08021355900.

44. Nagura and Yokoi, *Nichi-Ei heiki sangyōshi*, 379.

45. For the Tracey mission, see Asakawa, "Anglo-Japanese Military Relations, 1800–1900," 19–20, and Shinohara, *Nihon Kaigun oyatoi gaijin*, 92–100. More on the Douglas mission can be found in Gow, "The Douglas Mission," 144–57.

46. Quoted in Nihon Kaigun Kōkūshi Hensan Iinkai, *Nihon Kaigun kōkūshi 3*, 269.

47. Nihon Kaigun Kōkūshi Hensan Iinkai, *Nihon Kaigun kōkūshi 1*, 75.

48. US Military Attaché London, May 27, 1921, *U.S. Military Intelligence Reports Japan, 1918–1941*, reel 27.

49. Nihon Kaigun Kōkūshi Hensan Iinkai, *Nihon Kaigun kōkūshi 2*, 707–19.
50. Quoted in Till, *Air Power and the Royal Navy*, 64.
51. For a detailed account (which makes no reference to Japanese sources), see Ferris, "A British 'Unofficial' Aviation Mission," 419–39.
52. In January 1918, the Air Ministry was put in charge of the newly established Royal Air Force, putting the ministry's leaders at the same level as the Royal Navy's Admiralty Board and the British Army's Imperial General Staff.
53. Brackley, *Brackles*, 169.
54. National Archives Kew KV2-871, 61.
55. Forbes-Sempill, "The British Aviation Mission," 556.
56. Comment of the US military attaché in London on "Japanese Aeronautical Activities," *U.S. Military Intelligence Reports*, reel 27.
57. For a complete list, see Forbes-Sempill, "British Aviation Mission," 560.
58. *Asahi Tokyo*, April 17, 1921, evening edition, 2.
59. *Japan Gazette*, September 3, 1921.
60. Forbes-Sempill, "British Aviation Mission," 568.
61. Brackley, *Brackles*, 174.
62. Brackley, *Brackles*, 175.
63. Quoted in Hoare, "Ernest Cyril Comfort," 186.
64. Matsuoka, *Mitsubishi hikōki monogatari*, 36.
65. *Asahi Tokyo*, February 23, 1923, evening edition, 2.
66. Nozawa, *Nihon kōkūki sōshū: Mitsubishi hen*, 112.
67. Genda, *Kaigun Kōkūtai shimatsuki*, 1:33.
68. Kimura and Tanaka, *Nihon no meiki hyakusen*, 51.
69. Genda, *Kaigun Kōkūtai shimatsuki*, 1:34.
70. Brackley, *Brackles*, 195.
71. Brackley in a letter to General W. Caddell, in Brackley, *Brackles*, 177–78.
72. Forbes-Sempill, "British Aviation Mission," 584.
73. Ferris, "A British 'Unofficial' Aviation Mission," 428.
74. O'Brien, *The Anglo-Japanese Alliance*, 262.
75. Fearon, "Aircraft Manufacturing," 218.
76. For the British judgment of the Japanese Army, see Ferris, "'Worthy of Some Better Enemy?,'" 223–56.
77. Brackley, *Brackles*, 132.
78. Forbes-Sempill, "British Aviation Mission," 568.
79. Report, May 31, 1923, Air Ministry, quoted in Till, *Air Power and the Royal Navy*, 64.
80. Wada, *Kaigun kōkū shiwa*, 167.
81. Itō, *Daikaigun o omō*, 339.
82. British documents related to the Sempill spy case were released only in May 2002. The National Archives, Kew, "William Francis Forbes-Sempill," KV 2/871, 1921–1926.
83. Forbes-Sempill, "British Aviation Mission," 582.
84. Forbes-Sempill, "British Aviation Mission," 582.
85. Brackley, *Brackles*, 146–47.

86. Brackley, *Brackles*, 169.
87. National Archives Kew KV2-871, 158.
88. National Archives Kew KV2-871, 10.
89. National Archives Kew KV2-871, 171.
90. The Japanese turned down the offer to purchase the British carrier aircraft *Plover*. Mitsubishi imported the 385 horsepower Armstrong Siddeley Jaguar engine but installed it on only one single-passenger aircraft.

6. Japan's Naval Aviation Taking the Lead

1. Quoted in Nish, *Alliance in Decline*, 390.
2. Nish, *Alliance in Decline*, 207.
3. The US military attaché to London reported in May 1921: "The F-5 Boat is not now highly thought of and is obsolescent if not practically obsolete. It is of no great value except for training purposes as it takes too long to get off the water with anything like a full load." Lester (ed)., *U.S. Military Intelligence Reports*, reel 27, frame 140.
4. Budraß, "Rohrbach und Dornier," 210.
5. Takada Yoshimitsu, "Doitsu kōkū kanshi geppō" [Monthly German aviation inspection reports], 1920, JACAR ref. C04015195400.
6. Kuwabara, *Kaigun kōkū kaisōroku*, 197.
7. *Nihon no kōkū runesansu*, 111.
8. For details, see Ebert, "Rohrbach," 4–5.
9. Budraß, "Rohrbach und Dornier," 210.
10. Adolf Rohrbach, "Gründung und wirtschaftliche Entwicklung der Rohrbach-Metall-Flugzeugbau GmbH August 1922–Dezember 1923," 1924, BA-MA RH8-3606.
11. Mitsubishi Jūkōgyō KK, *Ōjibōbō*, 1:19.
12. For Hattori Jōji's vivid account of the humble beginnings of Mitsubishi's aircraft department, see Mitsubishi Jūkōgyō KK, *Ōjibōbō*, 1:28.
13. Moriya, *Meikōkōsakubu no senzensengoshi*, 17–19.
14. Adolf Rohrbach,"Gründung und wirtschaftliche Entwicklung."
15. Board meetings, July 1922 to December 1923, Archives at the Mitsubishi Economic Research Institute (MERI), Tokyo, MHI-00342.
16. Quoted in Budraß, "Rohrbach und Dornier," 211.
17. Adolf Rohrbach, "Ganzmetall-Grossflugboote" 1924, BA-MA RH8-3606.
18. "Close-Hauled," *Flight*, November 6, 1924, 714.
19. Multiple authors, "Mitsubishi zōsenjo ni oite Doitsu hikōki kōnyū no ken" [The purchase of a German aircraft by Mitsubishi Shipyard], 1924–25, JACAR ref. B07090278600, 0334.
20. Adolf Rohrbach, "Die Zukunft des Marineflugwesens. Vortrag vor der Admiralität der Kaiserlichen Japanischen Marine," 1925, BA-MA RH8-3606.
21. Percentage from BA-MA RH8-3606. Archives at the Mitsubishi Economic Research Institute (MERI), Tokyo, MHI 00220, gives a stock distribution of 100:100:500 among Adolf Rohrbach, Rohrbach Metallflugzeugbau, and Mitsubishi.
22. "Mitsubishi ni okeru kitaisekkei shisaku no hensen," 56.
23. Mitsubishi Nainenki, *R gata hikōtei zumenyō goi*.

24. Kuwabara, *Kaigun kōkū kaisōroku*, 197–98.
25. Nozawa, *Nihon kōkūki sōshū: Kawanishi Hiroshō hen*, 178.
26. Nihon Kaigun Kōkūshi Hensan Iinkai, *Nihon Kaigun kōkūshi 1*, 510–11.
27. Sanuki, *Sanuki Matao no hitorigoto*, 260.
28. Wagner quoted in Hirschel and Prem, *Aeronautical Research in Germany*, 270.
29. Former Vice-Admiral Funakoshi was chairman of the board of directors of the Mitsubishi-Rohrbach aircraft company. The quotation is taken from Mitsubishi Jūkōgyō KK, *Mitsubishi Jūkōgyō*, 640.
30. *Mitsubishi shashi*, 34:7139.
31. Von Kármán, *Wind and Beyond*, 131.
32. Ikari, *Saigo no nishiki taitei*, 41–42; *Sekkeisha no shōgen*, 2:241.
33. Comparable flying boats were the British Short Sunderland and the US Consolidated PB2Y Coronado. Although the speed and load capacity of these two planes were comparable to the H6K, their flight ranges (2,850 kilometers and 1,700 kilometers, respectively) were inferior.
34. Shin Meiwa Kōgyō Kabushiki Kaisha Shashi Henshū Iinkai, *Shashi*, 97.
35. Francillon, *Japanese Aircraft of the Pacific War*, 312.
36. "The navy's aircraft—building a new project for next year," *Yomiuri Shinbun*, February 19, 1923, 2.
37. The Washington Naval Treaty determined the "standard displacement" of a vessel as "displacement of the ship complete, fully manned, engined, and equipped ready for sea . . . but without fuel or reserve feed water on board." It therefore differs from a fully tanked ship's actual weight. To make things even more complicated, one "ton" is defined as 1,016 kilograms.
38. Notes on the Case of Squadron Leader Rutland, RAF National Archives Kew KV2-328, 221a.
39. Letter to MI1. National Archives Kew KV2-328, 223a.
40. Ikari, *Kaigun kūgishō*, 22–24.
41. National Archives Kew KV2-328, 227a.
42. Nihon Kaigun Kōkūshi Hensan Iinkai, *Nihon Kaigun kōkūshi 1*, 196–97.
43. Nihon Kaigun Kōkūshi Hensan Iinkai, *Nihon Kaigun kōkūshi 1*, 274.
44. BA-MA RH 8 I/3679.
45. Deutsches Museum Archiv FA 001 / 0821 07.1935–03.1939.
46. Matsuoka, *Mitsubishi hikōki monogatari*, 240.
47. Nakajima's designers incorporated features of the Boeing F2B and the Bristol Bulldog.
48. Nihon Kaigun Kōkūshi Hensan Iinkai, *Nihon Kaigun kōkūshi 1*, 39–47.
49. Nihon Kaigun Kōkūshi Hensan Iinkai, *Nihon Kaigun kōkūshi 1*, 202–29.
50. Okumiya and Horikoshi, *Zero!*, 74–75.
51. Nozawa, *Nihon kōkūki sōshū: Nakajima hen*, 144.
52. Peattie, *Sunburst*, 39.
53. Munson, *Bombers between the Wars*, 111–12.
54. Nozawa, *Nihon kōkūki sōshū: Aichi, Kūgishō hen*, 78–79.
55. Nozawa, *Nihon kōkūki sōshū: Nakajima hen*, 144.
56. Miki, "Aichi Tokei," 102.

57. Mizusawa, *Gunyōki no tanjō*, 61.

58. For the important role of dive bombers in the attack on Pearl Harbor, see Israel, *Marineflieger einst und jetzt*, 176.

59. Heinkel, *Stürmisches Leben*, 170.

60. JACAR ref. C04015838300.

61. Takeda, *Zerosen no ko*, 42.

62. Nihon Kaigun Kōkūshi Hensan Iinkai, *Nihon Kaigun kōkūshi 1*, 287.

63. JACAR ref. C11080535800.

64. JACAR ref. C11080535800.

65. Genda, *Kaigun kōkūtai shimatsuki*, 1:70.

66. Marder, *Old Friends, New Enemies*, 344.

67. Genda, "Evolution of Aircraft Carrier Tactics," 23–28.

68. JACAR ref. C13072074800.

69. Genda, *Kaigun kōkūtai shimatsuki*, 2:9–10.

70. John Ferris brings forward a similar argument, referring to "military ethnocentrism." See Ferris, "Double-Edged Estimates," 91–108.

71. "Handbook on the Air Services of Japan," 1939, in Lester (ed.), *U.S. Military Intelligence Reports, Japan, 1918–1941*, reel 31, 110a.

72. The aircraft bought from Great Britain after 1930 were the de Havilland D. H. 83 Fox Moth (1932) and the Airspeed Envoy (1935).

73. For a case study of the dive bomber project, see Nishiyama, *Engineering War and Peace*, 53–56.

74. Nihon Kaigun Kōkūgaishi Kankōkai, *Umiwashi no kōseki*, 39.

75. Till, *Air Power and the Royal Navy*, 93.

76. For an overview of British interwar aviation, see Edgerton, *England and the Aeroplane*, 28–59.

77. Göring quoted in Duppler and Forsmann, *Marineflieger*, 38.

7. US Know-How for Japanese Aircraft Makers

1. According to Bōeichō bōeikenshūjō senshi shitsu, *Kaigun kōkū gaishi*, 228, the night air raid on Singapore started on December 8, 1941, at 3:45 AM local time.

2. Informational Intelligence Summary no. 85: Flight Characteristics of the Japanese Zero Fighter (Washington, DC: Intelligence Service, US Army Air Forces, 1942), 2–3.

3. Simonson, "The Demand for Aircraft and the Aircraft Industry," 363.

4. Dickey, *The Liberty Engine*, 9, 66–67.

5. Pattillo, *Pushing the Envelope*, 32.

6. Hounshell, *From the American System to Mass Production*.

7. Army Aviation Headquarters, "American Military Aircraft and Engines 1928," JACAR ref. C01007438000, 1088.

8. I take the expression from Green's "A Little of What the World Thought of Lindbergh," in Lindbergh, *"We,"* 233–318.

9. In 1939, US engineers calculated that an increase in cruising speed of 1 mph to be worth US$2,500; a decrease of one pound in aircraft weight would save US$75. See Wright, "American Methods of Aircraft Production," 139.

10. For a detailed discussion of the "the mature propeller-driven airplane," see Anderson, *The Airplane*, 183–282.

11. Okamura, *Kōkū gijutsu no zenbō*, 1:34–35.

12. Ono, *Beikoku kōkūkōgyō ni tsuite*, 35.

13. For a detailed account of Tani's aerodynamic research, see Hashimoto, *Hikōki no tanjō*, 247–96. See also Mizusawa, *Gunyōki no tanjō*, 157–62.

14. "Report No. NY269, Report on Nakajima Aircraft Co. Purchases of Machine Tools and Aircraft Accessories in the U.S. 7," in *Beikoku Shihōshō*, 2:240.

15. Katō, *Sakuma Ichirō den*, 105–18.

16. Ikari, *Kaigun gijutsushatachi no Taiheiyō sensō*, 146.

17. US Military Intelligence Reports, reel 30, 817.

18. Nakagawa and Mizutani, *Nakajima hikōki enjin-shi*, 41.

19. The army named the engine Ha-8.

20. Okamura, *Kōkū gijutsu no zenbō*, 1:467.

21. Hallion, "Airplanes That Transformed Aviation."

22. "Nihon demo seisaku Dagurasu-ki no kenri 800000 en nari" [Even produced in Japan, the Douglas aircraft's production license costs ¥800,000], *Asahi Shinbun*, June 17, 1935.

23. "Report No. NY360, Confidential Report on Industrial Purchases in USA from 1935–41 by 588 Japanese Concerns," in *Beikoku Shihōshō*, 2:48–50.

24. US Military Intelligence Reports, reel 30, 787.

25. "Purchase and Experimentation of Foreign Transport Planes and Engine," US Military Intelligence Reports, reel 30, 957.

26. Francillon, *Japanese Aircraft of the Pacific War*, 499–501.

27. "Report No. NY269, Report on Nakajima Aircraft Co. Purchases of Machine Tools and Aircraft Accessories in the U.S.," in *Beikoku Shihōshō*, 2:300.

28. "Report No. NY272, Report on Japanese Army Arsenals," in *Beikoku Shihōshō*, 2:104.

29. The DC-4E's successor, the DC-4A (or DC-4), was a downsized version that together with its military version, the C-54 Skymaster, became a best-seller.

30. "Report No. NY269, Report on Nakajima Aircraft Co. Purchases of Machine Tools and Aircraft Accessories in the U.S.," in *Beikoku Shihōshō*, 2:304.

31. Anderson, *The Airplane*, 260.

32. *Sumitomo seimitsu kōgyō shashi*, 10–17.

33. US Military Intelligence Reports, reel 28, 218.

34. Nohara, "Kokusan puropera monogatari," 81–91.

35. Wright, "American Methods of Aircraft Production," 139. A summary of this article was published in *Nihon kōkūgakkaishi*, vol. 6., no. 51 (July 1939): 763.

36. Wright, "American Methods of Aircraft Production," 131–224, 144–45.

37. Sawai, "Amerikasei kōsakukikai no yunyū," 1.

38. Nelson, *Industrial Architecture of Albert Kahn*, 168–69.

39. Direct negotiations with the architectural firm Albert Kahn, Inc., which had designed the Pratt & Whitney plant, failed in 1937. For more on this point, see Fukao Junji, *Fukao's Counseling Reports*, 1937, Mitsubishi Heavy Industries Komaki Archive, not cataloged.

40. Mitsubishi Jūkōgyō KK, *Ōjibōbō*, 2:104.

41. Mitsubishi Jūkōgyō KK, *Ōjibōbō*, 1:472.

42. Kōno Hiroshi, *Ōbeikōkūkōgyō shisatsuhōkoku* [Inspection Report about the US and European Aviation Industry], 1939, Mitsubishi Heavy Industries Komaki Archive, not cataloged, 11.

43. Kōno Hiroshi, *Ōbeikōkūkōgyō shisatsuhōkoku*, 11.

44. In all likelihood, Kōno was alluding to President Franklin Roosevelt's 1938 announcement to build up an air force of 10,000 aircraft. See, for example, Kennedy, *American People in World War II*, 4.

45. Kōno Hiroshi, *Ōbeikōkūkōgyō shisatsuhōkoku*, 16.

46. Roosevelt's address of October 5, 1937, on the world political situation.

47. Tow, "The Great Bombing of Chongqing," 256–82. For Chinese air defense efforts, see Baumler, "Keep Calm and Carry On," 1–36.

48. American Committee for Non-Participation in Japanese Aggression, *America's Share in Japan's War Guilt*, 74.

49. Letter by Joseph Green, Office of Arms and Munitions Control to Okura & Company, dated July 1, 1938, reproduced in Ono, *Beikoku kōkūkōgyō*, 38–39, emphasis added.

50. Aviation weapons cooperative group, US, Beikokukōkūheiki kōbaidan gyōmu hōkoku sōfu no ken [Sending operation report of American ordnance purchase group], 1940, JACAR ref. C01004910000.

51. Files of US Consular Reports, Tokyo, to the State Department from 1937.

52. US Military Intelligence Reports, reel 31, 21–29, Memos: "Visit of Japanese Aviation Inspection Group to the US, May 13–Jul 26, 1937."

53. A May 16, 1938, *Asahi* article recognized Okada's distinguished service to the project.

54. "Report No. NY272, Report on Japanese Army Arsenals," in *Beikoku Shihōshō*, 2:163.

55. US Military Intelligence Reports, reel 28, 339–43, G-2 Report, "Expansion of Aviation Manufacturing Industry."

56. US Military Intelligence Reports, reel 31, 31.

57. Nozawa, *Nihon kōkūki sōshū: Yunyūki hen*, 148–49.

58. "Report No. NY272, Preliminary Examination of San Francisco Office Files of Mitsui and Mitsubishi," in *Beikoku Shihōshō*, 2:167; and "Report No. NY272, Report on Japanese Army Arsenals," in *Beikoku Shihōshō*, 2:165.

59. Aviation weapons cooperative group, US, Beikokukōkūheiki kōbaidan gyōmu hōkoku sōfu no ken [Sending operation report of American ordnance purchase group], 1940, JACAR ref. C01004910000, 503.

60. "Report No. NY272, Report on Japanese Army Arsenals," in *Beikoku Shihōshō*, 2:166–68.

61. US Military Intelligence Reports, reel 28, 175. There is some irony: as could be seen in the 1925 Mitchell trial, many US military officers were not aware of the "power of bombing," either.

62. US Military Intelligence Reports, reel 28, 362.

63. Gunkihogohō [Military Secrets Act], 1937, JACAR ref. A03022076900.

64. US Military Intelligence Reports, reel 27, 759, 770.
65. US Military Intelligence Reports, reel 27, 617–22.
66. US Military Intelligence Reports, reel 27, 752–61.
67. Yamawaki Masataka, "Wright Engineers at Nakajima," 1939, JACAR ref. C01007348500, 502.
68. Sutton and Parke, *Kōkūhatsudōki tairyōseisan ni kan suru kōshūroku* [A course on the mass production of aero-engines], Mitsubishi Heavy Industries Komaki Archive, not cataloged.
69. US Military Intelligence Reports, reel 31, 226–30 (June 23).
70. In addition to the aforementioned reports, see the magazine *Flying and Popular Aviation* that bluntly stated in January 1941: "Japan Is Not an Air Power"; and Lucien Zacharoff's article on "Japanese Air Power," *Aviation* (September 1941).
71. Quoted in Yamazaki, *Nippon ga nekkyō shita daikōkū jidai*, 132.
72. US Military Intelligence Reports, reel 29, 1026–27, report quoting a British aeronautical expert.
73. For more details on the *Kōkenki* project, especially the involvement of Tokyo Imperial University, see Tomizuka, *Kōkenki sekai kiroku juritsu e no kiseki.*
74. Morikawa, *Sora no eiyū.*
75. US Military Intelligence Reports, reel 31, 114.
76. US Military Intelligence Reports, reel 31, 164.
77. Quoted in Yamazaki, *Nippon ga nekkyō shita*, 169.
78. G3M bombers engaged in the December 1941 sinking of the British warships *Prince of Wales* and *Repulse*. In October 1939, the first prototype of the G3M's successor, the experimental G4M, was completed.
79. US Military Intelligence Reports, reel 31, 177.
80. US Military Intelligence Reports, reel 31, 186; and Mitsubishi Jūkōgyō KK, *Ōjibōbō*, 1:285. The *Nippon*'s Kinsei engine was made by Mitsubishi.
81. "Report No. NY-369," in *Beikoku Shihōshō* 1:1.
82. "Report No. SF146, Preliminary Examination of San Francisco Office Files of Mitsui and Mitsubishi," in *Beikoku Shihōshō*, 2:87.
83. "Report No. NY272, Report on Japanese Army Arsenals," in *Beikoku Shihōshō*, 2:165.
84. Coox, "Flawed Perception," 239–54; May, *Knowing One's Enemies*; Kotani, *Japanese Intelligence in World War II.*
85. For instance, there was a less than 15 percent error in the number of the US Navy's airplanes (estimated at 4,535 against an actual 5,291). *Japanese Military and Naval Intelligence Division, Japanese Intelligence Section, G-2. Reports. Pacific War* (Washington, DC: Government Printing Office, 1946), 46.
86. *Japanese Military and Naval Intelligence Division, Japanese Intelligence Section, G-2*, 48.
87. "Kaigai no America kōkūkai (jō)" [America on the opposite shore: Its aviation (1)], *Asahi*, October 26, 1941, 5.
88. The article's estimate for US annual aircraft production was astonishingly close to the actual number of about 12,800 aircraft. See, on this point, Pattillo, *Pushing the Envelope*, 125.

89. *Asahi*, November 10, 1941, 3; "Kaigai no America kagakukenkyū" [America on the opposite shore: Its scientific research], *Asahi*, November 8, 1941, 5.

90. May, *Knowing One's Enemies*, 446, 452.

91. *Japanese Military and Naval Intelligence Division, Japanese Intelligence Section, G-2*, 47.

92. *Japanese Military and Naval Intelligence Division, Japanese Intelligence Section, G-2*, 46.

93. "*Inakasumō ga yokozuna ni idonda yōna mono*," in Mitsubishi Jūkōgyō KK, *Ōjibōbō*, 3:94.

94. Katō, *Sakuma Ichirō*, 152.

8. Jet and Rocket Technology for Japan's Decisive Battle

1. Cohen, *Japan's Economy in War and Reconstruction*, 48–109.

2. Such a view is advanced, for instance, in *United States Strategic Bombing Survey, The Japanese Aircraft Industry*.

3. See, on this point, Matogawa, "Senzen no Nihon no roketto kenkyū"; and Grunden, *Secret Weapons and World War II*, 129.

4. Naitō, *Kimitsu heiki funryū*, 75.

5. US Naval Technical Mission to Japan, Japanese Metallurgy—Article 1, High Temperature Alloys for Gas Turbines, Rocket Nozzles and Lines (n.p., 1946), 7–10.

6. Okamura, *Kōkū gijutsu no zenbō*, 1:478.

7. In 1940, in the wake of a major reorganization, the Yokosuka Naval Air Arsenal (Kaigun Kōkūshō) changed its name to Naval Air Technical Arsenal (Kaigun Kōkū Gijutsushō).

8. Fujihira, *Kimitsuheiki no zenbō*, 52–54.

9. "Ne" stands for the first syllable of *nenshō funsha suishinki* (combustion jet propeller).

10. For a detailed analysis of early German jet and rocket technology, see Schabel, *Illusion der Wunderwaffen*, 35–62, and von Gersdorff and Grasmann, *Flugmotoren und Strahltriebwerke*, 181–244.

11. On April 26, 1939, a Messerschmitt Me 209 powered by a twelve-cylinder DB 601 piston engine had set a speed record of 755 kilometers per hour.

12. Okamura, *Kōkū gijutsu no zenbō*, 1:479.

13. For the German-Italian-Japanese project of a transcontinental air service see Herde, *Der Japanflug*.

14. "Abkommen über wirtschaftliche Zusammenarbeit," signed on January 20, 1943. For more on this topic, see Martin, *Deutschland und Japan im Zweiten Weltkrieg*, 152–71.

15. Braun, "Technologietransfer im Flugzeugbau zwischen Deutschland und Japan," 336.

16. The contract's title is a classic example of German officialese: *Vereinbarung über die gegenseitige Zurverfügungstellung von Nachbaurechten und Rohstoffen zwischen Deutschland und Japan*. For a reprint of the document, see Martin, *Deutschland und Japan*, 251.

17. *German Technical Aid to Japan* (Washington, DC: War Department, Military Intelligence Division, 1945), 33.
18. Iwaya, "Roketto ki Shūsui ni tsuite," 82–88.
19. Fujihira, *Kimitsuheiki no zenbō*, 15.
20. Nagamori, *Ichi chūdoku gijutsushikan no omoide*, 13.
21. Fujihira, *Kimitsuheiki no zenbō*, 43.
22. The original meaning of Shūsui is "a clear stream in autumn." In its figurative sense, it translates to "cold steel" or "polished sword."
23. Nihon Kōkū Gakujutsushi Henshū Iinkai, *Nihon kōkū gakujutsushi*, 26.
24. Makino, *Saishū kessen heiki Shūsui*, 17–18.
25. Both quotations from Nagoya Kōkūuchū Shisutemu Seisakujo, *Shūsui*, Mitsubishi Heavy Industries Komaki Archive, not cataloged, n.d.
26. Nihon Kōkū Gakujutsushi Henshū Iinkai, *Nihon kōkū gakujutsushi*, 26–27.
27. Shibata, *Yūjin roketto sentōki Shūsui*, 48.
28. Shibata, *Yūjin roketto sentōki Shūsui*, 33.
29. Iwaya, "Roketto ki Shūsui," 87.
30. Nihon Kōkū Gakujutsushi Henshū Iinkai, *Nihon kōkū gakujutsushi*, 26–27.
31. Okamura, *Kōkū gijutsu no zenbō*, 1:253–55.
32. Iwaya, "Roketto ki Shūsui," 86.
33. Iwaya, "Roketto ki Shūsui," 88.
34. *United States Strategic Bombing Survey, The Japanese Aircraft Industry*, 31, speculates on the "political implications" of these unrealistic production targets.
35. Okamura, *Kōkū gijutsu no zenbō*, 1:255; and *United States Strategic Bombing Survey, Nippon Airplane*, 6.
36. Okamura, *Kōkū gijutsu no zenbō*, 1:482.
37. Mitsubishi Jūkōgyō KK, *Ōjibōbō*, 2:63.
38. Mitsubishi Jūkōgyō KK, *Ōjibōbō*, 2:77–78.
39. Mitsubishi Jūkōgyō KK, *Ōjibōbō*, 2:80.
40. Mitsubishi Jūkōgyō KK, *Ōjibōbō*, 2:69; and *United States Strategic Bombing Survey, Mitsubishi*, 18.
41. Makino, *Saishū kessen heiki Shūsui*, 17–18.
42. Mitsubishi Jūkōgyō KK, *Ōjibōbō*, 2:66.
43. US Naval Technical Mission to Japan, *Target Report Miscellaneous Reports of Various Japanese Naval Research Activities* (n.p., 1946), 51.
44. Iwaya, "Roketto ki Shūsui," 84.
45. *German Technical Aid to Japan*, 46–64.
46. *Asahi Shinbun*, November 13, 1944.
47. US Naval Technical Mission to Japan, *Target Report Japanese Fuels and Lubricants, Research on Rocket Fuels of the Hydrogen Peroxide-Hydrazine Type* (n.p., 1946).
48. Okamura, *Kōkū gijutsu no zenbō*, 1:481.
49. Ishizawa, *Kikka*, 105.
50. Ishikawajima was the obvious choice. Its turbine branch was already established in 1936.
51. Fujihira, *Kimitsuheiki*, 62. For the dramatic shift of the navy's budget from vessels to aircraft, see Koyama, *Nihon gunji kōgyō no shiteki bunseki*, 300–309.

52. Fujihira, *Kimitsuheiki*, 61.

53. US Naval Technical Mission to Japan, *Target Report Miscellaneous Reports of Various Japanese Naval Research Activities* (n.p., 1946), 37ff.

54. Fujihira, *Kimitsuheiki*, 64–66.

55. Ishikawajima delivered its first Ne-20 jet engine to the Yokosuka arsenal on August 1, 1945. *Ishikawajima Jūkōgyō Kabushiki Kaisha 108-nenshi*, 885.

56. Okamura, *Kōkū gijutsu no zenbō*, 1:248–51.

57. Nohara, *Nihon gunyōki jiten: Kaigun hen, 1910–1945*, 162–63.

58. Okamura, *Kōkū gijutsu no zenbō*, 1:251.

59. Nihon Kaigun Kōkūshi Hensan Iinkai, *Nihon Kaigun kōkūshi 3*, 548–49.

60. Quoted in Nihon Kaigun Kōkūgaishi Kankōkai, *Umiwashi*, 260.

61. Nihon Kōkū Gakujutsushi Henshū Iinkai, *Nihon kōkū gakujutsushi*, 27.

62. Shibata, *Yūjin roketto sentōki Shūsui*, 72.

63. Mitsubishi Jūkōgyō KK, *Ōjibōbō*, 2:72; and Iwaya, "Roketto ki Shūsui," 87.

64. Air Technical Intelligence Group, "General Design and Development of Mitsubishi Aircraft Engines" (n.p., 1945), 5.

65. Nihon Kaigun Kōkūgaishi Kankōkai, *Umiwashi*, 266–69.

66. Quoted in Nihon Kaigun Kōkūgaishi Kankōkai, *Umiwashi*, 268–69.

67. Nagoya Kōkūuchū Shisutemu Seisakujo, *Shūsui*, n.d., Mitsubishi Heavy Industries Komaki Archive, not cataloged, 315.

68. For an account of these engineers' moral dilemma, see Nishiyama, *Engineering War and Peace*, 72–79.

69. Naitō, *Thunder Gods*, 48.

70. Naitō, *Thunder Gods*, 119.

71. *German Technical Aid to Japan*, 90.

72. Quoted in Naitō, *Ōka: Hijō no tokkō heiki*, 163.

73. US Naval Technical Mission, *Target Report Japanese Fuels* (n.p., 1946), 11.

74. Kimura, *Hikōki no hon*, 151. Kimura was in charge of designing the Ōka rocket-powered attack plane.

75. "On the day of defeat, we burned all classified documents," *Asahi Shinbun*, September 21, 2010.

76. Mitsubishi Jūkōgyō KK, *Ōjibōbō*, 2:75.

77. H. S. Tsien, *Technical Intelligence Supplement*, 1946, 160.

78. Mitsubishi Jūkōgyō KK, *Ōjibōbō*, 2:74; and Yokoyama, Yuyama, Akojima, and Moriya, "The Rocket Fighter Shusui," 274.

79. *German Technical Aid to Japan*, 6, 45.

80. ONI Special Activities Branch (IOP 16-Z), *Interrogation Nirschling, 1945*, n.p.

81. Butler, *War Prizes*, 239, 252, 255; Chambers, *Wings of the Rising Sun*, 264–71.

Epilogue

1. There seems to be no shortage of biographies in this field—some largely factual, some wildly (self-)aggrandizing. To name a few: de Havilland, *Sky Fever*; Doi, *Hikōki sekkei 50 nen no kaisō*; Dornier, *Aus meiner Ingenieurlaufbahn*; Freudenthal, *Flight into History*; Heinkel, *Stürmisches Leben*; Horikoshi, *Eagles of Mitsubishi*; Nowarra, *Flugzeuge*

des Alexander Baumann; Schmitt, Hofmann, and Hofmann, *Hugo Junkers und seine Flugzeuge*; Toyoda, *Hikōki-ō Nakajima Chikuhei*; Vogt, *Weltumspannende Memoiren*; von Kármán, *Wind and Beyond*; Weyl, *Fokker.*

2. I borrow this expression from Staudenmaier, *Technology's Storytellers*, 61–64.

3. Philosopher Joseph Agassi, as quoted in Staudenmaier, *Technology's Storytellers*, 81.

4. For more on the emergence of the "rational inventor" in the wake of the Second Industrial Revolution, see Mokyr, *Lever of Riches*, 113–48.

5. For more on the Japanese engineers' "self-assessment and soul-searching," see Maema, *Jetto enjin ni toritsukareta*, 180–99.

6. I took this and the previous quotation from Shiga Fujio in his preface to Fujihira, *Kimitsuheiki no zenbō.*

7. Mizuno argues in a similar way in *Science for the Empire*, 173.

8. By 1949, more than 10,000 scientists had joined the federation. See Nakayama, *Science, Technology and Society*, 18–20.

9. Quoted in Morris-Suzuki, *Technological Transformation of Japan*, 163.

10. SCAP (General Headquarters Supreme Commander for the Allied Powers), "SCAPIN 301 Commercial and Civil Aviation," Instruction Note 301, November 18, 1945.

11. For more details, see Koch, *Rüstungskonversion in Japan*, 75–82.

12. Nishiyama, *Engineering War and Peace*, 157–83.

13. Caspary, *Kooperation und Konkurrenz*, 47–112.

14. Quoted in Maema, *Jetto enjin*, 218.

15. Kawasaki joined the Nihon Jet Engine Company in 1956.

16. Shibata, "Chōsensensōgo no America no taigaienjo," 184.

17. Society of Japanese Aerospace Companies, *Nihon no kōkū uchū kōgyō*, 14–15.

18. Takeda, *Burūinparusu*, 106–7.

19. See the Japan Air Self-Defense Force's website at http://www.mod.go.jp/asdf/pr_report/paperplane/index39.html.

20. A turboprop is a jet engine that uses a turbine to drive a propeller.

21. Maema, *Kokusan ryokyakki MRJ hishō*, 16.

22. Caspary, *Kooperation und Konkurrenz*, 152.

23. See Samuels, *"Rich Nation, Strong Army,"* 210–14; Caspary, *Kooperation und Konkurrenz*, 125–82; Yokokura, *YS-11 Tobe!*; Maema, *YS-11 jō;* Maema, *YS-11 ge.*

24. Nihon Kōkū Kyōkai, *Nihon kōkūshi: Shōwa sengo hen*, 556.

25. Ann Crittenden, "C.I.A. Said to Have Known in 50's of Lockheed Bribes," *New York Times*, April 2, 1976.

26. Quoted in Itō and Itō, *Nihon kōkūshi nenpyō*, 254.

27. Society of Japanese Aerospace Companies, *Nihon no kōkū uchū kōgyō*, 34–36.

28. At that time, besides Japan, only the United States, the Soviet Union, Great Britain, France, and Sweden had developed supersonic aircraft.

29. For a comprehensive study on the FS-X project, see Lorell, *Troubled Partnership.*

30. Ishihara, *The Japan That Can Say NO*, 44.

31. The concept of technonationalism comes from Samuels, *"Rich Nation, Strong Army."*

32. Quoted in Caspary, *Kooperation und Konkurrenz*, 271.

33. Tochio, *Nihon no kōkūkisangyō*, 4.

34. Society of Japanese Aerospace Companies, *Nihon no kōkū uchū kōgyō*, 53–56.

35. Here is yet another example of the intriguing ambiguity of the Japanese language: using a different first logographic character with the same pronunciation, the expression for "quasi-national production," *junkokusan*, can be also understood as "purely domestically produced."

36. "MRJ no jūchū" [Accepting orders for the MRJ], *Nihon Keizai Shinbun*, October 31, 2017.

37. Fukunaga, *Gunyōkiseizō no sengo shi*, 92–95.

Bibliography

50 Jahre Dornier 1914–1964. Ein unvollständiges Bilderbuch zur Geschichte des Hauses Dornier. Friedrichshafen: Dornier-Werke, 1964.

Abe Hiroshi. "*Hatsukaze, Kochikaze* hōōhikō no omoide" [*Hatsukaze* and *Kochikaze*, recollections of the visit Europe flight]. In *Nihon minkan kōkūshiwa* [Stories of Japan's civil aviation], 69–72. Tokyo: Nihon Kōkū Kyōkai, 1966.

Abe Ichisuke. *Kawasaki Zōsenjo yonjūnenshi* [A 40-year history of Kawasaki Dockyard Company]. Kōbe: Kawasaki Zōsenjo, 1936.

Ader, Clément. *L'aviation militaire.* Paris: Berger-Levrault, 1911.

Aichi no kōkūshi [An aviation history of Aichi prefecture]. Nagoya: Chūnichi Shinbun Honsha, 1978.

Akimoto Minoru. *Kyojinki monogatari: Shirarezaru Nihon no kūchū yōsai* [The story of giant airplanes: Japan's unknown flying fortresses]. Tokyo: Kōjinsha, 2002.

———. *Nihon hikōsen monogatari: Kōkūkai no tokuina kōseki o tadoru* [A story of Japanese airships: Pursuing a unique track in the aviation world]. Tokyo: Kōjinsha, 2007.

American Committee for Non-participation in Japanese Aggression. *America's Share in Japan's War Guilt.* New York: Burland Print, 1938.

Anderson, John David. *The Airplane: A History of Its Technology.* Reston, VA: American Institute of Aeronautics and Astronautics, 2002.

Apollonio, Umbro, ed. *Futurist Manifestos.* New York: Viking Press, 1973.

Asada Sadao. *From Mahan to Pearl Harbor: The Imperial Japanese Navy and the United States.* Annapolis, MD: Naval Institute Press, 2006.

Asakawa, Michio. "Anglo-Japanese Military Relations, 1800–1900." In *The History of Anglo-Japanese Relations, 1600–2000*, edited by Ian Gow, Hirama Yōichi, and John Chapman, 3:13–34. Basingstoke: Palgrave Macmillan, 2003.

Bach, Martin. *Luftfahrtindustrie im Ersten Weltkrieg.* Allershausen: Nara, 2003.

Barnhart, Michael A. *Japan Prepares for Total War: The Search for Economic Security, 1919–1941.* Ithaca, NY: Cornell University Press, 1987.

Barth, Boris. *Dolchstoßlegenden und politische Desintegration. Das Trauma der deutschen Niederlage im Ersten Weltkrieg 1914–1933.* Düsseldorf: Droste, 2003.

Bartholomew, James R. *The Formation of Science in Japan: Building a Research Tradition*. New Haven, CT: Yale University Press, 1989.

Baumler, Alan. "Aviation and Asian Modernity 1900–1950." In *Oxford Research Encyclopedia of Asian History*. https://dx.doi.org/10.1093/acrefore/9780190277727.013.177. Accessed January 12, 2019.

———. "Keep Calm and Carry On: Airmindedness and Mass Mobilization during the War of Resistance." *Journal of Chinese Military History* 5, no. 1 (2016): 1–36.

Bayly, C. A. "AHR Conversation: On Transnational History." *American Historical Review* 111, no. 5 (2006): 1441–64.

Beikoku Shihōshō Senji Keizaikyoku tainichi chōsa shiryōshū / Economic Intelligence based on the Files of the Major Japanese Trading Companies in the United States: Selected Reports from the Japanese Files Research Project Prepared by the War Division of the Department of Justice, Economic Warfare Section between 1942 and 44. 5 vols. Edited by Miwa Munehiro. Tokyo: Kurosu Karuchā Shuppan, 2008.

Bijker, Wiebe E., Thomas P. Hughes, and Trevor Pinch. *The Social Construction of Technological Systems: New Directions in the Sociology and History of Technology*. Cambridge, MA: MIT Press, 2012.

Bōeichō bōeikenshūjō senshi shitsu. *Kaigun kōkū gaishi* [An outline of naval aviation]. Tokyo: Asagumo Shinbunsha, 1976.

———. *Rikugun kōkū heiki no kaihatsu seisan hokyū* [The development of the army's aerial weaponry, its production, and supply]. Tokyo: Asagumo Shinbunsha, 1975.

———. *Rikugun kōkū no gunbi to un'yō* [The army's aerial armament and operations]. 3 vols. Tokyo: Asagumo Shinbunsha, 1971.

Brackley, Frida H. *Brackles: Memoirs of a Pioneer of Civil Aviation*. Chatham: W. & J. Mackay, 1952.

Braun, Hans-Joachim. "Technologietransfer im Flugzeugbau zwischen Deutschland und Japan 1936–1945." In *Deutschland-Japan in der Zwischenkriegszeit*, edited by Erich Pauer, 324–40. Munich: Iudicium, 1992.

Budraß, Lutz. *Flugzeugindustrie und Luftrüstung in Deutschland 1918–1945*. Düsseldorf: Droste, 1998.

———. "Rohrbach und Dornier. Zwei Unternehmen aus dem Zeppelin-Flugzeugbau in der Weimarer Republik und im Nationalsozialismus." In *Zeppelins Flieger: Das Flugzeug im Zeppelin-Konzern und seinen Nachfolgebetrieben*, edited by Wolfgang Meighörner. Berlin: Wasmuth, 2006.

Burūinparusu paafekuto gaido [Blue Impulse perfect guide]. Tokyo: Ikaros, 2003.

Butler, Phil. *War Prizes: An Illustrated Survey of German, Italian and Japanese Aircraft Brought to Allied Countries during and after the Second World War*. Leicester: Midland Counties, 1994.

Caspary, Sigrun. *Kooperation und Konkurrenz: Die japanische Luftfahrtindustrie: Eine Studie zur japanischen Industriepolitik am Beispiel der Luftfahrtindustrie im Spannungsfeld von nationaler Strategie und oligopolem Markt*. Bonn: Bier'sche Verlagsanstalt, 1998.

Chambers, Mark. *Wings of the Rising Sun: Uncovering the Secrets of Japanese Fighters and Bombers of World War II*. Oxford: Osprey, 2018.

Cohen, Jerome Bernard. *Japan's Economy in War and Reconstruction*. Minneapolis: University of Minnesota Press, 1949.

Coox, Alvin D. "Flawed Perception and Its Effect upon Operational Thinking: The Case of the Japanese Army, 1937–41." *Intelligence and National Security* 5, no. 2 (1990): 239–54.

Corn, Joseph J. *The Winged Gospel: America's Romance with Aviation, 1900–1950*. New York: Oxford University Press, 1983.

Cox, Jafna L. "A Splendid Training Ground: The Importance to the Royal Air Force of Its Role in Iraq, 1919–32." *Journal of Imperial and Commonwealth History* 13, no. 2 (1985): 157–84.

Crawcour, Sydney. "Industrialization and Technological Change, 1885–1920." In John W. Hall et al., *The Cambridge History of Japan*, 6:385–450. New York: Cambridge University Press, 1988.

Daniels, Gordon. *A Guide to the Reports of the United States Strategic Bombing*. London: Royal Historical Society, 1981.

de Havilland, Sir Geoffrey. *Sky Fever: The Autobiography of Sir Geoffrey de Havilland*. Shrewsbury: Airlife, 1979.

Demetz, Peter. *Die Flugschau von Brescia*. Wien: Paul Zsolnay, 2002.

Dickey, Philip S. *The Liberty Engine, 1918–1942*. Washington, DC: Smithsonian Institution Press, 1968.

Dickinson, Frederick R. *War and National Reinvention: Japan in the Great War, 1914–1919*. Cambridge, MA: Harvard University Press, 1999.

Doi Takeo. *Hikōki sekkei 50 nen no kaisō* [Reflections on fifty years of aircraft design]. Tokyo: Kantōsha, 1989.

Dornier, Claude. *Aus meiner Ingenieurlaufbahn*. Zug: Author, 1966.

"Dornier Flugzeuge in Japan." *Dornier-Post*, no. 3 (1973): 54–55.

Drea, Edward J. *Japan's Imperial Army: Its Rise and Fall, 1853–1945*. Lawrence: University Press of Kansas, 2009.

Duppler, Jörg, and Heinrich Forsmann. *Marineflieger: Von der Marineluftschiffabteilung zur Marinefliegerdivision*. Herford: Mittler, 1988.

Ebert, Hans Joachim. "Rohrbach." In *Neue deutsche Biographie*, edited by Otto zu Stolberg-Wernigerode. Berlin: Rohmer-Schinkel, 2005.

Edagawa, Ryūsuke. *Aichi Tokei Denki 85-nenshi* [An 85-year history of Aichi Tokei Denki]. Nagoya: Aichi Tokei Denki Kabushiki Kaisha, 1984.

Edgerton, David. *England and the Aeroplane: Militarism, Modernity and Machines*. London: Penguin, 2013.

Engel, Leonard. "Japan Is Not an Air Power." *Flying and Popular Aviation* 28, no. 1 (1941): 12–14, 70–72.

Etherton, P. T., and Tillman Hubert Hessell. *Manchuria, the Cockpit of Asia*. New York: Frederick A. Stokes, 1932.

Evans, David C., and Mark R. Peattie. *Kaigun: Strategy, Tactics, and Technology in the Imperial Japanese Navy, 1887–1941*. Annapolis, MD: Naval Institute Press, 1997.

Fearon, Peter. "Aircraft Manufacturing." In *British Industry between the Wars: Instability and Industrial Development 1919–1939*, edited by N. K. Buxton and Derek Howard Aldcroft. London: Scolar Press, 1979.

———. "The Formative Years of the British Aircraft Industry, 1913–1924." *Business History Review* 43, no. 4 (1969): 476–95.

Ferris, John. "Armaments and Allies: The Anglo-Japanese Strategic Relationship, 1911–1921." In *The Anglo-Japanese Alliance, 1902–1922*, edited by Phillips Payson O'Brien, 249–66. New York: Routledge Curzon, 2004.

———. "A British 'Unofficial' Aviation Mission and Japanese Naval Developments, 1919–1929." *Journal of Strategic Studies* 5, no. 3 (1982): 419–39.

———. "Double-Edged Estimates: Japan in the Eyes of the British Army and the Royal Air Force, 1900–1939." In *The History of Anglo-Japanese Relations, 1600–2000*, edited by Ian Gow, Hirama Yōichi, and John Chapman, 3:91–108. Basingstoke: Palgrave Macmillan, 2003.

———. "'Worthy of Some Better Enemy?': The British Estimate of the Imperial Japanese Army, 1919–41, and the Fall of Singapore." *Canadian Journal of History* 28 (August 1993): 223–56.

Forbes-Sempill, William. "The British Aviation Mission to the Imperial Japanese Navy." *Journal of the Royal Aeronautical Society* 28, no. 165 (1924): 553–84.

Francillon, René J. *Japanese Aircraft of the Pacific War.* Annapolis, MD: Naval Institute Press, 1987.

Freudenthal, Elsbeth Estelle. *Flight into History: The Wright Brothers and the Air Age.* Norman: University of Oklahoma Press, 1949.

Fritzsche, Peter A. *Nation of Fliers: German Aviation and the Popular Imagination.* Cambridge, MA: Harvard University Press, 1992.

Fujihira Ukon. *Kimitsuheiki no zenbō* [An overall picture of secret weapons]. Tokyo: Hara Shobō, 1976.

Fujii Tadayoshi. "Shōwa shoki sensōkaishiji ni okeru taishūteki gunjishien kyanpeen no ichitenkei: Gunjōki (Aikokugō, Hōkokugō) ken'nōundō no katei ni tsuite" [A typical campaign for wartime mass support at the beginning of the war in early Shōwa Japan: The donation campaign of the warplanes Aikokugo and Hōkokugo]. *Surugadai University Studies* 6 (1992): 39–55.

———. *Zaigō Gunjinkai: Ryōhei ryōmin kara akagami, gyokusai e* [The Military Reserve Association: From "good soldiers, good citizens" to draft notices to dying an honorable death]. Tokyo: Iwanami Shoten, 2009.

Fujitani, Takashi. *Splendid Monarchy: Power and Pageantry in Modern Japan.* Berkeley: University of California Press, 1996.

Fukunaga Akihiko. *Gunyōkiseizō no sengo shi: Sengo kūhaku ki kara senshin gijyutsu jisshōki made* [Postwar history of the production of military aircraft: From the postwar vacuum to the demonstration of advanced technology]. Tokyo: Fuyō shobō shuppan, 2016.

Garon, Sheldon. "Transnational History and Japan's 'Comparative Advantage.'" *Journal of Japanese Studies* 43, no. 1 (2017): 65–92.

Genda Minoru. "Evolution of Aircraft Carrier Tactics of the Imperial Japanese Navy." In *Air Raid, Pearl Harbor!*, edited by Paul Stillwell, 23–28. Annapolis, MD: Naval Institute Press, 1981.

———. *Kaigun Kōkūtai shimatsuki* [The beginning and the end of the naval air forces]. 2 vols. Tokyo: Bungei Shunjū Shinsha, 1961.

Giyū zaidan kaibō gikai no mokuteki shimei [The Maritime Defense Volunteer Association's purpose and mission]. Tokyo: Kaibō Gikai, 1927.

Goldman, Emily O., and Leslie C. Eliason. *The Diffusion of Military Technology and Ideas*. Stanford, CA: Stanford University Press, 2003.

Gow, Ian. "The Douglas Mission (1873–79) and Meiji Naval Education." In James Hoare, *Britain & Japan: Biographical Portraits*, 3:144–57. Richmond, UK: Japan Library, 1999.

Grunden, Walter E. *Secret Weapons and World War II: Japan in the Shadow of Big Science*. Lawrence: University Press of Kansas, 2005.

Haddow, G. W., and Peter M. Grosz. *The German Giants: The Story of the R-Planes, 1914–1919*. London: Putnam, 1969.

Hallion, Richard P. "Airplanes That Transformed Aviation." In *Air & Space Smithsonian* (July 2008). https://www.airspacemag.com/history-of-flight/airplanes-that-transformed-aviation-46502830/. Accessed April 26, 2019.

Hartmann, Gérard. "Les premiers Farman." In *Dossiers historiques et techniques aéronautique française*. http://www.hydroretro.net/etudegh/index.php. Accessed January 12, 2019.

Hashimoto, Takehiko. *Hikōki no tanjō to kūki rikigaku no keisei: kokkateki kenkyū kaihatsu no kigen o motomete* [The invention of the airplane, the emergence of aerodynamics, and the formation of national systems of research and development]. Tokyo: Tōkyō daigaku shuppankai, 2012.

Hata Shunroku, Takashi Itō, and Takeshi Hara. *Gensui Hata Shunroku kaikōroku* [The memoirs of Marshal Hata Shunroku]. Tokyo: Kinseisha, 2009.

Hayashi Teisuke. "Aru enjin sekkeigijutsusha no kaisō" [Memoirs of an engine designer]. *Nihon kikai gakkaishi* 86, no. 776 (July 5, 1983): 776–81.

Heinkel, Ernst. *Stürmisches Leben*. Stuttgart: Mundus, 1953.

Herde, Peter. *Der Japanflug: Planungen und Verwirklichung einer Flugverbindung zwischen den Achsenmächten und Japan 1942–1945*. Stuttgart: F. Steiner, 2000.

Higgins, George J. "The Characteristics of the N.A.C.A. M-12 Airfoil Section." *Technical Notes National Advisory Committee for Aeronautics* 343 (1926): 1–7.

Hiraki Kunio. *Baron Shigeno no shōgai* [The life of Baron Shigeno]. Tokyo: Bungei Shunjū, 1990.

Hirschel, Ernst-Heinrich, and Horst Prem. *Aeronautical Research in Germany: From Lilienthal until Today*. New York: Springer, 2004.

Hoare, J. E. "Ernest Cyril Comfort: The Other British Aviation Mission and Mitsubishi 1921–24." In *Britain & Japan: Biographical Portraits*, edited by Hugh Cortazzi, 6:182–90. Folkestone, UK: Global Oriental, 2007.

Horikoshi, Jiro. *Eagles of Mitsubishi: The Story of the Zero Fighter*. Seattle: University of Washington Press, 1981.

Hounshell, David A. *From the American System to Mass Production, 1800–1932: The Development of Manufacturing Technology in the United States*. Baltimore, MD: Johns Hopkins University Press, 1984.

Humphreys, Leonard A. *The Way of the Heavenly Sword: The Japanese Army in the 1920's*. Stanford, CA: Stanford University Press, 1995.

Ikari Yoshirō. *Kaigun gijutsushatachi no Taiheiyō sensō* [The Pacific War of the naval engineers]. Tokyo: Kōjinsha, 1989.

———. *Kaigun kūgishō: Hokori takaki zunōshūdan no eikō to shuppatsu* [The Naval Air Arsenal: Glory and beginnings of a proud think tank]. Tokyo: Kōjinsha, 1996.

———. *Saigo no nishiki taitei: Kaigun hikōtei no kiroku* [The last Type 2 giant flying boat: A chronicle of the navy's flying boats]. Tokyo: Kōjinsha, 1994.

Inoue Ikutarō. "Kōkūnihon no sōsho" [The beginnings of Japanese aviation]. In Nihon Kōkū Kyōkai, *Nihon kōkūshi* [The history of Japanese aviation], 1:248–58. Tokyo: Nihon Kōkū Kyōkai, 1936.

Inoue Ikutarō Kankōkai. *Inoue Ikutarō den* [The life of Inoue Ikutarō]. Tokyo: Kōbun Shoin, 1966.

Ishida Keigo and Hamada Hide. "Kyūnihongun ni okeru jinji hyōkaseido" [The evaluation of personnel matters in the former Japanese military]. *Bōeikenkyūjokiyō* 9 (2006): 43–82.

Ishihara, Shintaro. *The Japan That Can Say NO*. New York: Simon & Schuster, 1989.

Ishikawajima Jūkōgyō Kabushiki Kaisha 108-nenshi [A 108-year history of Ishikawajima Heavy Industries]. Tokyo: Ishikawajima Harima Jūkōgyō, 1961.

Ishizawa Kazuhiko. *Kikka: Nihon hatsu no jetto enjin Ne-20 no gijutsu kenshō: Kaigun tokushu kōgekiki* [Kikka: Japan's first jet engine: A review of the Ne-20: The navy's special attack plane]. Tokyo: Miki Shobō, 2006.

Israel, Ulrich. *Marineflieger einst und jetzt*. Berlin: Brandenburgisches Verlagshaus, 1991.

Itō Masanori. *Daikaigun o omō* [Thinking of the navy]. Tokyo: Bungei Shunjū Shinsha, 1956.

Itō Ryōhei and Itō Hitoshi. *Nihon kōkūshi nenpyō: Shōgen to shashin de tsuzuru 70-nen* [A chronology of Japanese aviation: 70 years pieced together by testimonies and photographs]. Tokyo: Nihon Kōkū Kyōkai, 1981.

Iwaya Eiichi. "Roketto ki Shūsui ni tsuite" [The rocket-powered aircraft Shūsui]. *Sekai no Kōkūki* 5 (January 1952): 82–88.

Iwaya Sazanami. *Hikō shōnen otogi ebanashi* [The young aviator, an illustrated picture story]. Tokyo: Bunundō, 1911.

Izawa Yasuho. *Nihon Rikugun jūbakutai* [The heavy bomber squad of the Japanese Army]. Tokyo: Gendaishi Shuppankai, 1982.

Jauneaud, Marcel. *L'aviation militaire et la guerre aérienne*. Paris: Flammarion, 1923.

Johnson, Chalmers A. *MITI and the Japanese Miracle: The Growth of Industrial Policy, 1925–1975*. Stanford, CA: Stanford University Press, 1982.

Jōhō Yoshio. *Rikugunshō gunmukyoku shi, Gekan (Shōwa)* [The history of the Bureau of Military Affairs, vol. 2 (Shōwa)]. Tokyo: Fuyō Shobō Shuppan, 2002.

———. *Rikugunshō gunmukyoku shi, Jōkan (Meiji Taishō)* [The history of the Bureau of Military Affairs, vol. 1 (Meiji and Taishō)]. Tokyo: Fuyō shobō shuppan, 2002.

Joint Chiefs of Staff. *Department of Defense Dictionary of Military and Associated Terms*. Ft. Belvoir, VA: Defense Technical Information Center, 2010.

Jones, Hazel J. *Live Machines: Hired Foreigners and Meiji Japan*. Vancouver: University of British Columbia Press, 1980.

Kaigun kōkū gaishi [An outline of the history of naval aviation]. Tokyo: Asagumo Shinbunsha, 1976.

Kakamigaharashi kyōiku iinkai. *Kakamigaharashi shi tsūshi hen kinsei kindai gendai* [The complete early-modern, modern, and present history of Kakamigahara]. Kakamigahara: Kakamigahara City, 1987.

Kariya Masai. *Nihon rikugun shisakuki monogatari* [About the Japanese Army's experimental machines]. Tokyo: Kōjinsha, 2007.

Kasza, Gregory James. *The State and the Mass Media in Japan, 1918–1945*. Berkeley: University of California Press, 1988.

Katō Isamu. *Sakuma Ichirō den* [The life of Sakuma Ichirō]. Yokosuka: Sakuma Ichirō den kai, 1977.

Katsuragi Yōji. *Rekishi no naka no Nakajima hikōki* [A history of Nakajima's aircraft]. Tokyo: Guranpurishuppan, 2002.

Kawasaki Heavy Industries. *Kawasaki Jūkō Gifu kōjō no omoide* [Kawasaki Heavy Industries: Recollections of the Gifu factory]. Kakamigahara: Kawasaki Jūkōgyō, 1988.

Kawasaki Jūkō Gifu kōjō gojūnen no ayumi [Kawasaki's Gifu factory: 50 years of development]. Kakamigahara: Kawasaki Jūkōgyō kōkūkijigyōhonbu, 1987.

Kawasaki Jūkōgyō KK. *Kawasaki Jūkōgyō kabushiki gaisha hyakunenshi: Yume o katachi ni: 1896–1996* [100 years of Kawasaki Heavy Industries: Like a dream: 1896–1996]. Kobe: Kawasaki Jūkōgyō, 1997.

Kawasaki Kōkūkikōgyō KK. *Kōkūkiseisaku enkaku kitai no bu* [The development of aircraft production—airframes]. Kakamigahara: Kawasaki kōkūkikōgyō, 1946.

Kennedy, David M. *The American People in World War II: Freedom from Fear, Part Two.* New York: Oxford University Press, 1999.

Kern, Stephen. *The Culture of Time and Space, 1880–1918.* Cambridge, MA: Harvard University Press, 2003.

Kimura Hidemasa. *Hikōki no hon* [The aircraft book]. Tokyo: Shinchōsha, 1962.

Kimura Hidemasa and Tanaka Shōichi. *Nihon no meiki hyakusen* [A selection of one hundred famous Japanese airplanes]. Nagoya: Chūnichi Shinbunsha, 1985.

Koch, Matthias. *Rüstungskonversion in Japan nach dem Zweiten Weltkrieg: von der Kriegswirtschaft zu einer Weltwirtschaftsmacht.* Munich: Iudicium, 1998.

Koiso Kuniaki. *Katsuzan kōsō* [Arrowhead mountain, wild goose claw]. Tokyo: Koiso Kuniaki Jijoden Kankōkai, 1963.

Koiso Kuniaki and Musha Kinkichi. *Kōkū no genjō to shōrai* [The present and future of aviation]. Tokyo: Bunmei Kyōkai, 1928.

Kōri Katsu. *Aireview's the Fifty Years of Japanese Aviation, 1910–1960: A Picture History with 910 Photographs.* Tokyo: Kantōsha, 1961.

Kotani Ken. *Japanese Intelligence in World War II.* Oxford: Osprey, 2009.

Koyama, Hirotake. *Nihon gunji kōgyō no shiteki bunseki Nihon shihon shugi no hattenkōzō to no kankei ni oite* [A historical analysis of Japan's military industry and its relation to the structure of the development of Japan's capitalism]. Tokyo: Ochanomizu Shobō, 1972.

Kudō Akira, Tajima Nobuo, and Erich Pauer. *Japan and Germany: Two Latecomers to the World Stage, 1890–1945.* 3 vols. Folkestone, UK: Global Oriental, 2009.

Kusakari Shirō. *Taiō nikki* [The diary of my stay in Europe]. N.p.: privately published by Shinozaki Masaru, 2008.

Kuwabara Torao. *Kaigun kōkū kaisōroku* [Recollections of the navy air force]. Tokyo: Kōkū Shinbunsha, 1964.

Launius, Roger D. *Innovation and the Development of Flight*. College Station: Texas A & M University Press, 1999.

Law, John. *Aircraft Stories: Decentering the Object in Technoscience*. Durham, NC: Duke University Press, 2002.

Le Corbusier. *Aircraft: The New Vision*. 1935; New York: Universe, 1988.

Lester, Robert, ed. *US Military Intelligence Reports, Japan, 1918–1941*. Frederick, MD: University Publications of America, 1987.

Lindbergh, Charles A. *"We."* New York: Putnam's, 1927.

Lorell, Mark A. *Troubled Partnership: A History of U.S. Japan Collaboration on the FS-X Fighter*. Santa Monica, CA: Rand, 1995.

Maema Takanori. *Asahi Shinbun hōō daihikō: Ge, senko no miyako Rōma e* [The Asahi Newspaper's Great Visit-Europe-Flight, vol. 2: Towards the Eternal City of Rome]. Tokyo: Kōdansha, 2004.

———. *Asahi Shinbun hōō daihikō: Jō, Shiberia ōdan* [The Asahi Newspaper's Great Visit-Europe-Flight, vol. 1: Crossing Siberia]. Tokyo: Kōdansha, 2004.

———. *Jetto enjin ni toritsukareta otoko* [Men obsessed with jet engines]. Tokyo: Kōdansha, 1992.

———. *Kokusan ryokyakki MRJ hishō* [The flight of the made-in-Japan passenger aircraft MRJ]. Tokyo: Yamato Shobō, 2008.

———. *Man mashin no shōwa densetsu* [A Shōwa legend of men and machines]. 2 vols. Tokyo: Kōdansha, 1996.

———. *Shinkansen o kōkūki ni kaeta otokotachi* [The men who changed the Shinkansen into an airplane]. Tokyo: Sakurasha, 2014.

———. *YS-11 ge: Kunan no hatsu hikō to meiki no unmei* [YS-11 vol. 2: The maiden flight's hardships and the destiny of a masterpiece]. Tokyo: Kōdansha, 1999.

———. *YS-11 jō: Kokusan ryokyakki wo tsukutta otokotachi* [YS-11 vol. 1: The men who built a made-in-Japan airliner]. Tokyo: Kōdansha, 1999.

Mainichi Shinbun sha. *Nihon kōkū shi* [The history of Japanese aviation]. Tokyo: Mainichi Shinbun sha, 1979.

Makino Ikuo. *Saishū kessen heiki Shūsui sekkeisha no kaisō: Mihappyō shiryō ni yori kaimeisuru kyūkyoku no mekanizumu* [Recollections of the designers of the Shūsui, the weapon for the last decisive battle: Clarification of the final mechanism by unpublished sources]. Tokyo: Kōjinsha, 2006.

Marder, Arthur Jacob. *Old Friends, New Enemies: The Royal Navy and the Imperial Japanese Navy*. Vol. 1, *Strategic Illusions, 1936–1941*. Oxford: Clarendon, 1981.

Markus, Andrew L. "The Carnival of Edo: Misemono Spectacles from Contemporary Accounts." *Harvard Journal of Asiatic Studies* 45, no. 2 (December 1985): 499–541.

Martin, Bernd. *Deutschland und Japan im Zweiten Weltkrieg: 1940–1945*. Hamburg: Nikol, 2001.

Matogawa Yasunori. "Senzen no Nihon no roketto kenkyū" [Japan's prewar rocket research]. *ISAS News*, No. 239 (January 2001).

Matsuoka Hisamitsu. *Mitsubishi hikōki monogatari* [The story of Mitsubishi aircraft]. Tokyo: Atene Shobō, 1993.

May, Ernest R. *Knowing One's Enemies: Intelligence Assessment before the Two World Wars.* Princeton, NJ: Princeton University Press, 1984.

Meilinger, Phillip S. "Giulio Douhet and the Origins of Airpower Theory." In *The Paths of Heaven: The Evolution of Airpower Theory*, edited by Phillip S. Meilinger, 1–41. Maxwell AFB, AL: Air University Press, 1997.

Melzer, Jürgen. "Warfare 1914–1918 (Japan)." In *1914–1918-online. International Encyclopedia of the First World War*, edited by Ute Daniel, Peter Gatrell, Oliver Janz, Heather Jones, Jennifer Keene, Alan Kramer, and Bill Nasson. Freie Universität Berlin. https://doi.org/10.15463/ie1418.11172. Accessed January 12, 2019.

———. "'We Must Learn from Germany': Gliders and Model Airplanes as Tools for Japan's Mass Mobilization." *Contemporary Japan* 26, no. 1 (2014): 1–27.

Mikesh, Robert, and Shorzoe Abe. *Japanese Aircraft: 1910–1941.* London: Putnam, 1990.

Miki Tetsuo. "Aichi Tokei Denki no hikōki" [The aircraft of Aichi Tokei Denki]. In *Nihon minkan kōkūshiwa* [Stories of Japan's civil aviation], 101–3. Tokyo: Nihon Kōkū Kyōkai, 1966.

Millet, Allan A. "Patterns of Military Innovation." In *Military Innovation in the Interwar Period*, edited by Williamson Murray et al., 329–67. New York: Cambridge University Press, 1996.

Mimura, Janis. *Planning for Empire: Reform Bureaucrats and the Japanese Wartime State.* Ithaca, NY: Cornell University Press, 2011.

Mitsubishi Jūkōgyō KK. *Mitsubishi Jūkogyō Kabushiki Kaisha shi* [The history of Mitsubishi Heavy Industries Ltd.]. Tokyo: Mitsubishi Jūkōgyō Kabushiki Kaisha, 1956.

Mitsubishi Jūkōgyō KK. *Ōjibōbō: Mitsubishi Jūkō Nagoya 50 nen no kaiko* [Memories of the past: 50 years of recollections of Mitsubishi Nagoya]. 3 vols. Nagoya: Ryōkōkai, 1970.

Mitsubishi Nainenki. *R gata hikōtei zumenyō goi* [A dictionary for blueprints for the R-type flying boat]. N.p., n.d.

"Mitsubishi ni okeru kitaisekkei shisaku no hensen" [Changes in Mitsubishi's aircraft body design and trial manufacture]. In *Minkan kōkūki kōgyōshi* [The history of civil aircraft manufacturing], edited by Kōkū kōgyōshi hensan iinkai, 152–56. N.p.: 1948.

Mitsubishi shashi [History of the Mitsubishi company]. 41 vols. Tokyo: Tōkyō Daigaku Shuppankai, 1979–2016.

Miyamoto Mataji and Yoshio Togai. *Sōgō shōsha no keieishi* [A history of the management of general trading companies]. Tokyo: Tōyō Keizai Shinpōsha, 1976.

Mizuno, Hiromi. *Science for the Empire: Scientific Nationalism in Modern Japan.* Stanford, CA: Stanford University Press, 2009.

Mizusawa Hikari. *Gunyōki no tanjō: Nihongun no kōkū senryaku to gijyutsu kaihatsu* [The birth of military aircraft: The development of the Japanese military's strategy and technology]. Tokyo: Yoshikawa kōbunkan, 2017.

Mokyr, Joel. *The Lever of Riches: Technological Creativity and Economic Progress.* New York: Oxford University Press, 1990.

Morikawa Hajime. *Sora no eiyū Fujita Yūzō chūsa* [Wing Commander Fujita Yūzō, the hero of the skies]. Tokyo: Shōwa Shobō, 1939.

Moriya Gakuji. *Meikōkōsakubu no senzensengoshi* [The pre- and postwar history of aircraft construction at the Nagoya factory]. Nagoya: Mitsubishi Jūkōgyō KK Nagoya Kōkūkiseisakujo, 1988.

Morris-Suzuki, Tessa. *The Technological Transformation of Japan: From the Seventeenth to the Twenty-First Century.* New York: Cambridge University Press, 1994.

Munson, Kenneth. *Bombers between the Wars, 1919–39.* New York: Macmillan, 1970.

Muramatsu, Teijiro. *Westerners in the Modernization of Japan.* Tokyo: Hitachi, 1995.

Nagamori Yoshio. *Ichi chūdoku gijutsushikan no omoide* [The memoirs of one technical officer stationed in Germany]. N.p.: Kōsai insatsu, 1980.

Nagura Bunji and Yokoi Katsuhiko. *Nichi-Ei heiki sangyōshi: Buki iten no keizaishiteki kenkyū* [The Japanese-British arms industry: Research on the economic history of weapons transfer]. Tokyo: Nihon Keizai Hyōronsha, 2005.

Naitō Hatsuho. *Kimitsu heiki funryū* [The top-secret weapon Funryū]. Tokyo: Tosho Shuppansha, 1979.

———. *Ōka: Hijō no tokkō heiki* [Ōka: A merciless special attack weapon]. Tokyo: Bungei Shunjū, 1982.

———. *Thunder Gods: The Kamikaze Pilots Tell Their Story.* New York: Kodansha International, 1989.

Nakagawa Ryōichi and Mizutani Sōtarō. *Nakajima hikōki enjin-shi: Wakai gijutsusha shūdan no katsuyaku* [The history of Nakajima's aircraft engines: The activities of a group of young engineers]. Tokyo: Kantōsha, 1985.

Nakayama Shigeru. *Science, Technology and Society in Postwar Japan.* London: Kegan Paul, 1991.

Nelson, George. *Industrial Architecture of Albert Kahn, Inc.* New York: Architectural Book Publishing, 1939.

Newdick, Thomas. *Japanese Aircraft of World War II 1937–1945.* London: Amber Books, 2017.

Nihon Kaigun Kōkūgaishi Kankōkai. *Umiwashi no kōseki: Nihon Kaigun kōkū gaishi* [The wake of the sea eagles: An unofficial history of Japanese naval aviation]. Tokyo: Hara Shobō, 1982.

Nihon Kaigun Kōkūshi Hensan Iinkai. *Nihon Kaigun kōkūshi 1 yōhei hen* [The history of Japanese naval aviation 1: Strategy]. Tokyo: Jiji Tsūshinsha, 1969.

———. *Nihon Kaigun kōkūshi 2 gunbi hen* [The history of Japanese naval aviation 2: Armaments]. Tokyo: Jiji Tsūshinsha, 1969.

———. *Nihon Kaigun kōkūshi 3 gijutsu seido hen* [The history of Japanese naval aviation 3: Technology]. Tokyo: Jiji Tsūshinsha, 1969.

Nihon Kōkū Gakujutsushi Henshū Iinkai. *Nihon kōkū gakujutsushi (1910–1945)* [Aeronautical research in Japan (1910–1945)]. Tokyo: Hatsubaijo Maruzen, 1990.

Nihon Kōkū Kyōkai. *Nihon kōkūshi* [The history of Japanese aviation]. 2 vols. Tokyo: Nihon Kōkū Kyōkai, 1936.

———. *Nihon kōkū shi Meiji Taishō* [The history of Japanese aviation in the Meiji and Taishō periods], vol. 1. Tokyo: Nihon Kōkū Kyōkai, 1956.

———. *Nihon kōkūshi Shōwa sengo hen* [The history of Japanese aviation in the postwar Shōwa period], vol. 3. Tokyo: Nihon Kōkūshi Hensan Iinkai, 1992.

———. *Nihon kōkūshi Shōwa zenkihen* [The history of Japanese aviation in the early Shōwa period], vol. 2. Tokyo: Nihon Kōkū Kyōkai, 1956.

Nihon minkan kōkūshiwa [Stories of Japan's civil aviation]. Tokyo: Nihon Kōkū Kyōkai, 1966.

Nihon no kōkū runesansu [The renaissance of Japan's aviation]. Tokyo: Kantōsha, 2000.

Nippon sekai isshū daihikō [The great around-the-world flight of the *Nippon*]. Osaka: Osaka Mainichi Shinbunsha, 1940.

Nish, Ian Hill. *Alliance in Decline: A Study in Anglo-Japanese Relations 1908–23*. London: Athlone, 1972.

———. "Japan and the Outbreak of War in 1914." In *Collected Writings of Ian Nish, Part 1*, 173–87. Richmond, UK: Japan Library, 2001.

Nishiyama, Takashi. *Engineering War and Peace in Modern Japan, 1868–1964*. Baltimore, MD: Johns Hopkins University Press, 2014.

Nohara Shigeru. "Kokusan puropera monogatari" [The story of the made-in-Japan propeller]. *Rekishi Gunzō* 81 (February 2007): 81–91.

———. *Nihon gun'yōki jiten: Kaigun hen, 1910–1945* [Encyclopedia of Japanese warplanes (Navy) 1910–1945]. Tokyo: Ikarosu Shuppan, 2005.

———. *Nihon gun'yōki jiten: Rikugun hen, 1910–1945* [Encyclopedia of Japanese warplanes (Army), 1910–1945]. Tokyo: Ikarosu Shuppan, 2005.

———. *Zukai sekai no gun'yōkishi: 5, Nihon kaigun gun'yōkishū* [Illustrated warplane history vol. 5: Warplanes of the Japanese Navy]. Tokyo: Guriinarō Shuppansha, 1994.

———. *Zukai sekai no gun'yōkishi: 6, Nihon rikugun gun'yōkishū* [An illustrated history of the world's warplanes vol. 6: Warplanes of the Japanese Army]. Tokyo: Guriinarō Shuppansha, 1997.

———. *Zukai sekai no gun'yōkishi: 8, Nihon rikukaigun shisaku keikakuki 1924–1945* [An illustrated history of the world's warplanes vol. 8: Experimental planes of the Japanese Army and Navy, 1924–1945]. Tokyo: Guriinarō Shuppansha, 1999.

Nowarra, Heinz J. *Die Flugzeuge des Alexander Baumann*. Friedberg: Podzun-Pallas, 1982.

Nozawa Tadashi. *Nihon kōkūki sōshū: Aichi, Kūgishō hen* [Encyclopedia of Japanese aircraft: Aichi, Navy aviation technology arsenal]. Tokyo: Shuppankyōdōsha, 1959.

———. *Nihon kōkūki sōshū: Kawanishi Hiroshō hen* [Encyclopedia of Japanese aircraft: Kawanishi and Hiro Navy Arsenal]. Tokyo: Shuppankyōdōsha, 1959.

———. *Nihon kōkūki sōshū: Kawasaki hen* [Encyclopedia of Japanese aircraft: Kawasaki]. Tokyo: Shuppankyōdōsha, 1960.

———. *Nihon kōkūki sōshū: Kyūshū • Hitachi • Shōwa • Nippi • shosha hen* [Encyclopedia of Japanese aircraft: Kyūshū, Hitachi, Shōwa, Nippi, others]. Tokyo: Shuppankyōdōsha, 1980.

———. *Nihon kōkūki sōshū: Mitsubishi hen* [Encyclopedia of Japanese aircraft: Mitsubishi]. Tokyo: Shuppankyōdōsha, 1961.

———. *Nihon kōkūki sōshū: Nakajima hen* [Encyclopedia of Japanese aircraft: Nakajima]. Tokyo: Shuppankyōdōsha, 1963.

———. *Nihon kōkūki sōshū: Tachikawa • rikugun kōkūkōshō • Manpi • Nikkoku hen* [Encyclopedia of Japanese aircraft: Tachikawa, army aviation arsenal, Manpi, Nikkoku]. Tokyo: Shuppankyōdōsha, 1980.

———. *Nihon kōkūki sōshū: Yunyūki hen* [Encyclopedia of Japanese aircraft: Imported aircraft]. Tokyo: Shuppankyōdōsha, 1972.

Oba Yahei. *Warera no kūgun* [Our air force]. Tokyo: Dai Nihon Yūbenkai Kōdansha, 1937.
O'Brien, Phillips Payson. *The Anglo-Japanese Alliance, 1902–1922*. New York: Routledge Curzon, 2004.
Ogasawara Kazuo. *Kōkū senjutsu kōjuroku* [Lecture notes on air strategy]. Tokyo: Rikugundaigakkō shōkō shūkaijo, 1922.
Ogawa Toshihiko. *Nihon kōkūki daizukan: 1910–1945* [An illustrated guide to Japanese aircraft: 1910–1945]. 3 vols. Tokyo: Kokusho Kankōkai, 1993.
Okamura Jun. *Kōkū gijutsu no zenbō* [An overall picture of aviation technology]. 2 vols. Tokyo: Hara Shobō, 1976.
Okumiya Masatake and Horikoshi Jirō. *Zero! The Story of Japan's Air War in the Pacific: 1941–45*. New York: Dutton, 1956.
Ono Eisuke. *Beikoku kōkūkōgyō ni tsuite* [The US aviation industry]. Yokohama: Yokohama Specie Bank, 1938.
Ono Shōzō. *Kūchū hikōki no genzai oyobi shōrai* [The present and future of aircraft]. Tokyo: Senryūdō, 1910.
Ōsaka Honsha Hanbai Hyakunenshi Henshū Iinkai. *Asahi Shinbun hanbai hyakunen shi: Ōsaka hen* [A 100-year sales history of the *Asahi Shinbun* (Osaka edition)]. Osaka: Asahi Shinbun Ōsaka Honsha, 1979.
Ōshū hōmon dai hikō kinen gahō [The Great Visit-Europe Flight, a memorial picture report]. 2 vols. Osaka: Osaka Asahi Shinbun, 1925.
Ōtani Shinichi. "Kyōto to Shimazu Genzō fushi" [Kyoto and Shimazu Genzō and his son]. *Kagaku to kyōiku* 44, no. 1 (1996): 18–19.
Ōura Genzaburō. *Saikin sekai no hikōsen* [The world's latest airships]. Tokyo: Hakubunkan, 1909.

Palmer, Scott W. *Dictatorship of the Air: Aviation Culture and the Fate of Modern Russia*. New York: Cambridge University Press, 2006.
Pattillo, Donald M. *Pushing the Envelope: The American Aircraft Industry*. Ann Arbor: University of Michigan Press, 1998.
Peattie, Mark R. *Sunburst: The Rise of Japanese Naval Air Power, 1909–1941*. Annapolis, MD: Naval Institute Press, 2001.
Plüschow, Gunther. *Chintao kara tobidashite* [Flying out of Tsingtau]. Tokyo: Rakuyōdō, 1918.
———. *Die Abenteuer des Fliegers von Tsingtau, meine Erlebnisse in drei Erdteilen*. Berlin: Ullstein, 1916.
———. *My Escape from Donington Hall preceded by an account of the siege of Kiao-Chow in 1914*. London: John Lane, 1922.
Polak, Christian Philippe. *Fude to katana: Nihon no naka no mō hitotsu no Furansu 1872–1960* [The brush and the sword: Yet another France in Japan, 1872–1960]. Tokyo: Zainichi Furansu shōkō kaigisho, 2005.
———. *Kinu to hikari: Shirarezaru Nichi-Futsu kōryū 100-nen no rekishi (Edo jidai-1950-nendai)* [Silk and radiance: The unknown 100-year history of Japanese-French exchange (from the Edo period to the 1950s)]. Tokyo: Ashetto Fujin Gahōsha, 2002.

———. "La mission militaire française de l'aéronautique au Japon (1919–1921)." *Ebisu*, no. 51 (2014): 163–83.

Porte, Rémy. "L'échec de la Mission militaire française d'Aéronautique au Japon 1918–1920." *Revue historique des armées*, no. 236 (2004): 88–96.

Presseisen, Ernst Leopold. *Before Aggression: Europeans Prepare the Japanese Army*. Tucson: University of Arizona Press, 1965.

Roach, Edward J. *The Wright Company: From Invention to Industry*. Athens: Ohio University Press, 2014.

Roland, Alex. *Model Research: The National Advisory Committee for Aeronautics, 1915–1958*. 2 vols. Washington, DC: Scientific and Technical Information Branch, National Aeronautics and Space Administration, 1985.

Ruttan, Vernon W., and Yujiro Hayami. "Technology Transfer and Agricultural Development." *Technology and Culture* 14 (1973): 119–51.

Saaler, Sven. "The Imperial Japanese Army and Germany." In *Japanese-German Relations 1895–1945. War, Diplomacy and Public Opinion*, edited by Christian W. Spang and Rolf-Harald Wippich, 21–41. New York: Routledge, 2006.

Samuels, Richard J. *"Rich Nation, Strong Army": National Security and the Technological Transformation of Japan*. Ithaca, NY: Cornell University Press, 1994.

Sanuki Matao. *Sanuki Matao no hitorigoto: Kōkū no sekai o horiokoshite hakkutsu shita jijitsu* [Sanuki Matao's monologue: Facts unearthed from the aviation world]. Tokyo: Gurīn arōshuppansha, 1997.

Sawai, Minoru. "Amerikasei kōsakukikai no yunyū to shōsha katsudō: 1930–1965" [The import of US machine tools and trading companies from 1930 to 1965]. *Ōsaka Daigaku keizaigaku* 45, no. 2 (1995): 1–25.

———. *Kindai Nihon no kenkyū kaihatsu taisei* [The completion of the development of research in modern Japan]. Nagoya: Nagoya Daigaku Shuppankai, 2012.

Schabel, Ralf. *Die Illusion der Wunderwaffen: die Rolle der Düsenflugzeuge und Flugabwehrraketen in der Rüstungspolitik des Dritten Reiches*. Munich: R. Oldenbourg, 1994.

Schatzberg, Eric. *Wings of Wood, Wings of Metal: Culture and Technical Choice in American Airplane Materials, 1914–1945*. Princeton, NJ: Princeton University Press, 1999.

Schencking, J. Charles. *Making Waves: Politics, Propaganda, and the Emergence of the Imperial Japanese Navy, 1868–1922*. Stanford, CA: Stanford University Press, 2005.

Schivelbusch, Wolfgang. *The Railway Journey: The Industrialization of Time and Space in the 19th Century*. Berkeley: University of California Press, 1986.

Schmitt, Günter, Angelika Hofmann, and Thomas Hofmann. *Hugo Junkers und seine Flugzeuge*. Stuttgart: Motorbuch, 1986.

Sekigawa, Eiichiro. *Pictorial History of Japanese Military Aviation*. London: Allan, 1974.

"Sekkei wa aato ofu conpuromaizu" [Design is the art of compromise]. *Japan Airlines Flight Crew News*, 91 (January 1992): 6–17.

Sekkeisha no shōgen: Nihon kessakuki kaihatsu dokyumento [The testimony of designers: Documents of the development of Japanese technological masterpieces]. 2 vols. Tokyo: Kantōsha, 1994.

Senba Tadashi. *Hikōki ni miserarete* [Fascinated by aircraft]. Gifu: Chūbu senkōkai, 2000.

Shibata Kazuya. *Yūjin roketto sentōki Shūsui: Kaigun dai312 kōkūtai Shūsuitai shashinshi* [The manned rocket fighter Shūsui: A photo history of the Navy's 312th Air Corps]. Tokyo: Dainihon Kaiga, 2005.

Shibata Shigeki. "Chōsensensōgo no America no taigaienjo (MSA) to Nihon no kōkūkikōgyō" [US foreign assistance to Japan (MSA) and the Japanese aircraft industry after the Korean war]. *Shakai-keizai-gaku* 67, no. 2 (2001): 169–90.

Shibuya Atsushi. *Hino Kumazō den* [The life of Hino Kumazō]. Kumamoto: Tamakina Shuppansha, 2006.

Shigeno Kiyotake and Kuwano Tōka. *Tsūzoku hikōki no hanashi* [A popular (nonspecialist) story about aircraft]. Tokyo: Nittōdō shoten, 1913.

Shin Meiwa Kōgyō Kabushiki Kaisha Shashi Henshū Iinkai. *Shashi* [A company's history]. Nishinomiya: Dōsha, 1984.

Shinohara Hiroshi. *Nihon Kaigun oyatoi gaijin: bakumatsu kara Nichi-Ro Sensō made* [The navy's foreign advisers: from the closing days of the Tokugawa bakufu to the Russo-Japanese War]. Tokyo: Chūō Kōronsha, 1988.

Shōnen kokubōkai. *Warera no kūgun* [Our airforce]. Tokyo: Kimura Shobō, 1932.

Simonson, G. R. "The Demand for Aircraft and the Aircraft Industry, 1907–1958." *Journal of Economic History* 20, no. 3 (1960): 361–82.

Smith, Merritt Roe, and Leo Marx, eds. *Does Technology Drive History?: The Dilemma of Technological Determinism*. Cambridge, MA: MIT Press, 1994.

Society of Japanese Aerospace Companies. *Nihon no kōkū uchū kōgyō 50nen no ayumi* [The 50-year progress of Japanese aerospace companies]. Tokyo: Society of Japanese Aerospace Companies, 2003.

Staudenmaier, John M. *Technology's Storytellers: Reweaving the Human Fabric*. Cambridge, MA: Society for the History of Technology and MIT Press, 1985.

Stillwell, Paul, ed. *Air Raid, Pearl Harbor!* Annapolis, MD: Naval Institute Press, 1981.

Sugimoto Kaname. *Tobe, MRJ: Sekai no kōkūki shijō ni nozomu Hinomaru jetto* [Fly MRJ, the Hinomaru jet that challenges the international aircraft market]. Tokyo: Nikkan Kōgyō Shinbun Sha, 2015.

Sugita Ichiji. *Nihon no seisenryaku to kyōkun: Washington Kaigi kara shūsen made* [Japan's political war strategy and lessons: From the Washington Conference to the end of the war]. Tokyo: Hara Shobō, 1983.

Suikōkai. *Kaisō no Nihon Kaigun* [Recollections of the Japanese Navy]. Tokyo: Hara Shobō, 1985.

Sumitomo Seimitsu Kōgyō shashi [A history of Sumitomo Precision Products]. Amagasaki: Sumitomo Seimitsu Kōgyō Kabushiki Kaisha Shashi, Henshū Iinkai, 1981.

Takano Tatsumi. *Hino Tokugawa ryō taii hikōki daikyōsō shōka* [The song of the big contest between the two Captains Hino and Tokugawa]. Tokyo: Hotta Wataru Morikan, 1911.

Takeda Yorimasa. *Burūinparusu: Ōzora o kakeru samurai tachi* [Blue Impulse: The samurai flying through the wide open sky]. Tokyo: Bungeishunjū, 2014.

———. *Zerosen no ko: Densetsu no mōshō Kamei Yoshio to sono kyōdai* [A child of the Zero: The legendary brave warrior Kamei Yoshio and his brothers]. Tokyo: Bungeishunju, 2014.

Till, Geoffrey. *Air Power and the Royal Navy, 1914–1945: A Historical Survey*. London: Macdonald and Jane's, 1979.

Tochio Takako. *Nihon no kōkūkisangyō: Genjō to kongo no kadai* [Japan's aviation industry: Its present condition and its future issues]. Tokyo: National Diet Library, 2016.

Tokorozawa kōkūshiryō chōsa shūshū suru kai. *Yūhi sora no makuake Tokorozawa* [Launching out: Tokorozawa and the opening of the skies]. Tokorozawa: Daiichi insatsu, 2005.

Tokugawa Yoshitoshi. *Nihon kōkū koto hajime* [The beginnings of Japanese aviation]. Tokyo: Shuppankyōdōsha, 1964.

Tomizuka Kiyoshi. *Kōkenki sekai kiroku juritsu e no kiseki* [*Kōkenki*: The drive for setting a world record]. Tokyo: Mikishobō, 1998.

Torikai Tsuruo, ed. *Shirarezaru gunyōki kaihatsu* [The development of unknown military aircraft]. 2 vols. Tokyo: Kantōsha, 1999.

Tow, Edna. "The Great Bombing of Chongqing and the Anti-Japanese War, 1937–1945." In *The Battle for China: Essays On the Military History of the Sino-Japanese War of 1937–1945*, edited by Mark R. Peattie, Edward J. Drea, and Hans van de Ven, 256–82. Stanford, CA: Stanford University Press, 2011.

Toyoda Jō. *Hikōkiō Nakajima Chikuhei* [Nakajima Chikuhei, the king of aircraft]. Tokyo: Kōdansha, 1992.

Tsuji Hideo and Kurita Masatada. *Hondo bōkū sakusen* [Strategy for the air defense of the mainland]. Tokyo: Asagumo Shinbunsha, 1968.

Ugaki Kazushige. "Kokka sōdōin ni sakuō suru Teikoku Rikugun no shinshisetsu" [The Imperial Army's new facilities and their connection to national mobilization]. In *Kokka sōdōin no igi* [The meaning of national mobilization], edited by Sawamoto Mōko, 263–84. Tokyo: Aoyama Shoin, 1926.

United States Strategic Bombing Survey, Mitsubishi Heavy Industries. Washington, DC: United States Strategic Bombing Survey, Aircraft Division, 1947.

United States Strategic Bombing Survey, Nippon Airplane (Nippon Hikoki KK). Washington, DC: United States Strategic Bombing Survey, Aircraft Division, 1945, 1946.

United States Strategic Bombing Survey, The Japanese Aircraft Industry. Washington, DC: United States Strategic Bombing Survey, Aircraft Division, 1947.

van Creveld, Martin. *Technology and War: From 2000 B.C. to the Present*. New York: Free Press, 1989.

van der Mey, Michiel. *Dornier Wal: "A Light Coming Over the Sea."* Vicchio, Italy: LoGisma, 2005.

Van Vleck, Jenifer L. *Empire of the Air: Aviation and the American Ascendancy*. Cambridge, MA: Harvard University Press, 2013.

Vogt, Richard. *Weltumspannende Memoiren eines Flugzeug-Konstrukteurs*. Steinebach: Luftfahrt-Verlag W. Zuerl, 1976.

von Gersdorff, Kyrill, and Kurt Grasmann. *Flugmotoren und Strahltriebwerke.* Munich: Bernard & Graefe, 1981.
von Kármán, Theodore. *The Wind and Beyond; Theodore von Kármán, Pioneer in Aviation and Pathfinder in Space.* Boston: Little, Brown, 1967.

Wachtel, Joachim. *Claude Dornier: ein Leben für die Luftfahrt.* Planegg: Aviatic, 1989.
Wada Hideho. *Kaigun kōkū shiwa* [Historical episodes of naval aviation]. Tokyo: Meiji Shoin, 1944.
Wagner, Wolfgang. *Hugo Junkers, Pionier der Luftfahrt—seine Flugzeuge.* Bonn: Bernard & Graefe, 1996.
Watabe Kazuhide. *Nihon no hikōkiō Nakajima Chikuhei: Nihon kōkūkai no ichidai senkakusha no shōgai* [Nakajima Chikuhei, the king of aircraft: the life of Japan's number one aviation pioneer]. Tokyo: Kōjinsha, 1997.
Weckbach, Hubert. *Heilbronner Köpfe.* Heilbronn: Archiv der Stadt Heilbronn, 1998.
Weyl, Alfred R. *Fokker: The Creative Years.* New York: Funk and Wagnalls, 1968.
Williamson, Murray. "Innovation: Past and Future." In *Military Innovation in the Interwar Period,* edited by Murray Williamson et al., 300–328. New York: Cambridge University Press, 1996.
"Wind Tunnels in Japan: Leading Particulars and Scale Drawings of Six Tunnels Used in Aerodynamic Investigations." *Aircraft Engineering and Aerospace Technology* 3, no. 4 (1931): 79.
Wohl, Robert. *A Passion for Wings: Aviation and the Western Imagination, 1908–1918.* New Haven, CT: Yale University Press, 1994.
———. *The Spectacle of Flight: Aviation and the Western Imagination, 1920–1950.* Carlton: Melbourne University Press, 2005.
Wright, T. P. "American Methods of Aircraft Production." *Journal of the Royal Aeronautical Society* 43 (March 1939): 131–224.

Yamazaki Akio. *Nippon ga nekkyō shita daikōkū jidai* [The era of Japan's wild aviation enthusiasm]. Tokyo: Ei Shuppansha, 2007.
Yokokura Jun. *YS-11 tobe! Sekai o tonda nihon no tsubasa* [Fly YS-11! Japanese wings that flew in the world]. Tokyo: Shōgakkan, 2004.
Yokota Junya. *Kumo no ue kara mita Meiji* [The Meiji era seen from above the clouds]. Tokyo: Gakuyō shobō, 1999.
Yokoyama Hisayuki. "Military Technological Strategy and Armaments Concepts of Japanese Imperial Army—Around the Post-WWI Period." *NIDS Security Reports,* no. 2 (March 2001): 116–59.
Yokoyama T., K. Yuyama, I. Akojima, and S. Moriya. "The Rocket Fighter Shusui—As Re-developed from Incomplete and Vague Me163B Data." *Transactions of the Newcomen Society for the Study of the History of Engineering and Technology* 70 (1998–99): 257–76.
Young, Edward M. *Aerial Nationalism: A History of Aviation in Thailand.* Washington, DC: Smithsonian, 1994.
Young, Louise. *Japan's Total Empire: Manchuria and the Culture of Wartime Imperialism.* Berkeley: University of California Press, 1998.

Index

Page numbers for maps, figures, and tables are in italics.

Abe Hiroshi, 88, 93, 294n76
acrobatic flight, xx, 125, 277–78, *278*, 283
Ader, Clément, 25
advertising, 89, 222–27, 269
aerial armament, 5–7, 37, 102, 183, 264, 293n53; and the British Aviation Mission to Japan, 154, 157; and international relations, 270
Aeronautical Research Institute, 3, 112, 299n4
Aichi Aircraft Company, 195–96; D1A1, 190; D1A2, 190; D3A, 191, 207
Aichi Tokei, 80, 145, 148–49, 171, 186–87, 190–91. *See also* Aichi Aircraft Company
Aikokugō, 128–30
aircraft, carrier, 6, 168–69, 185, 187, 192, 197; A6M1, 212; and the British Aviation Mission to Japan, 158, 162, 165–66; D1A1, 190; D1A2, 190; D3A1, 191; early years of, 149–51, *150*; Experimental Kūshō 6-Shi Special Bomber, 189; F8C, 189; He 66, 188; He 70 Blitz, 190–91; and hydroplanes, 176; and launching technology 148; Mitsubishi Type 10, 153, 161, 186; Nakajima Type 97, 201; Plover, 302n90; and preemptive air strikes, 188–91; second generation of, 186–88; Suisei, 195; Zero-sen fighter, 107, 120, 201, 212, 218, 251, 255, 277
aircraft, civil, 56, 80, 130, 135, 196, 281–82; and British influence, 152, 157; and Junkers, 116; and technology transfer, 266; and US aviation, 204, 207–8
aircraft, reconnaissance, 7, 102, 116, 126–27, 135; balloons, 14, 19–20, 23, 38; and British influence, 152, 154; De Havilland D. H. 9, *159*; Do N, 100–101, *115*; floatplanes, 44; and flying boats, 173–74; Hansa Brandenburg, 145, *146*; Rumpler Taube, 47–48, *48*, 77; Salmson 2A2, 64, *134*; Tobi, 104–6, *106*; Type 88, 108–9, *134*; and US aviation, 206, 208, 229
aircraft, water-based combat, 177–82
aircraft carriers, 6, 193, 283; *Akagi*, 183, 185–86, 195; *Ark Royal*, 197; and the British Aviation Mission to Japan, 154–55, *158*, 160, 166; expansion of aircraft carrier force, 183–88; and flying boats, 176; *Furious*, 143, 149–51, 169, 184; *Hōshō*, 149, 151–53, 161–62,

aircraft carriers (*continued*)
168, 183–85; *Kaga*, 183, 185–86, 192, 195; *Langley*, 149, 151
air doctrine, 80–82, 93, 126–27
airfields, *xxvi*, 29, 87, 144, 149, 170, 221; Issy-les-Moulineaux, 42; and jet and rocket technology, 251, 254; Johannisthal, 29, 50, 264; Kakamigahara, 60, 97, 120–21, *122*, 132; on Luzon, 116; Oppama, 253; near Stuttgart, 107; Tachikawa, 89, 113–14, 128; Tokorozawa, 34, 36, 54–55, 60–61, 77–78; Yokosuka, 192
airframes, 102, 104, 109, 184, 201, 281; and early German influence, 73; and flying boats, 171, 174; and jet and rocket technology, 235, 241–44, 248, 261, 263; and Junkers, 120; Kawasaki Type 87 bomber, *101*; Kikka, 251–52, 254; and US aviation, 203, 212, 225, 229
airliners: Boeing 222, 281; Boeing 787, 281, 283; Calcutta, 181; Douglas DC-1, 207; Douglas DC-2, 207–8, 227; Douglas DC-3, 208, 210, 227; Douglas DC-4E, 210, *211*; German Junkers in Japan, 114–19; and license negotiations, 119–21; and Mitsubishi, 124–25; Type 92 Superheavy Bomber, 114–15, 121–24, *122*, 130–32; US airliners, 207–10; YS-11, 278–79, 281–83
air-mindedness, 12, 24, 34, 37, 53, 60, 66, 70, 71, 86, 92, 93, 94, 264, 269, 277, 286n2
air raids, 6–7, 46, 94, 135–36, 201, 270; and carrier-based aircraft, 149; and jet and rocket technology, 235–236, 243, 245, 250–51, 257; and Junkers, 121; and the Manchurian crisis, 127; and US machine tools, 216–17
airships, 23, *28*, 36, 53, 78, 151; Yamada Airship no. 2, 21–22, *21*, 24; Zeppelin LZ 4, 53, 289n32
air shows, 5, 34, 277
air strikes, 6–7, 132, 149, 201, 268; preemptive, 188–91
Air Technical Research Institute (Rikugun Kōkūgijutsu Kenkyūjo), 135, 196
All Nippon Airways, 278
Andō Nario, 114
Anti-Comintern Pact, 190
arms reduction. *See* disarmament
Army Aviation Headquarters, 96, 114, 116, 126, 132, 135, 221, 223
around-the-world trips, *228*, 269
Asahi Shinbun, 47–48, 60, 78, 112, 269; and Junkers, 123; *Kamikaze*, 223–25, *224*; and the Manchurian crisis, 128; and US aviation, 207, 231; and the "Visit Europe Flight," 86–93, *91–92*
A6M1, 212
Atsuta Arsenal, 61
autonomous diplomacy, 6
Aviator of Qingdao. *See* Plüschow, Gunther

Baldwin, Thomas Scott, 19
Balloon Committee. *See* Provisional Committee for Military Balloon Research
balloon launches, 5, 268; Japan's first balloon launch, 1, 13–17, *16*, 86; Shimazu Genzō's balloon launch, 17–18, *18*
balloons, 11–13, 37–38, 57, 62, 223, 272; balloon fever, 17–19; early balloons, 13–14; reconnaissance balloons, 286n7; in the Russo-Japanese War, 19–22; and technology transfer, 265; Yamada Airship, 19–22, *21*; *See also* balloon launches; Provisional Committee for Military Balloon Research
Battle of Britain, 193
battleships, 6, 24, 141–42, 168, 183, 195–96; and the British Aviation Mission to Japan, 158, 160; and carrier-based aircraft, 149–51, *150*; and flying boats, 170; and ship-based floatplanes, 144–47; and technology transfer, 267; and US aviation, 203
Baumann, Alexander: at Mitsubishi, 102–7, 111, 133; Tobi reconnaissance aircraft, *106*; and Vogt, 107, 109
Berthaut, Henri, 57

biplanes, 48, 101, 268; Avro 504K, *158*, *159*; Calcutta, 181; Farman biplane, 26, *28*, 30–31, *31*, *36*, 45, 97; H3K1, 181; Keystone B-6A, 203; Sopwith Camel, 149–52; Sopwith Pup, 143, 147, 152; Type 10 carrier fighter, 153, 161, 186; Vogt and, 109, 112; *Wakadori*, 41–42, *42*
Blackburn, 159, *159*, 162, 192
Blériot, Louis, 25–26, 29, 37
Blue Impulse, 277–78, *278*, 283
BMW, 72, 112; 003 jet engine, 240, 248, *249*
board games: Great Visit Europe Flight as, 92
Boeing, 282; B-17, *115*, 210, 227; B-29, 7, 14, 235, 245, 261; F2B, 303n47; 767, 281; 777, 281; 787, 281, 283; YB-9, 102, 295n19
bombers, *115*; B5N, 218; B-17 Flying Fortress, *115*, 210, 227; B-29, 7, 114, 136, 235, 243, 245, 261; Blackburn Swift, *159*, 162; Do N (Type 87), 100–102, *101*, 115, *115*, 126, *134*; Farman biplanes, *36*; F.60 Goliath, 97, *115*, *134*; G10N Fugaku, 299n125; heavy bombers, *100*, 107, 128; Ki-20, *134*, 174; Ki-91, 136; Shinzan, 210; Sopwith Cuckoo, 159, *159*; superbombers, 123–25, 130–32, 135–36; Type 92, 114–15, *115*, 120–23, *122*, 130, 132; Type 93 Heavy Bomber, 125; Type 93 Light Bomber, 125; ultra-heavy bomber, 116, *117*
bombing, 40, 60, 65, 70, 193, 196; and the British Aviation Mission to Japan, 165, 170; and disarmament, 85; Douhet and, 81, 127, 293n49; and Japanese aviation after 1945, 279; and Japanese observers in Britain during World War I, 143; and jet and rocket technology, 260; and Junkers, 118; and the Manchurian crisis, 127; and a new air doctrine, 80–82; and the Qingdao air war, 44–46, 49; and US aviation, 208, 214, 220, 229. *See also* air raids; air strikes; bombers; bombs
bombs, 46, 49, 65, 94, 104, 142; and the British Aviation Mission to Japan, 166; and disarmament, 84; and early German influence, 75, 77; and flying boats, 178; jet-propelled, 7, 258; and Junkers, 118; and a new air doctrine, 82; and preemptive air strikes, 189–90; and ship-based floatplanes, 144; and US machine tools, 216–17. *See also* bombers; torpedoes
Brackley, Herbert, 157, 161–66
British Aviation Mission, 154–58, *158*, 164, 168, 192, 265
British Gloucestershire Aircraft Company, 187
"buy Japanese," 278

catapult, 7, 147, 149, 170, 176, 260, 300n31, 300n32
Chappel, R. W., 192–93
China, 96, 220; air raids on, 6–7, 216; Manchurian Incident, 125–32, 135; and the Mutual Security Agreement, 276; provisional balloon unit and, 20; and the Qingdao air war, 44–49, 154; return of Shandong province 83; and stealth technology, 282; and US air bases, 235; and US embargoes, 216, 232
Cold War, 274, 282
Comfort, Ernest Cyril, 152–53
Costruzioni Meccaniche Aeronautiche, 98
Crowe, Sir Edward, 163
Curtiss, Glenn, 25, 141
Curtiss Aeroplane and Motor Company, 189
Curtiss P-6 Hawk, 112
Curtiss-Wright, 205–6, 213, 221; Cyclone, 206, 208

Daikō plant, *214*, 245
Daimler, 54. *See also* Daimler-Benz
Daimler-Benz, 139
Dai Nippon Kōkū KK (Greater Japan Airlines Co., Ltd.), 208, 210
D'Annunzio, Gabriele, 36

De Villaret, Étienne, 57
Defense Agency, 274, 275, 279, 280
diplomacy, 79, 83, 88, 93, 169, 184, 269–70; aerial armament and, 5–7
disarmament, 6, 70, 83, 188, 270; Ugaki, 83–85, 94, 131; Yamanashi, 83
Dittmar, Heini 238
dive bombing, 139, 169, 189–91, 195–97, 206–7, 268
Doi Takeo, 109, 112, 114, 278
Do N, 100–101, *100*, *115*
donation campaigns, 5, 53, 128. *See also* "Donation Fever"
"Donation Fever," 127–30, 134, 269
Dornier, Claude, xxi, 72–73; Do H Falke, 111, 296n48; Dornier Merkur transport aircraft, 129; and independent aircraft design, 134; and Kawasaki, 97–102, 108; and Rohrbach, 173; and Vogt, 107–9, 111, 133; Wal, 98–99
Dornier-Kawasaki, 97–100, *100*, 108–9, *108*, 133; Ki-5, 114; Type 87 Heavy Bomber, 101–2, *101*, 126
Douglas, Archibald, 155
Douglas Aircraft Corporation, 207–11, 218; DC-2, 207–8, 227; DC-3, 208, 210, 227; DC-4E, 210–11, *211*; F-4 Phantom, 279–80
Douhet, Giulio, 52, 81, 127, 142, 293n49
Driant, Émile, 36
Dunning, Edwin, 143
duralumin, 73, 104, 112, 171–74, *172*, 291n14

Ely, Eugene, 299n14
embargoes, 7, 202, 216–19, 232–33, 270
English Channel, crossing of, 25
enthusiasm. *See* popular aviation enthusiasm

Farman, Henri, 25–26; École de Pilotage de Henri Farman, 27. *See also* Farman aircraft
Farman, Maurice, 288n8
Farman aircraft, 26, *28*, 29–30, *31*, 35–37, *36*; bombers, 46–48, 65–66, 97, *115*, *134*; seaplanes *45*, 142
Faure, Jacques-Paul, 57–64, 66, 80–81, 290n58; triumphal arch for, *59*
Fiala von Fernbrugg, Benno, 120–21
fighter aircraft, 7, 64, 94–98, 102, 134–35, 169; A5M, 187–88; Army Type 92, *134*; A6M1, 212; and Britain's influence, 151–53, 193–95; Buzzard, *159*; Chance Vought V-143, 218; competition for, 110–13; and early German influence, 72; F-1, 280; F-16, 280; F-22 Raptor, 282; F-86, 276–77; F-104 Starfighter, 279; and the French Aeronautical Mission to Japan, 56–57, 60; FS-X, 280; Gloster, *159*, 192; Hayabusa, 111–12, 218; Hien, 114, 278; J8M2, 257; and jet and rocket technology, 239, 241–42, 247, 257–58, 262; and Junkers, 118; Kawasaki KDA-5, 112–13, *134*; Ki-5, 114; and the Manchurian crisis, 125–26; Martinsyde F.4, *159*; Me 262, 240, 251; Nakajima Army Type 91, *134*, 182; Navy Type 10, 153, 161, 186; Ne-20 engine, 250, *250*; and a new air doctrine, 82, 85; Nieuport 29, *134*; S. E.5 A, *159*; and ship-based floatplanes, *134*, 144; 608; Sopwith Camel, 149–50, 152; Spad type 13; Taka, 187; and technology transfer, 267–68, 350, 385, US intelligence assessments of, 233; Zero-sen fighter, 107, 120, 201, 212, 218, 251, 255, 277
flight simulators, 60–61, *61*
floatplanes, 45, 149–50; Farman, 45, 142; Hansa Brandenburg W.29, 145; Heinkel's, *150*, 168, 170, 195; Junkers, 146; ship-based, 144–47; Type Mo, 40, *45*
flying boats, 168, 169–71, 176–79, 181–83, 195–96, 303n33; and bombers, 97–98; and the British Aviation Mission to Japan, *159*, 160; Experimental R-1, 175–76; H3K1, 181; H6K, 182, 303n33;

H8K, 182, 277; *Iris*, 167; Ro II, 175; Short F.5, *159*, 160, 164, 171, 195; and technology transfer, 267; Type 90-1, 179; Type 90-2, 181–82; Type 91, *134*, 182; Type R, *177*; Wal, 98–99
Fokker, 173, 279
Forbes-Sempill, Sir William Francis, 157–58, *158*, 165–67
Fowler, F. B., 157
French Aeronautical Mission, 40, 56–64, 80, 157, 265
Friedrichshafen, 99, 107
fuel, 13, 87, 170, 219, 235–36, 270; fuel-dumping device, 187; fuel-efficiency, 118, 204, 225, 250, 281; fuel tanks, 120, 182; gas turbulence, 238; rocket fuel, 241–42, 244–47, 252–55, *253*, 257–58, 261
Fuji, 274, 285
Fuji Heavy Industries, 263
Fuji Hikōki, 243–44
Fujimoto Akio, 215
Fuji Precision Industry, 275
Fukao Junji, 213–14
Funakoshi Kajishirō, 179, 303n29
Futurism, 4

Genda Minoru, 192–94, 277
General Dynamics, 280
German-Japanese Anti-Comintern Pact. *See* Anti-Comintern Pact
Gifu Station, *59*
Gloster, *159*, 192, 262
goodwill flights, 5, 225
Göring, Hermann, 197, 239
Grade, Hans, 29–31, *31*
"Great Visit Europe Flight," 70, 86–94, 223, 269; *Asahi*'s Memorial Picture Report on, *91*; as a board game, *92*
Grey, Sir Edward, 44
Grumman, 279
gun turrets, 118, 123, 147–50, *150*

Hamamatsu, *xxvi*, 62, 102, 132
Hamilton Standard, 208, 212, 225
Hansa Brandenburg W.29, 145–46; made-in-Japan version of, *146*
Hayabusa, 111–12, 218
H8K flying boat, 182, 277
Heinkel, Ernst, 72, 107, 145–50, 168, 187, 190–91, 195; D3A1, 191; floatplane, *150*; HD 23, 187; He 70 Blitz, 190–91; He 118, 139; He 176, 238; He 178, 238. *See also* Hansa Brandenburg W.29
Hibi Yoshitarō, 237
Hino Kumazō, 26–27, 29–33, *31*, 35, 37–38, 41
Hiro Arsenal, *xxvi*, 171, 176, 179
Hirth, Hellmuth, 107
Hitachi, 218, 237, 280
Horikoshi Jirō, 107, 120, 233, 277
Hōshō (flying phoenix), 151–53, 161–62, 168, 183–85
Hundred Days Offensive, 58

Immelmann, Max, *52*
Imperial Defense Policy, 115, 131, 193, 270
Imperial Japanese Army Academy (Rikugun Shikan Gakkō), 57
Imperial Rescript on the Termination of the War, 257
Imperial Way faction, 97, 131, 132
industrialists, 6, 69, 97–98, 231, 264, 278; Michelin, 288n1; Shimazu Genzō, 17–18, *18*
Inoue Ikutarō, 54–56, 66, 110, 131; and the French Aeronautical Mission to Japan, 58, 60, 63–64; and Junkers 116, 118–19
Inter-Allied Aeronautical Commission of Control, 71, 72, 82, 97, 107, 291n6
internationalism, 86–93, *91–92*, 283; technointernationalism, 281
international relations, 5–7, 269–70. *See also* diplomacy
interservice rivalry, 12, 23, 139, 241, 267, 274
Inuzuka Toyohiko, 242, 253–54
Ishihara Shintarō, 280

Ishikawajima, 104, 145, 248, 257, 274, 309n50; Ishikawajima-Harima, 263, 279; T-2, 109
Ishikawajima Heavy Industries, 275. *See also* Ne-20
Iwaya Eiichi, 240–41, 244, 246, 248
Iwaya Sazanami, 35–36, *36*

Japan Air Transport (Nippon Kōkū Yusō KK), 181, 208–9, *209*
Japanese Files Research Project, 227–31, *230*
Japanese spirit, 84
JASDF (Japan Air Self-Defense Force), xvi, 277
Jauneaud, Marcel, 81–82, 293n53
jet and rocket technology, 1, 7, 234–36, 260–63, 283; BMW 003, 240, 248–49, *249*; Boeing 767, 281; commercial jet aircraft, 282; early Japanese experiments, 236–38; and German technology, 234–40; German Walter-type rocket engine, 246; He 176, 238; He 178, 238; Ishikawajima-Harima J3, 263; Japan's first rocket aircraft, 241–47; Japan's postwar jet technology, 275–76; J8M2, 257; jet airplanes, 248–52; JO-1, 275; Kikka, 251–52, 254–57, *256*, 261–63; Me 163, 236, 238, 240–42, 262; Me 262, 240, 251; Mitsubishi Regional Jet (MRJ), 282; Ne-10, 238; Ne-12, 238; Ne-20, 249–51, *250*, 257, 260–63, 275, 310n55; Ne-130, 257; Ne-230, 257; Ne-330, 257; Ōka, 258–61, *259*, 277; Shūsui, 241–44, 252–254, *253*, 257, 261–63; Toku-Ro, 244–46; T1, 263, 275; T1B, 263; T-33, 276; TR-10, *237*, *250*
Jordan, William, 161, 169
Junkers, Hugo, 73, 108, 118, 120–21, *134*. *See also* Junkers aircraft
Junkers aircraft, 114–17, 133, 163, 173; CL I, 78; double wing design, *122*; float planes, 146; G 38, 118–19, 123, 297n81; K 37, 125, 129; and Mitsubishi, 124–25
Just, Erich, 99, 112

Kahn, Albert, 213
Kakamigahara Airfield, *xxvi*, 60, 97, 120, 132; flight simulator at, *61*
Kamikaze aircraft, 223–25, *224*
Kanda Minoru, 218–19, 231
Kaneko Yōzō, 142–44
Kantō earthquake, 83, 139–40
Kasai Taijirō, 246
Kasumigaura Air Base, 78, 99, 140, 164, 185
Katagiri Shōhei, 88
Katō Shigeo, 237
Kawanishi, 179–82, 196, 215, 218, 229, 277; K-7A seaplane, *180*
Kawasaki, 80, 96, 136, 145, 274; and Boeing, 281; Do N, 100–101, *100*, *115*, *134*; and FS-X, 280; and interceptors, 110; KDA-2, *134*; KDA-3, 111; KDA-5, 112–13, *134*; and the Mutual Security Agreement, 276; and supersonic aircraft, 279–81; Type 88, 109; Type Otsu-I, 104; and US aviation, 203, 229; and Vogt, 100, 107–9, 133. *See also* Dornier-Kawasaki
Kawata Akiharu, 22–25
Kayser, Erich, 179
Kikka jet aircraft, 261–63; building the airframe, 251–52; maiden flight, 254–57, *256*
Kikuhara Shizuo, 181, 277
Kimura Hidemasa, 258, 277
Kofukuda Mitsugi, 252
Koiso Kuniaki, 72–73, 84, 116, 118, 131
Kōkenki, 217, 225–26, *226*
kokusanka, 3, 69
Kōno Hiroshi, 215
Kōno report, 216
Korean War, 274, 276
Kriegsmarine, 197
Kumazawa Toshikazu, 238
Kure Naval Arsenal, 24, 185
Kusakari Shirō, 43, 50–54, *52*, 58, 66, 290n58
Kwantung Army, 125–26, 131

Lachmann, Gustav, 109
landing, xx, 34, 144, 151, 153, 223; carrier-deck landings, 160–62, 165, 169, 176, 185–87, 190, 192, 195, 299n14; emergency, 54, 121, 186–87, 254; and jet and rocket technology 254–55; and Junkers, 120–23; landing gear, 207–8, 218; landing skids, 42; landing speed, 120, *122*, 162; landing strips, 144; on water, 45, 121, 186–87
launching technology, 71, 147–49
Law for the Promotion of the Aircraft Manufacturing Industry, 278, 280
Law for the Protection of Military Secrets (*Gunkihogohō*), 221
League of Nations, 6, 171, 270
Le Corbusier, 4
light cruisers, 6, 188
Lindbergh, Charles, xx, 118, 204
Lockheed: Electra, 227; F-22 Raptor, 282; F-104 Starfighter, 279; T-33 jet trainer, 276
London Naval Treaty, 6, 188, 270
Lorraine engines, 65, 176, 290n54
Ludwig, Paul, 175
Luftwaffe, 119, 193, 197

MacArthur, Douglas, 272
Ma Chan-shan, 298n106
machine tools, 62, 98, 133, 207, 213–19, 233
"made-in-Japan," 37, 85, 109, 112, 133, 161; and carrier planes, 187; doctrine, 283; and flying boats, 176–77; and gas balloons, 14, 20; Hamilton propeller, 212, 224–25; and jet and rocket technology, 263; and Junkers, 121, 123; *Kōkenki*, 217, 225–26, *226*; Navy Type Hansa reconnaissance seaplanes, *146*; after 1945, 276, 280–81; policy 282
Mahan, Alfred Thayer, 141
Mainichi Shinbun, 225, *228*
Makino Ikuo, 246
Manchuria, 6, 20, 97, 113, 125–32, 134–35; and international relations, 269–70; and US aviation, 202. *See also* Manchurian Airlines; Manshū Hikōki Seizō KK
Manchurian Airlines, 208
Manila, 116–17, *117*
Manshū Hikōki Seizō KK (Manchurian Airplane Manufacturing Company), 215
Marie, André, 111
Marinetti, Filippo: Futurist Manifesto, 4
Marquerie, Charles Antoine, 57
Martin, Rudolf, 36
mass media, 5, 268–69, 283; and the French Aeronautical Mission to Japan, 59–60; and gas balloons, 14–15, 17; and jet and rocket technology, 247; and Junkers, 123; and Kusakari Shirō, 53; and the Manchurian crisis, 127–29; and the Qingdao air war, 46; and US aviation, 223–24, 231; and the "Visit Europe Flight," 86–94, *91–92*. *See also* advertising *and individual newspapers by name*
Maybach, 77, 289n27
media. *See* mass media
Messerschmitt, Willy, 238
Michelin, André and Édouard, 288n1
Milch, Erhard, 239
Mitchell, William, 142, 203, 306n61
MITI (Ministry of International Trade and Industry), 274
Mitsubishi, 96, 110, 133, 139, 194, 196; A5M, 187–88; Baumann at, 102–7; Boeing 767, 281; and British influence, 152; Daikō plant, 214, *214*, 245; and flying boats, 178–79; F-1, 280; F-86 Sabre, 276–77; FS-X, 280; G3M, 207, 225, 307n78; Hayabusa, 111–12, 218; and jet and rocket technology, 237, 241–42, 246; and Junkers, 119–21, 123; Ki-15, 223; Ki 20, *134*, 174; Kinsei Type 4, 207; and the media, 269; Ne-330, 257; after 1945, 274–76, 279; Regional Jet (MRJ), 282; and Rohrbach, 174–77; Shūsui, 241–44, 252–55, *253*, 257,

Mitsubishi (*continued*)
261–63; Tobi, 105–6, *106*, 109; Toku-Ro, 244–46; Type 10, 153, 161, 186; Type 92, 114–15, 121–25, *122*, 130–32; Type R flying boat, *177*; and US aviation, 203, 205–6, 213, 215, 224; Zero-sen fighter, 107, 120, 201, 212, 218, 251, 255, 277
Mitsubishi Shōji, 77, 184, 206, 229
Mitsui Bussan, 76–77, 79, 206–12, 218
Mochida Yūkichi, 244–45
monoplanes, 26, 101–2, 110–11, 124, 203; *Dreadnought*, 166; Grade, 29–32, *31*; Hino Type 1, 27; Staaken, 173; Taube, 47–48, *48*, 291n4; W.29, 145–46, *146*

NACA, 112, 205, 296n54
Nagano Osamu, 248
Nakaguchi Hiroshi, 241, 252
Nakajima Aircraft Company, 110–11, 113, 139, 196, 205–7; and Andō Nario, 114; Army Type 91, 134, *134*; AT-2, *209*; carrier-based dive bomber, 186–87, 189–90; DC-3, 210; Experimental LB-2 Long-Range Attack-Bomber, 208–10; and Fuji Heavy Industries, 263; Hayabusa fighter, 218; and the Kikka, 251, 254; light bomber, 104; long-range bombers, 299n125; Ne-130, 257; Ne-230, 257; Ogikubo aircraft factory, 192; Type 97, 201; and US assessments of Japanese air power, 222, 229; and the "Visit Europe Flight," 89; W.29, 145
Nakajima Chikuhei, 141–42, 233
Nakao Sumitoshi, 111
Nakata Nobushirō, 111
nationalism, 36–37, 53, 94, 135; and the media, 268–69; national identity, 4–5, 269; national prestige, 86–93, *91–92*; national pride, 5, 15, 33, 87, 113, 123, 269
National Order of the Legion of Honor, 43
Naval Air Technical Arsenal (Kaigun Kōkū Gijutsushō), 238, 242, 244, 248, 251, 260
Ne-20, 249–51, 257, 260–61, 263, 275, 310n55; functional diagram of, *250*
Neutrality Act, 216
Nieuport, 43, 46–47, 51, 60–61, 64, *134*
Nihon Airplane Manufacturing Company (NAMC), 278–79
Nihon Jet Engine Company (NJE), 275
Nippon aircraft, 227, *228*, 269
Nippon Hikōki, 243–44
Nippon Kōkū airline, 179–81; Kawanishi K-7A seaplane, *180*
North American Aviation: F-86 Sabre, 276–77
Northrop, 218

Ogawa Taichirō, 258
Oguri Masaya, 245
oitsuki, oikose (catch up and surpass), 3
Ōka suicide attacker, 258–61, *259*, 277
Ōkuma Shigenbu, 44
Olympics, 277–78, *278*
Ōmiya Fuji Industry, 275
Oota Shōichi, 258
Oppama, 245, 253
Osaka Asahi newspaper, 59–60
Osaka Mainichi newspaper, 46, *228*
Ōshima Hiroshi, 239
Ōta Minoru, 277
oyatoi gaikokujin (honorable hired foreigners), 3

Pacific War, 4, 7, 207, 227, 234, 260
Paget, A., 163
Panay incident, 190, 216
Parseval, 289n27, 291n4
patriotism, 5, 38, 53, 69, 128–29, 278–79
Peace Treaty of San Francisco, 274
Pearl Harbor attack, 7, 185, 191, 194, 201–2, 276–77; and US aviation, 218, 229, 231–33
plants, production, 175–76, 206–8, 251–52; Daikō, 214, 245; Messerschmitt, 240; Nagoya, 119, *214*, 244; Pratt & Whitney, 213–14, *214*

Plüschow, Gunther, 49; and Taube monoplane, 47–49, *48*
policy, 96; air doctrine, 80–82, 93, 126–27; aviation, 5, 281; Aviation Bureau, 85; Balloon Committee, 29; big-ship, big-gun, 141–42, 195, 267; British armament, 154; disarmament, 83, 94; Excess Profit Tax, 152; Imperial Defense Policy, 115–16, 270; "made-in-Japan," 282; military, 132, 155; National Defense Policy, 193; procurement, 3, 164; Soviet Union, 276; US, 212, 229, 231, 274, 282
Popular Aviation: "DC-4 Off for Japan," *211*
popular aviation enthusiasm, 4–5, 11–12, 17–19, 34–42, 70–71; and the French Aeronautical Mission to Japan, 58–60, *59*; and Kusakari Shirō, 50, 53; and the "Visit Europe Flight," 86–92
Prandtl, Ludwig, 140
Pratt & Whitney, 205, 213–15, *214*; R-1689 Hornet, 207
prototype competition, 96, 103, 104, 108, 113
Provisional Committee for Military Balloon Research (Rinji Gunyō Kikyū Kenkyū Kai), 23–24, 26–30, 37, 50–51
public, the. *See* popular aviation enthusiasm

Qingdao air war, 44–50, 66, 145, 154, 268

Railway Technology Research Institute (Tetsudō gijutsu kenkyūjo), 272, 275
reconnaissance, 40, 46, 100, 148, 195; and aircraft carriers, 183; and the British Aviation Mission to Japan, 160, 165; and disarmament, 85; and early German influence, 75; and flying boats, 170; and the French Aeronautical Mission to Japan, 56–57, 60; and Japanese observers in Britain during World War I, 143–44; and a new air doctrine, 82; and US aviation, 223. *See also* aircraft, reconnaissance
Regnault, Eugène, 58
research airplanes: *Kōkenki*, 217, 225–26, *226*
Robin, Maxime, 111
rocket interceptors, 235–36, 238, 240, 246; Shūsui, 241–44, 252–55, *253*, 257, 261–63. *See also* jet and rocket technology
rocket technology. *See* jet and rocket technology
Rohrbach, Adolf, 73, 80, 108, 171–74, 186; legacy of, 177–82; and Mitsubishi, 174–79, *177*
Roux, Charles, 41
Rumpler Taube monoplane, 47–48, *48*, 77, 291n4
Russo-Japanese War, 38, 50, 87, 141, 145, 265; and gas balloons 19–23
Rutland, Frederick Joseph, 147, 183–85, 192

Sanuki Matao, 178
seaplanes, 24, 139, 142, 145, 147–48, 160, 170; Hansa, *146*; K-7A, *180*; launched from cannon turrets, 191; Short 320, 144; Sopwith, 144; Type Mo, 45, *45*; and Wada, 173. *See also* floatplanes; flying boats
self-sufficiency, 6, 69, 85, 108, 195
Sempill mission, 159, *159*, 165, 184, 194
Senba Tadashi, 124
Shidehara Kijūrō, 83
Shigeno Kiyotake, 41–43, 51; and *Wakadori*, *42*
Shimazu Genzō, 17, *18*
Shin Meiwa, 274, 281
Shinohara Shunichirō, 88
Shirakawa Yoshinori, 119
Short Brothers company, 144, *159*, 164, 171, 173, 181
Shōwa Aeroplane Company, 210
Shūsui rocket interceptor, 241–44, 261–63; maiden flight, 252–54, *253*, 257
Siege of Port Arthur, 20
Siemens Scandal, 44, 288n7

Sino-Japanese War: First, 141; Second, 7, 190, 194, 213, 220, 270
Smith, Herbert, 152–54, 161, 186, 194
Solf, Wilhelm, 99
Sopwith Aviation Company: Camel biplane, 149, 152; Cuckoo, 159, *159*; Pup biplane, 143, 147, 152; Schneider plane, 144
Spad, 43, 51, 64, 290n54
Sparrowhawk, 159, *159*
spectacle, 5, 17, 30–34, 37, 70, 268
Spencer, Percival, 19
Sperry, 208
Spring Offensive, 57
stealth technology, 282–83
stressed-skin structure, 108–9, 133, 178
submarines, 2, 6, 103, 146–47, 188, 270; and jet and rocket technology, 239–40, 247
Sugiyama Gen, 97–98
suicide attackers, 7, 223; Ōka, 258–61, *259*, 277
Sumitomo Metal Industries, 212, 237
supersonic technology, 279–80, 283, 311n28
Supreme War Council (Gunji Sangikan), 131, 291n6
Suzuki Kantarō, 143

Tachikawa, *xxvi*, 89, 113–14, 128
Taiwan, 56, 116–17, *117*
Takaoka Susumu, 254–57, 263
Takezaki Tomokichi, 97–99, 112
Tamaru Nario, 237
Tanakadate Aikitsu, 224
Tanaka Giichi, 55–56, 60, 75, 83
Tanaka Kanbei, 112–13
Tanegashima Tokiyasu, 236–38, 248
Tani Ichirō, 196, 205, 258
Tateyama air corps, 192
Taube monoplane. *See* Rumpler Taube monoplane
technointernationalism, 281
technology transfer, 2, 11, 96, 169, 195, 264–68, 283; and aircraft carriers, 183; and flying boats, 171; and Japanese aviation after 1945, 276, 280; and jet and rocket technology, 239, 260; and US machine tools, 213
technonationalism, 281
Temporary Committee for Aviation Technique and Practice, 58
Terauchi Masatake, 20, 22, 23
Tobi reconnaissance aircraft, 105–6, *106*, 109
Tokorozawa, *xxvi*, 34, 36, 54–55, 60–61, 77–78
Tokorozawa Flight School, 3, 65, 96
Tokugawa Yoshinobu, 57
Tokugawa Yoshitoshi, 26–31, 33–38, 41–42; in a Farman biplane, *28*, *31*
Tokyo Army Arsenal, 46
Tokyo Imperial University. *See* Aeronautical Research Institute
torpedoes, 35, 75, 142, 144, 193, 197; and British influence, 154, 159, 160, 162, 165; and jet and rocket technology, 236, 247, 251, 258; and US machine tools, 217–18. *See also* bombers
Tosu Tamaki, 143–44
Transocean Program, 118, 297n81
Treaty of Versailles, 6, 69–72, 76–80, 94, 97–99, 168–70, 269; and flying boats, 175; and launching technology, 148; and ship-based floatplanes, 145
triumphal arch, *59*
Tsukiji: Aircraft Test Laboratory at, 139, 299n4; first balloon launch at, 14–16, *16*
turbojet engines, 234–35; TR-10, *237*. *See also* Ne-20

U-boats, 197
Ugaki Disarmament, 83–84, 131–32
Ugaki Kazushige, 70, 83–85, 94, 99, 110
US and Japan Mutual Defense Assistance Agreement, 274, 276

variable-pitch propeller, 208, 210–12
Vietnam War, 279

"Visit Europe Flight." *See* "Great Visit Europe Flight"
Vogt, Richard, 100, 107–9, *107*, 111–12, 133–34; legacy of, 113–14
Voisin, 43
Von Kármán, Theodore, 179
Von Ribbentrop, Joachim, 239
Von Zeppelin, Count Ferdinand, 28–29

Wada Hideho, 164
Wada Misao, 173, 176, 179, 260
Wakadori, 41–42, *42*
war trophies, 74–78
Washington Naval Treaty, 6, 83, 115–16, 146, 299n8, 303n37; and aircraft carriers, 183; and the British Aviation Mission to Japan, 162–63; and international relations, 270
Watanabe Jōtarō, 131
Watanabe Kōtarō, 75, 82
Wells, H. G., 36
Wieselsberger, Carl, 140, 299n4
wind tunnels, 4, 73, 110, 140, 143, 299n4; and flying boats, 179, 182; and Junkers, 120; and NACA 112, 296n54
World War I, 1–3, 6, 37, 40, 69–71, 93–94; and battleship-first proponents, 141; Baumann and, 103; and British influence, 151–52, 154–56, 159, 162, 167; and disarmament, 84–85; and early German influence, 71–72, 77; and flying boats, 170–71; and the French Aeronautical Mission to Japan, 58; and Inoue Ikutarō, 54; Japanese observers in Britain during, 142–44; and Junkers, 120; and Kusakari Shirō, 50, 52; and long-range bombers, 65; and the media, 268; and a new air doctrine, 80–81; and Shigeno Kiyotake, 43–44; and ship-based floatplanes, 144–45, 147; and technology transfer, 265; and US aviation, 203
World War II, 1, 77, 195, 197, 282–83; Japanese aviation after, 271–80; and jet and rocket technology, 242, 263
Wright, Orville and Wilbur, 25–26, 30, 202
Wright, T. P., 213
Wright Company, 215, 218
Wunderwaffen (miracle weapons), 239

X-2 Shinshin, 282

Yamada Isaburō, 19–22, 24, 38; Yamada Airship no. 2, *21*
Yamana Masao, 195, 258
Yamanashi Disarmament, 83
Yamanashi Hanzō, 79, 83
Yamashita Kamesaburō, 144
Yokosuka Naval Air Arsenal (Kaigun Kōkūshō), 24, *45*, 140, 151, 195–96, 308n7; and aircraft carriers, 185; and flying boats, 175–76; and jet and rocket technology, 237, 242–43, 254; and ship-based floatplanes, 146–47. *See also* Naval Air Technical Arsenal
Yomiuri Shinbun, 14, 46, 113, 123, 128, 130
Yoshida Shigeru, 274
Yoyogi flight, 29–36, 38, 66
Yoyogi Parade Ground, 89, 130, 185, 264. *See also* Yoyogi flight
YS-11 project, 278–79, 281–83

Zeppelin-Staaken R.XV, 77
Zero-sen fighter, 107, 120, 201, 212, 218, 251, 255, 277

Harvard East Asian Monographs

(most recent titles)

406. Michal Daliot-Bul and Nissim Otmazgin, *The Anime Boom in the United States: Lessons for Global Creative Industries*
407. Nathan Hopson, *Ennobling the Savage Northeast: Tōhoku as Japanese Postwar Thought, 1945–2011*
408. Michael Fuller, *An Introduction to Chinese Poetry: From the Canon of Poetry to the Lyrics of the Song Dynasty*
409. Tie Xiao, *Revolutionary Waves: The Crowd in Modern China*
410. Anne Reinhardt, *Navigating Semi-colonialism: Shipping, Sovereignty, and Nation-Building in China, 1860–1937*
411. Jennifer E. Altehenger, *Legal Lessons: Popularizing Laws in the People's Republic of China, 1949–1989*
412. Halle O'Neal, *Word Embodied: The Jeweled Pagoda Mandalas in Japanese Buddhist Art*
413. Maren A. Ehlers, *Give and Take: Poverty and the Status Order in Early Modern Japan*
414. Chieko Nakajima, *Body, Society, and Nation: The Creation of Public Health and Urban Culture in Shanghai*
415. Pu Wang, *The Translatability of Revolution: Guo Moruo and Twentieth-Century Chinese Culture*
416. Hwansoo Kim, *The Korean Buddhist Empire: A Transnational History (1910–1945)*
417. Joshua Hill, *Voting as a Rite: A History of Elections in Modern China*
418. Kirsten L. Ziomek, *Lost Histories: Recovering the Lives of Japan's Colonial Peoples*
419. Claudine Ang, *Poetic Transformations: Eighteenth-Century Cultural Projects on the Mekong Plains*
420. Evan N. Dawley, *Becoming Taiwanese: Ethnogenesis in a Colonial City, 1880s–1950s*
421. James McMullen, *The Worship of Confucius in Japan*
422. Nobuko Toyosawa, *Imaginative Mapping: Landscape and Japanese Identity in the Tokugawa and Meiji Eras*
423. Pierre Fuller, *Famine Relief in Warlord China*
424. Diane Wei Lewis, *Powers of the Real: Cinema, Gender, and Emotion in Interwar Japan*
425. Maram Epstein, *Orthodox Passions: Narrating Filial Love during the High Qing*
426. Margaret Wan, *Regional Literature and the Transmission of Culture: Chinese Drum Ballads, 1800-1937*
427. Takeshi Watanabe, *Flowering Tales: Women Exorcising History in Heian Japan*
428. Jürgen P. Melzer, *Wings for the Rising Sun: A Transnational History of Japanese Aviation*
429. Edith Sarra, *Unreal Houses: Character, Gender, and Genealogy in the Tale of Genji*
430. Yi Gu, *Chinese Ways of Seeing and Open-Air Painting*
431. Robert Cliver, *Red Silk: Class, Gender, and Revolution in China's Yangzi Delta Silk Industry*
432. Kenneth J. Ruoff, *Japan's Imperial House in the Postwar Era, 1945-2019*